Sustainable Development Goals Series

The **Sustainable Development Goals Series** is Springer Nature's inaugural cross-imprint book series that addresses and supports the United Nations' seventeen Sustainable Development Goals. The series fosters comprehensive research focused on these global targets and endeavours to address some of society's greatest grand challenges. The SDGs are inherently multidisciplinary, and they bring people working across different fields together and working towards a common goal. In this spirit, the Sustainable Development Goals series is the first at Springer Nature to publish books under both the Springer and Palgrave Macmillan imprints, bringing the strengths of our imprints together.

The Sustainable Development Goals Series is organized into eighteen subseries: one subseries based around each of the seventeen respective Sustainable Development Goals, and an eighteenth subseries, "Connecting the Goals", which serves as a home for volumes addressing multiple goals or studying the SDGs as a whole. Each subseries is guided by an expert Subseries Advisor with years or decades of experience studying and addressing core components of their respective Goal.

The SDG Series has a remit as broad as the SDGs themselves, and contributions are welcome from scientists, academics, policymakers, and researchers working in fields related to any of the seventeen goals. If you are interested in contributing a monograph or curated volume to the series, please contact the Publishers: Zachary Romano [Springer; zachary.romano@springer.com] and Rachael Ballard [Palgrave Macmillan; rachael.ballard@palgrave.com].

Shantanu Ghosh · Tarak Nath Sahu ·
Girijasankar Mallik

Financial Inclusion

A Drive Towards Achieving the Sustainable Development Goals

palgrave
macmillan

Shantanu Ghosh
Department of Commerce
Mugberia Gangadhar Mahavidyalaya
Bhupatinagar, West Bengal, India

Tarak Nath Sahu
Department of Commerce
Vidyasagar University
Midnapore, West Bengal, India

Girijasankar Mallik
School of Business
Western Sydney University
Parramatta, NSW, Australia

ISSN 2523-3084 ISSN 2523-3092 (electronic)
Sustainable Development Goals Series
ISBN 978-981-95-3384-8 ISBN 978-981-95-3385-5 (eBook)
https://doi.org/10.1007/978-981-95-3385-5

Cover credit: Nicolay Pandev gettyimages

This Palgrave Macmillan imprint is published by the registered company Springer Nature Singapore Pte Ltd.
The registered company address is: 152 Beach Road, #21-01/04 Gateway East, Singapore 189721, Singapore

Preface

In the landscape of global development, financial inclusion stands as a beacon of hope and opportunity for billions around the world. It is a concept that transcends mere access to financial services; it is a catalyst for empowerment, resilience, and sustainable progress. As we approach a new era marked by the Sustainable Development Goals (SDGs), the need to promote financial inclusion is more urgent than ever.

This book, "Financial Inclusion—A Drive Towards Achieving the Sustainable Development Goals," emerges at a pivotal moment in history, where the intersection of finance and development holds the promise of transformative change. Within these pages, we commence a voyage to investigate the multifaceted dimensions of financial inclusion and its profound implications for realizing the aspirations encapsulated in the SDGs.

This book aims to clarify the crucial importance of financial inclusion in promoting the global agenda for sustainable development by conducting thorough analysis, examining empirical evidence, and presenting real-world case studies. The chapters in this volume attempts to characterize and subsequently compare the achievements in financial inclusion. Further, it also explores how inclusive finance plays a crucial role in attaining the SDGs, including reducing poverty and hunger, and more importantly in advancing economic development.

Yet, this book is not merely a compendium of theoretical frameworks and policy prescriptions. It is a testament to the collective efforts of policymakers, practitioners, researchers, and advocates who are tirelessly working towards a more inclusive and equitable world. It is a celebration of innovation, resilience, and the indomitable human spirit that drives us towards progress, even in the face of daunting challenges.

As we deal with the intricacies of the contemporary world, it is our responsibility to acknowledge that financial inclusion is not only a social and economic necessity but also a moral necessity. It is a fundamental human right that lies at the heart of our quest for dignity, equality, and justice.

I extend my heartfelt gratitude to all those who have contributed to this endeavour—the authors, the editors, the reviewers, and the countless individuals and organizations working tirelessly to advance the cause of financial inclusion.

I earnestly aspire for this book to function as a wellspring of motivation, wisdom, and authority for individuals who endeavour to construct a globe that is both comprehensive and enduring.

Bhupatinagar, India — Shantanu Ghosh
Midnapore, India — Tarak Nath Sahu
Sydney, Australia — Girijasankar Mallik

Contents

Abbreviations

ADF test	Augmented Dickey Fuller test
AePS	Aadhar Enabled Payment Services
AFI	Alliance for Financial Inclusion
AIDS	Acquired Immune Deficiency Syndrome
ANBC	Adjusted Net Bank Credit
ANOVA	Analysis of Variance
APY	Atal Pension Yojana
ARDL	Autoregressive Distributed Lags
ARMA	Auto Regressive Moving Average
ASEAN	Association of Southeast Asian Nations
ATM	Automated Teller Machine
ATM_A	ATM's per 1000 km^2
ATM_D	ATMs of Commercial Banks scaled by the Demographics
ATM_P	ATM's per 100000 adults
BC	Business Correspondent
BDG	Bank's Deposit to GDP
BF	Business Facilitator
BLUE	Best Linear Unbiased Estimator
BP	Breusch Pagan statistics
BRAN_A	Branches of Commercial Bank's per 1000 km^2
BRAN_D	Branches of Commercial Bank's per 100000 adults
BRAN_P	Branches of Commercial Banks scaled by the Demographics
BRI	Bank Rakyat Indonesia
BZ	Bank's Z-score
CAGR	Compounded Annual Growth Rate
CCC	Certified Credit Counsellors

CDR	Bank's Credit to Bank's Deposit
CEAOBE	Credit Equivalent Amount of Off-Balance-Sheet Exposure
CFA	Common Factor Analysis
CFL	Centre for Financial Literacy
CGAP	Consultative Group to Assist the Poor
CII	Depth of Credit Information Index
CoC_estimate	Estimate of Control of Corruption
CPC	Communist Party of China
CPI	Consumer Price Index
CR	Liquid Assets to Deposits and Short-term Funding
CRISIL	Credit Rating Information Services of India Limited
CRR	Cash Reserve Ratio
CRWA	Bank's Regulatory Capital to Risk-weighted Assets
CTA	Bank's Capital to Total Assets
CV	Coefficient of Variation
DEPO	Depositors with Commercial Banks (per 1000 adults)
DMBAG	Deposit Money Bank's Assets to GDP
DOLS	Dynamic Ordinary Least Squares
DSDG	Division for Sustainable Development Goals
EC	Error Correction
ECM	Error Correction Mechanism
FA	Factor Analysis
FAO	Food and Agricultural Organization
FE model	Fixed Effect model
FI	Financial Inclusion
FIAC	Financial Inclusion Advisory Committee
FIP	Financial Inclusion Plans
FMOLS	Fully Modified Ordinary Least Squares
FPI	Food Production Index
FPI_C	Food Price Index at Ceiling Prices
FSC	Food Supply Chain
GCC	Gulf Cooperation Council
GDP	Gross Domestic Product
GE_estimate	Estimate of Government Efficiency
GFC	Global Financial Crisis
GFCF	Gross Fixed Capital Formation
GFDD	Global Financial Development Database
GMM	Generalized Method of Moments
GNI	Gross National Income
GNIpc	Gross National Income per capita
GPFI	Global Partnership for Financial Inclusion
GSE	Government Social Expenditure
HDI	Human Development Index

HEL	Governments Health Expenditure to Total Expenditure (in %)
HLPF	High-Level Political Forum
HSD	Honestly Significant Difference
ICDS	Integrated Child Development Services
IFC	International Finance Corporation
IFI	Index of Financial Inclusion
IMF	International Monetary Fund
IMPS	Immediate Payment Service
INF	Inflation
INFE	International Network for Financial Education
IV estimator	Instrumental Variable estimator
JAM	Jan-Dhan Aadhar and Mobile
JB	Jarque Bera statistics
JLG	Joint Liability Groups
JSY	Janani Suraksha Yojana
KCC	Kisan Credit Card
KMO test	Kaiser-Meyer-Olkin's test
KPI	Key Performance Indicator
LDC	Least Developed Countries
LLC test	Levin-Lin-Chu test
LLDC	Landlocked Developing Countries
LLG	Liquid Liabilities to GDP
LOAN	Outstanding Loans with Commercial Banks (per cent of GDP)
LOS	Level of Significance
LPG	Liquified Petroleum Gas
LRI	Strength of Legal Rights Index
MDG	Millennium Development Goals
MGNREGS	Mahatma Gandhi National Rural Employment Guarantee Scheme
MoSPI	Ministry of Statistics and Programme Implementation
MRS	Marginal Rate of Substitution
MSE	Micro and Small Enterprises
MSME	Micro, Small and Medium Enterprises
NAMCABS	National Mission for Capacity Building of Bankers
NAPCC	National Action Plan for Climate Change
NCFE	National Centre for Financial Education
NIM	Bank's Net Interest MArgin
NITI Aayog	National Institution for Transforming India Aayog
NMSA	National Mission on Sustainable Agriculture
NPLG	Bank's Non-performing Loans to Gross Loans
NSA	Number of Savings Account
NSDP	Net State Domestic Product

NSFE	National Strategy for Financial Education
NSFI	National Strategy for Financial Inclusion
NSSO	National Sample Survey Organization
OCTA	Bank's Overhead Costs to Total Assets
ODA	Official Development Assistance
OECD	Organization for Economic Cooperation and Development
OFC	Overseas Financial Centres
OIC	Organization of Islamic Cooperation
OLS	Ordinary Least Square
OWG	Open Working Group
PCA	Principal Component Analysis
PMJDY	Pradhan Mantri Jan Dhan Yojana
PMJJBY	Pradhan Mantri Jeevan Jyoti Bima Yojana
PMMY	Pradhan Mantri MUDRA Yojana
PMSBY	Pradhan Mantri Suraksha Bima Yojana
POP	Population
PoU	Prevalence of Undernourishment
POV	Poverty
PP test	Phillips Perron test
PRIVY	Private Credit by Deposit Money Bank's to GDP
PSAV_estimate	Estimate of Political Stability and Absence of Violence
RBI	Reserve Bank of India
RealIP	Industrial Production in Real terms (Inflation adjusted)
R-E-G-W-Q	Ryan-Einot-Gabriel-Welsch
RNSDP	Real Net State Domestic Product
ROA	Return on Assets
RRB	Regional Rural Bank
SAARC	South Asian Association for Regional Cooperation
SCB	Scheduled Commercial Bank
SDG	Sustainable Development Goals
SDG 1	End poverty in all its forms everywhere
SDG 2	End hunger, achieve food security and improved nutrition and promote sustainable agriculture
SDG 5	Achieve gender equality and empower all women and girls
SDG 8	Promote sustained, inclusive and sustainable economic growth, full and productive employment and decent work for all
SDG 9	Build resilient infrastructure, promote inclusive and sustainable industrialization and foster innovation
SDG 10	Reduce inequality within and among countries
SDG 16	Promote peaceful and inclusive societies for sustainable development, provide access to justice for all and build effective, accountable and inclusive institutions at all levels

SHG	Self Help Group
SIDS	Small Island Developing States
SLR	Statutory Liquidity Ratio
SME	Small and Medium Enterprises
UN	United Nations
UNDP	United Nations Development Program
UNEMP	Unemployment
UNGA	United Nations General Assembly
UPI	Unified Payments Interface
USA	United States of America
USD	United States Dollar
UT	Union Territory
VAR	Vector Autoregression
VECM	Vector Error Correction Model
VND	Vietnamese Dong
VNR	Voluntary National Review

List of Figures

List of Tables

CHAPTER 1

Introduction

1.1 Overview

That historical period is well past when financial inclusion was scoffed at. Financial inclusion as the foundation of economic empowerment covers a broad range of characteristics that contribute towards sustainable development. It is one of the primary engines recognized among development studies to build and sustain a fair and balanced economic growth expectedly creating impressive results. Poverty in its many expressions is a big concern across the globe. With the devastation of the whole world with the COVID-19 pandemic, so much more is at stake. It gives extremely effective of a responding point against the challenges associated with poverty and its related complexities such as generation of income, improvement in employability, equitable resource allocation, financial literacy, gender equality, women's empowerment and health and nutrition. It is crucial to have financial inclusion and financial sustainability as it offers a balance for both individual and socio-economic growth. Initially, the definition of financial inclusion comprises access to and usage of financial services, cost and appropriateness of the person's ability to pay for the service, financial literacy and consumer protection. It is more than access; it is having use of services such as savings, credit, insurance and payment mechanisms. Such a facing goes not to financial institutions but assesses how far access and usage of such resources make possible an individual's economic prosperity.

S. Ghosh et al., *Financial Inclusion*, Sustainable Development Goals Series, https://doi.org/10.1007/978-981-95-3385-5_1

This book would also include assessing financial inclusion utilizing a number of measures that represent how the individuals and businesses are included in the formal financial system. The major factors remain account ownership, use of banking services, availability of credit, insurance penetration and financial literacy level. These measures would provide insight into the depth, the breadth and the accessibility of the financial services for certain demography.

Diverse paths highlighted by comparative analysis throughout the Indian states and union territories, across a sample of 26 Asian countries and with 12 European and 12 Asian countries, stress the necessity for context-specific solutions. A bigger analysis including 46 countries located within Asia, Europe, Africa, North America and South America finds a patchwork of development and obstacles, reflecting varied economic, social and institutional situations. Such comparisons would further introspect into solutions to and provide valuable insights into the difference in regulatory frameworks, the infrastructural development and socio-economic elements that have been instrumental in those discrepancies. Yet, amidst this variation, a braided thread of financial inclusion running through all is its contribution towards reaching the Sustainable Development Goals (SDGs). It is a prerequisite for the success of numerous SDGs, including SDG 1 (eradicating poverty), SDG 2 (ending hunger) and SDG 8 (decent work and economic growth). Financial inclusion in itself has offered economic strength, resilience, and income transformation through many forms of financial resources, including credit, savings and insurance. Apart from this, financial inclusion has strengthened the underprivileged parts of society.

Financial inclusion to SDG 1 enables every person and household the freedom and opportunity to save and access their assets. It further enables access to a loan for productive investments, regulate consumption at times of financial trouble and provide wider coverage across poverty reduction and economic inclusion. This, for instance, goes a long way in supporting entrepreneurship and income-generating activities.

Furthermore, financial inclusion addresses the causes of food security and malnutrition (SDG 2). Financial services availability encourages farmers to invest in inputs and apply new ways of agriculture, therefore lowering climate change risks. On the other side, microfinance and savings products support vulnerable households in the need for food and income shocks that lead to improved access to food security and improved nutritional status.

Financial inclusion further encourages job creation as a type of development. Thus, it comes neatly within the purview of SDG 8. By enabling small to medium-sized firms (SMEs) access to sources of credit, it enhances activities on opening entrepreneurship among excluded populations and thus, economic growth, innovation and job creation. It also enhances access to financing, fortifying households against economic shocks and produces more stable and inclusive economies.

Financial inclusion appears as one of the accelerators for accomplishing several Sustainable Development Goals. It can also alleviate poverty, reduce hunger and boost economic prosperity. This report offers several methods that governments might expand financial access to financial goods and services and construct inclusive financial systems to tap into the transformative ability of money to accomplish the universal development goal and create a just and sustainable future for all. Therefore, this may be of good benefit to scholastics and aspiring academics interested in asserting theoretical ideas in empirical data around the development economics environment.

1.2 About the Book

In the landscape of global development, financial inclusion stands as a beacon of hope and opportunity for billions around the world. It is a concept that transcends mere access to financial services; it is a catalyst for empowerment, resilience, and sustainable progress. As we approach a new era marked by the Sustainable Development Goals (SDGs), the need to promote financial inclusion is more urgent than ever.

This book, "Financial Inclusion—A Drive Towards Achieving the Sustainable Development Goals," emerges at a pivotal moment in history, where the intersection of finance and development holds the promise of transformative change. Within these pages, the authors commence a voyage to investigate the multifaceted dimensions of financial inclusion and its profound implications for realizing the aspirations encapsulated in the SDGs.

This book aims to clarify the crucial importance of financial inclusion in promoting the global agenda for sustainable development by conducting thorough analysis, examining empirical evidence, and presenting real-world case studies. The chapters in this volume attempts to characterize and subsequently compare the achievements in financial inclusion. Further, it also explores how inclusive finance plays a crucial role in attaining the SDGs, including reduction of poverty and hunger, and more importantly in advancing economic development.

Yet, this book is not merely a compendium of theoretical frameworks and policy prescriptions. It is a testament to the collective efforts of policymakers, practitioners, researchers, and advocates who are tirelessly working towards a more inclusive and equitable world. It is a celebration of innovation, resilience, and the indomitable human spirit that drives us towards progress, even in the face of daunting challenges.

As we deal with the intricacies of the contemporary world, it is our responsibility to acknowledge that financial inclusion is not only a social and economic necessity but also a moral necessity. It is a fundamental human right that lies at the heart of our quest for dignity, equality, and justice.

I earnestly aspire for this book to function as a wellspring of motivation, wisdom, and authority for individuals who endeavor to construct a globe that is both comprehensive and enduring.

1.3 Brief Outline of Chapters

The study encompasses the context of this subject through ten sections and appropriate sub-sections as needed. Initially it characterizes financial inclusion and the developmental goals that are targeted to be achieved by the end of 2030. This research work attempts to measure and subsequently compare the achievements in financial inclusion across the four different samples. Further, it aims to empirically observe the theoretical influence that financial inclusion may have on certain indicators of development. A brief description of each chapter is as follows:

Chapter I: Introduction

The formal introduction of Financial Inclusion and its associated scope, as well as the discussion of the sustainable development goals that are intended to be achieved by 2030, are the main topics of the first chapter. It also provides a thorough overview of the current situation with regard to the attainment of the sustainable development goals and financial inclusion in both Indian and foreign provinces. In addition, this chapter provides a thorough explanation of the purpose, scope, and, in the conclusion, the chapter plan.

Chapter II: Financial Inclusion: Trends and Progress

With the help of empirical data, the second chapter offers a thorough examination of India's financial inclusion history as well as relevant national experiences. It also looks at the varied steps that nations and international organizations—like the IMF, CGAP, GPFI, and others—have made to promote financial inclusion globally. It seeks to highlight FI's achievements both domestically in India's various states and abroad.

Chapter III: Sustainable Development: Achievements and Strategic Roadmap

In the third chapter, the sustainable developmental goals are presented as targets to be accomplished by the year 2030. It focuses on the progression of objectives and the corresponding accomplishments achieved over time in India and globally.

Chapter IV: Assessing Financial Inclusion for Inclusive Growth and Development

The fourth chapter attempts to evaluate the achievements in financial inclusion. This is done by first describing the phenomena and then engaging in a thorough discussion of the current methods used to measure the notion. Moreover, it aims to assess the level of financial inclusion specifically in the Indian states and Union Territories, the Asian continent, the combined Asian and European continents, and finally all five continents (Asia, Europe, Africa, North and South America) collectively. This evaluation is conducted by utilizing the existing resources to verify the theoretical propositions across the Indian provinces and internationally.

Chapter V: Comparative Analysis of Financial Inclusion Across Regions

In the fifth chapter, a comparative analysis is provided for the different units of evaluation. The objective is to emphasize the advancements in the levels of financial inclusion during the study period. Additionally, an analysis has been conducted to compare the accomplishments of different regions and socio-economic levels. This chapter has also focused on the evaluation of continental comparisons.

Chapter VI: Financial Inclusion and Sustainable Development Goals—Unfolding the Association

The sixth chapter attempts to reduce the dimensionality of the association and emphasize on Social Targets (Goal 1–6), Economic Goals (Goal 7–11) and Environmental Sustainability (Goal 12–15). It attempts to find the inter-linkages while searching for the driving forces that ensure synergic benefits.

Chapter VII: Financial Inclusion and Economic Growth (SDG 8)

The seventh chapter focuses on the correlation between financial inclusion and the economic development of the region. The thoroughly researched 'Finance-led-Growth argument' has been analyzed in conjunction with the examination of the classical 'Reverse Causality scenario'.

Chapter VIII: Financial Inclusion and Poverty Alleviation (SDG 1)

The purpose of the eighth chapter is to examine the traditional concept of reducing poverty based on income, which is facilitated through programs promoting financial inclusion.

Chapter IX: Financial Inclusion and Zero Hunger (SDG 2)

The ninth chapter is dedicated to examining the objective of attaining zero hunger through financial inclusion initiatives. In the pursuit of SDG 2, the connections between undernourishment (which signifies hunger) and factors such as food prices, availability, and the ability to buy healthy or nutritious food have been effectively managed.

Chapter X: Conclusion and Policy Recommendations

The tenth chapter concludes with a final remark and provides policy recommendations. The summary encompasses the entirety of the research and aims to provide significant insights for policymakers. These insights will enable policymakers to assess and tailor policies accordingly. Finally, this study also delineates the extent to which subsequent research can be pursued.

These sections were followed by a bibliography containing alphabetically arranged references.

1.4 Methodology

One of the severe challenges development strategies are often faced with is the quantification of its parameters and subsequently tracking its progress. Additionally, the characterization of the phenomena largely varies across the cross-sectional units (be it across the states and union territories of a country or countries from different continents). This study attempts to characterize financial inclusion in the most comprehensive way through engaging both the data from the demand as well as supply side and becomes the first academic endeavor that considers as many as possible countries (categorized on the basis of continents and income groups) over a longer period of time. The findings of the study offer empirical support to the theoretical propositions that adds the typical flavor often studies lack to serve.

Competition in any form creates a sense of excitement. This study offers comparisons across the continents, income groups, regional categorizations and more importantly across the period of time with respect to the average achievement scores in financial inclusion. Contemporary literature along with the previous literary evidences has not offered such a broader perspective of competition that might attract reader's attention.

The inter-linkages between the developmental goals are expected to have synergic benefits. Development economists particularly the scholars and potential academicians are genuinely in search of these kinds of contents those exhibit empirical validations of proposed/expected economic theory. This study is a classic exhibition of the strategies aimed to accelerate economic development (SDG 8), reduce poverty (SDG 1) and malnutrition (SDG 2) through improving the provision of financial services, especially in a developing economy set-up.

1.5 Scope of the Book

The research on financial inclusion spans over a broader spectrum, which emphasizes further on the potential of exploring novel ideas on the subject. Finance is a vital component of economic development, which further influences a variety of other developmental indicators. Therefore, it goes without saying that finance appears to be an important driver that either directly or indirectly contributes to development. As a result, it becomes fair to first guarantee that a certain level of penetration has been achieved by the regulated financial system before formulating plans to expand its services and, ultimately, fulfill the aim of development. Keeping this in mind, the research will seek to capture the degree of penetration across nations from five different continents, as well as across the Indian states and UTs, while also considering the possibility for future service expansions. A variety of studies with comparable aims have attempted to depict the amount of penetration obtained by the formal financial system, but they vary in important ways. Taking into account a distinct set of characteristics while also using various approaches, studies so far have sufficiently reflected the degree of performance obtained by nations. Though there has been a lack of agreement on what the factors should be or even on the acceptable approach that adequately measures the financial inclusion performances. The lack of unanimity is essentially the basis for the research to be carried out. Furthermore, the study presents a thorough picture and also seeks to fill the gaps either highlighted by existing literature as an additional topic of investigation or altogether missed in identifying the specific subject. Aside from emphasizing on the identification of significant variables and changes in the existing methodologies used to compute the degree of achievement, the current study has some additional characteristics those makes it a unique one. With a wide number of variables accessible for nineteen consecutive years, the study aims to identify a trend and further analyze whether there has been an increase or decline in terms of the accomplishment value in financial inclusion throughout the years. Further, the study attempts to perform a comparison based on the earning potential by categorizing the nations in terms of their income. Furthermore, considering countries from the continent of Africa, Asia, Europe, North and South America as well as different states and UTs of India belonging to the five different regions broadens the scope for further comparisons between the continents and the regions of India.

Because financial inclusion is theoretically expected to impact development, the current research seeks to validate the causal linkages between financial inclusion and specific development indicators (i.e., SDG 1, SDG 2, and SDG 8) among the 17 main objectives to be achieved by 2030. It basically proposes to evaluate (both the magnitude and the direction of impact) how financial inclusion reduces poverty (SDG 1), undernourishment (SDG 2) and promotes growth in the economy (SDG 8) through reducing the dimensionality of the association and emphasize on Social Targets (Goal 1–6), Economic Goals (Goal 7–11) and Environmental Sustainability (Goal 12–15). It attempts to find the inter-linkages while searching for the driving forces that ensure synergic benefits.

CHAPTER 2

Financial Inclusion: Trends and Progress

2.1 Background of Financial Inclusion in India

There is not much debate that the financial system has a considerable impact on access to banking services. A robust banking system that covers both urban and rural regions is critical for enabling commerce, increasing savings rates, and speeding economic development. India's progress towards financial inclusion was accelerated first by the nationalization of the Life Insurance Corporation of India in late 1956. Following this, banks were nationalized in 1969 and 1980, respectively, while non-life insurers were nationalized in 1972. India has been an active member of global forums that include the "Global Partnership for Financial Inclusion" (GPFI) and the "Organization for Economic Cooperation and Development" (OECD) in addition to supporting the concept of financial inclusion. Furthermore, it has served as a co-chair in GPFI's Regulation and Standard Setting Bodies, with Indonesia and the United Kingdom, and has been actively developing important policy recommendations on digitalization and regulation of financial inclusion. The RBI is also an essential member of the OECD's International Network for Financial Education (INFE), actively driving the operations of the four working groups. The National Strategy for Financial Inclusion (NSFI) adopted by the Financial Inclusion Advisory Committee (FIAC) in India serves as a model for the INFE. Indian policymakers have shown an unshakable commitment to inclusive development, culminating in the Pradhan

S. Ghosh et al., *Financial Inclusion*, Sustainable Development Goals Series, https://doi.org/10.1007/978-981-95-3385-5_2

Mantri Jan Dhan Yojana, the state-of-the-art national mission for incorporating the excluded sections within the ambit of formal finance (PMJDY), which went into effect in August 2014. The major goal was to offer basic financial services to all families by using the existing huge banking network and technical infrastructure.

2.1.1 Financial Inclusion—Indian Experiences

In only five years, the PMJDY managed to create 34.01 crore bank accounts, garnering a deposit of ₹89,257 crore (NSFI). The weekly opening of bank accounts reached 18,096,130, earning a position in the Guinness Book of World Records and providing a slew of services such as a ₹100,000 overdraft facility, accidental death and disability insurance, term-life insurance, and an old-age pension. To supplement the expansion of inclusivity, schemes such as PMSBY and PMJJBY have been created to cover the risks associated with unforeseeable future occurrences for a person as part of a full package with PMJDY. Pradhan Mantri Suraksha Bima Yojana (PMSBY) extends accidental fatalities and permanent disabilities insurance on a yearly renewing basis at an annual premium of ₹12. Pradhan Mantri Jeevan Jyoti Bima Yojana (PMJJBY), on the other hand, provides an annual life insurance policy for just ₹330. Subscriptions to these insurance offerings have been carefully priced to meet the requirements of the most vulnerable groups. Furthermore, to meet the financial requirements of the marginalized sectors in old age, programs such as the Atal Pension Yojana (APY) provide a monthly pension ranging between ₹1000 to ₹5000 for individuals beyond the age of 60.

Aside from measures aimed at individuals or households as economic units, sector-specific initiatives are also in existence. Capacity development programs, such as the National Mission for Capacity Building of Bankers for funding the MSME Sector (NAMCABS), focus on molding bankers' perspectives on funding and encouraging such promising economic sectors. A Certified Credit Counsellors (CCC) program was introduced to impart a professional perspective in creating financial plans and project reports. This plan would aid banks in analyzing and making more informed lending choices. Furthermore, the Pradhan Mantri Mudra Yojana (PMMY) plan, with a cap of ₹10 lakh, focuses on the financial needs of small business organizations. Credit accessibility to individuals is further encouraged through dedicated online websites such as 'Udyami Mitra' and 'psbloanin59minutes', both of which were established in April

2015. Among the notable efforts focusing on the development engine (i.e. the MSME sector) are interest subvention programs to minimize the cost of their borrowing and TReDS to address the issue of late payments.

Agriculture, as a major contributor to the Indian economy accounts for 15% of GDP, 11% of exports, and providing a living for half of the country's population, has been identified as a priority area for formal financing. Further, the scheduled commercial banks along with the foreign banks (having at least 20 branches) have been set out for achieving a lending target of 18% of their Adjusted Net Bank Credit (ANBC)[1] or Credit Equivalent Amount of Off-Balance-Sheet Exposure (CEAOBE), whichever is larger. This focus is reinforced by channeling formal credit of 8% of ANBC or CEAOBE, whichever is more, to marginal as well as small farmers. With a view to accommodate for emergency cash banks have been recommended to provide collateral-free lending to small and marginal farmers to a maximum of 1.6 lacs. Schemes such as the Kisan Credit Card (KCC) offer revolving cash credits for a period of 5 years on an annual renewal basis to provide appropriate and timely financial assistance for agriculture and other requirements. The KCC plan has been expanded to include animal husbandry and fisheries.

2.1.2 *FI in India—Some Empirical Evidences*

Financial inclusion is among the critical policy objectives that contribute to the overall aim of economic growth and social well-being. It is prominent in policies aimed at ensuring fair distribution of opportunities and, therefore, addressing inequality to some degree in both developed and developing nations. As part of a larger picture (i.e. social inclusion), the provision of financial services provides a safe refuge for savings while also bridging the gap between the deficit and excess of cash via loan extension. Such an arbitrage encourages capital creation and consequently economic growth, as well as poverty reduction by boosting the borrower's efficiency as an economic unit. Starting with the establishment of cooperatives, financial inclusion initiatives have developed to be compatible with shifting demographics, economic situations, and social requirements.

[1] Usually ANBC refers to a broader measure of credit extended by commercial banks (including loans, investments and other assets). The Reserve Bank of India uses the ANBC to emphasize (monitor and evaluate) on the lending to Priority Sectors by commercial banks on a quarterly and annual basis.

In light of recent improvements in financial inclusion, an entry point might be having initially access to a bank account to further avail a wide range of financial services. To further the aim for financial inclusion, the RBI has allowed for specialized licensing for Small Finance Banks and Payment Banks, which would better meet the usual demands of customers. However, in terms of fundamental banking operations, certain data might give a more promising representation of the current landscape of the available services. According to the recent "Financial Inclusion & Future of Financial Services in India - Vision 2030" summit, SCBs would operate 1.59 lakh branches by June 2022, with a combined deposit of well over 170 lakh billion rupees. This equates to roughly fifteen outlets for every 100,000 people. This is complemented by the presence of 2.17 lakh Automated Teller Machines, 47% of whom are located in rural and semi-urban areas, as well as being close to 32 million banks' Business Correspondents. Furthermore, as of 2021, 78% of Indians (attaining an age of 15 and above) own a bank account, up from 53% in 2014, following the World Bank's Global Findex Database. Banking access has been provided to almost every hamlet within a 5-kilometer radius in 25 states and 7 UTs, accounting for 99.94% of all identifiable settlements. Similar data from the census in 2001 and 2011 shows an improvement in household access to financial services. Households using total banking services climbed by 112% between 2001 and 2011, at a compounded annual growth rate (CAGR) of 7.8%. The CAGR for rural and urban families using banking services remained at 8.2 and 7.2% each year, respectively. In 2011, the proportion of households using banking services remained at 58.7% (compared to 35% in 2001), and the split between rural and urban areas remained at less than one-third and less than half, respectively, compared to 54.44 and 67.77% in 2001. However, according to NSSO statistics, an astounding 560 million individuals are still excluded from formal finance. Review of Fig. 2.1 shows the worldwide concentration of the unbanked population, where India carries a stake of 11%. It further highlights that such incidence is more prominent across the income groups those are more economically challenged (Ghosh & Sahu, 2022).

Thus, rather of focusing just on formal finance, we must also include aspects such as 'Usage' and 'Quality'. In this regard, creating an index that provides a thorough overview of the advances in this area has been common practice. The RBI's FI index emphasizes three dimensions: "access," "usage," and "quality." The CRISIL Inclusix that covers

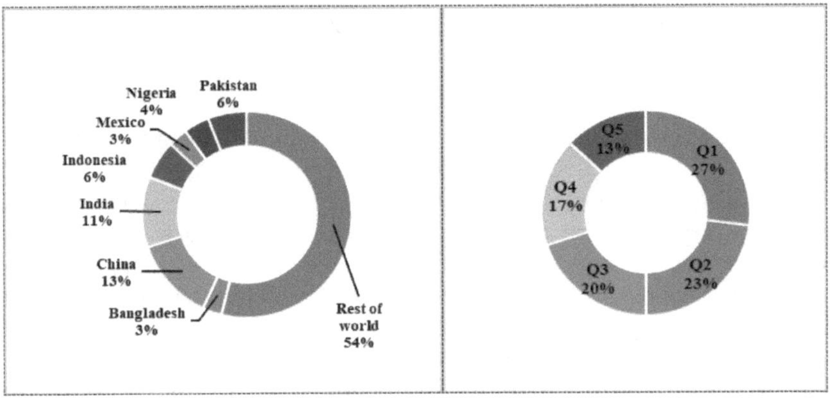

Fig. 2.1 Concentration of Unbanked Population (Left Image) and Unbanked Among the Poor (Right Image) Worldwide (Group Q1 to Q5 represents the unbanked population categorically from lowest income to highest income. *Source* Presented by Authors)

roughly 666 districts throughout India has produced an index on a scale of 0-100 incorporating variables into important criteria of 'Branch Penetration', 'Deposit Penetration', 'Credit Penetration', and 'Insurance Penetration'. However, the development of such an index can be traced back to Sarma's (2008) pioneering work, which focused on the three fundamental characteristics of financial inclusion (i.e. Availability, Accessibility, and Usage).

The following section seeks to offer a quick overview of the existing literature on the formation related to a financial inclusion index. It also introduces the likely factors that are theoretically significant for constructing such an index. In the first phase, we offer the most relevant layoff pertaining to the modulations for the establishment of a measure representing the most suitable achievement scores in financial inclusion. The section then presents several theoretically significant variables and attempts to explain their context in order to build a relationship between the variables and the eventual index to be produced.

2.2 Financial Inclusion—The Global Perspective

Financial landscape development is seen as a fundamental driver of economic progress (Demirguc-Kunt, 2006). However, empirical research on finance-growth proposition appears to be vague and mixed (positive, negative, or absent) (Bist, 2018). There is also discussion of the unidirectional and bidirectional causal linkages involving financial inclusion and economic expansions in the literature. There is still a lack of agreement on the indicators that correctly indicate the state of the financial sector. Access to low-cost financial services promotes formal saving and capital development, allowing the population as a whole to thrive (Abu-Bader & Abu-Qarn, 2008; Bittencourt, 2012; Pal & Vaidya, 2009). Furthermore, a wider client base increases the economy's inherent capacity to sustain shocks of unprecedented magnitudes while stabilizing the service provider (Levine, 1997; Yang & Yi, 2008). Service providers can grow more efficiently by decreasing expenses and offering a wide range of solutions to their customers at a reasonable or even reduced price. This is accomplished by seeing the neglected area as an opportunity to go further.

The notion of financial inclusion—alternatively, involving an omitted element within the framework of the official financial system—turned popular while dealing with the problem of restricted usage of branches of banks following financial sector liberalization in the early 1990s (Leyshon & Thrift, 1995). The primary goal was to avoid financial marginalization, which is often exploitative as a result of predatory loan practices.

Financial inclusion is also an effective strategy in light of the global agenda's aim of long-term economic prosperity. The central banks across developing and affluent countries have used a variety of ways to address the problem of exclusion, particularly involuntary exclusion. The United States of America's adoption of the legislation known as the *Community Reinvestment Act* (1977) ensured that basic banking services were offered to a greater audience. Another prominent endeavor addressing the particular issue of financial exclusion is a voluntary agreement established in 1996 by the *German Bankers' Association*. It is also worth noting the South African Banking Association's efforts to provide an inexpensive account with a bank known as *Mzansi*, as well as the French Government's introduction of the Law on Exclusion (1998) to broaden the reach of banking services (Maity, 2018). Similarly, a Financial Inclusion

Task Force was established in 2005 to investigate the state of financial inclusion in Britain as a whole.

Globally, 1.2 billion individuals acquired access to a bank account between 2011 and 2017, signifying considerable progress towards financial inclusion. By the year 2017, 69% of the population worldwide had an account. Digital means of accessing financial services majorly through the mobile phones have been a common practice of around 80 countries around the world. However, the divide between high-income and developing countries appears significant so far as usage of formal savings channels are concerned (Fig. 2.2).

A review of G20 nations on an average further show that less than half of the adults (48%) were able to answer 70% of the queries focusing to understand their financial knowledge. The understandability of the concepts of diversification and compounding was even lesser. Astoundingly it was found that 40% of the respondents do not understand diversification and a major chunk was unaware about their financial security after being retired. Except for the three countries, the divide between men and women was also significant. Men on an average were found to have almost 11-percentage point high financial knowledge compared to the women. The report reveals that on an average only three out of five (60%) uses a proper budget whereas 25% of them do not consider their ability to afford before a purchase decision. The awareness and usage of a particular financial product was also starkly different across the nations.

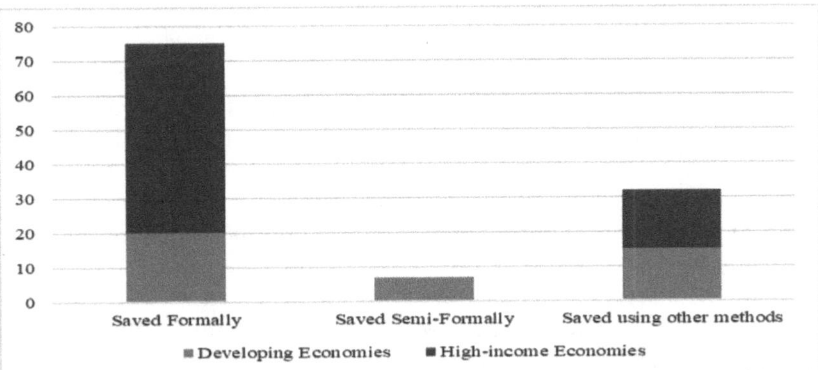

Fig. 2.2 Adults Savings Behavior (*Source* Global Findex Report, 2017)

Across the G20 members, it was observed that only two-thirds on an average were having accessibility to a payment product which had a formal recognition. So far as the usage of the same was concerned, it came down to 10% on an average, where standalone positions for Brazil were 8% and Indonesia at 2%. Further, observing the accessibility to informal savings and credit arrangements, it was observed that on an average 19% of the people from G20 member nations turned to family and friends which remarkably stood at 54% for India.

The World Bank Group with IMF during the spring meeting in 2015 with its collaborations (both public and private) committed to enable 1 billion people to give access to basic financial services under the Universal Financial Access (UFA) 2020 Plan. Along with 30 partners that aims to ensure private sector engagement in financial inclusion with 25 priority countries (mostly the LDCs) the aim was to foster the following issues:

1. Ensuring access to transaction accounts through a robust regulatory framework
2. Enlarging the network of point of accessibility
3. Developing financial independence
4. Increasing the outreach to disadvantaged sections (e.g. Women, Rural dwellers and like).
5. Ensuring social transfers to those accounts to make sure they do not become dormant.
6. Encouraging people to take advantage of safe and efficient digitized financial services.

Following these six broad targets under the UFA, 2020 several countries came up with customizations of their existing strategies to promote financial inclusion. In alignment with these major plans, India through launching of the Pradhan Mantri Jan Jhan Yojana aimed to initially provide every household with at least one bank account, focusing on the poor and marginalized. Further, through promotion of the cashless transactions by way of allowing rewards and incentives on using the Unified Payment Interface (UPI) and mobile wallets, it aimed to foster a digital payment ecosystem. Similarly, Kenya especially across its rural settlements through leveraging on its available technological infrastructure promoted the expansion of its mobile money services (e.g. M-Pesa and like), thus encouraging cashless transactions. Further, the government there in collaboration with the NGOs and corporate from the telecommunication industry has organized several campaigns on financial awareness and financial literacy to educate the end user on managing finances and using mobile banking. While mainstreaming the marginalized population into the formal financial system, the government of Brazil through relaxing the micro-credit regulations enables microfinance institutions to extend credit to low-income individuals and small businesses. Further, while addressing the percentage of adults without access to financial services in Nigeria, the government under the National Financial Inclusion Strategy (NFIS) outlined a comprehensive framework. Under this framework, banking services were offered to underserved areas through agents, which typically had the advantage of greater outreach without involving substantial capital investments. In line with the UFA 2020, Bangladesh aimed to reduce unemployment and consequently boost economic development through offering microcredit especially to the most vulnerable sections of the society (particularly women and poor). Further, to improve the access to financial services, the government encouraged mobile banking and digital wallets to facilitate transactions and savings electronically. In South Africa, following the UFA 2020, the National Financial Literacy Strategy (NFLS) was designed to escalate the financial knowledge of the population especially across the young and underserved groups. Dedicated policies were also into effect aiming to include the workers especially those working in the informal sectors into the formal financial system. Alike India, Philippines adopted the National Strategy for Financial Inclusion (NSFI) as a comprehensive framework promoting digital payments, microfinance, and insurance products. Additionally, the government of Philippines allowed the fintechs to innovate while keeping the regulatory sandbox active to ensure consumer protection and financial stability across the nation. Among others, the Mexican government through promoting the programs like 'Bienestar' allowed transfer of funds digitally. Further, the government played an active role to strengthen the regulatory environment while encouraging the participation of fintechs and microfinance institutions. The Egyptian experience is also in no way different. Access to conventional banking services was emphasized for the women and the youth, especially across the rural areas. Additionally, digital transactions were promoted to reach the underserved populations.

The cross-country initiatives commonly exhibits a commitment for collaboration between the conventional banks, fintech companies, NGOs and local governments to maximize the outreach and offering universal access to available resources. Strategic public-private partnerships along with relevant amendments have been made to keep a balance between innovation and consumer protection. Further, through continuous monitoring and evaluation of the initiatives taken, the regulatory bodies often keep the track on whether the need of the target population is met.

In the following section, we highlight briefly certain experiences of some countries with regards to their present status of achieving financial inclusion.

Bangladesh

Under the UFA 2020 vision, the landscape of financial services accessibility has drastically changed for Bangladesh. Following Findex 2021, it is observed that among 120.63 million population, the country had 56.94 million adults which were unbanked. However, share of adults with a transaction account has grown from 32% in 2011 to 53% in 2021. This acceleration has magnified especially when transactions taken place electronically is looked upon. From 34% in 2017, it has risen to 45.3% in 2021. Yet the gender gap is shrinking, since, the transaction accounts held by women has grown to 43.5% in 2021 from 36% in 2017. For men, the same proportion has fallen from 65% in 2017 to 63% in 2021. With a comprehensive microfinance network and emphasis especially on the rural belt, the Government of Bangladesh has significantly improved accessibility to formal credit which fosters employment.

India

Following the data from the World Bank in 2021, in India, 229.3 million out of a 1019 million (i.e. almost 23%) are unbanked. However, over the recent decade or so it has witnessed tremendous growth in the percentage of adults with a transaction account as well as the transactions those are taking place electronically. The number of transaction account grew from 35% in 2011 to 78% in 2021, albeit a fall was observed during 2017–21. The transactions taking place electronically grew more than two and a half times from 13% in 2011 to 35% in 2021. Female account holders grew from less than 26% in 2011 to 78% in 2021, whereas for the male this growth was from 44% in 2011 to 78% in 2021, albeit a fall was observed during 2017–21 for the male population. This growth was majorly backed by the aggressive policy initiatives by the government and subsequent implementation by the commercial banks under the regulatory guidelines of the Reserve Bank of India (The Central Bank of the country). Among some of the promising initiatives the MGNREGA scheme, the Pradhan Mantri Jan Dhan Yojana (PMJDY), Business Correspondent (BC) and Business Facilitator (BF) models, Financing through the Self-Help Groups, Organizing Financial Literacy programs are noteworthy (Datta & Sahu, 2023).

Indonesia

The position of Indonesia is no different. Among a population of 202.6 million, 97.3 million are unbanked as of 2021 survey. From a 20% of adults with a transaction account in 2011, it rose up to 52% in 2021. The percentage of population performing transactions electronically rose from 12% in 2011 to 37% in 2021. The proportion of male and female account holders in 2021 stood at 51% and 52% respectively. Although being from the leading nations apart from the spread of microfinance networks the approach from commercial banks somehow discourages the efforts towards financial inclusion (Datta & Sahu, 2023; Rosengard & Prasetyantoko, 2011).

China

Research of Chinese experiences found that boosting financial inclusion was formally stated in 2013 during the Third Plenary Session of the Communist Party of China's (CPC) 18th Central Committee. Further, the State Council of China on 31st December, 2015 introduced the *Plan for Advancing the Development of Financial Inclusion* covering a period from 2016–2020. The Plan outlines a strategy for increasing financial product and service mobility, satisfaction, and effectiveness, with a focus on MSEs, rural residents, economically disadvantaged urban residents, the poor, the differently-abled, and the elderly as an intended recipient of financial inclusion. To lower the cost of operations and communication costs of servicing consumers and to drive technological and product innovation, a wide variety of regulatory measures, involving monetary, tax, and supervisory constraints, have been imposed. The People's Bank of China has implemented a number of steps to motivate financial institutions to cater better services to MSEs, the farming industry, agriculturalists, and rural regions, including revised necessary reserve ratios, loan restructuring, and rediscounted lending. Since 2018, a steady drop in commercial banks' necessary reserve ratios has lowered lending rates to meet the demands of neglected market groups. The gradual reduction of these reserve percentages increases the available cash in banks' hands, which may better support the financing of agriculture, microenterprises, and rediscounting loans. This had resulted into a huge leap from 64% in 2011 to 89% in 2021 in terms of owing a transaction account. The male-female participation rates have also shown similar growth. The growth in having a transaction account for male from 2011 to 2021 stood at 68–90% whereas for the female counterpart it was at 60–87%.

Philippines

The number of unbanked populations across Philippines has been 37.3 million out of a total of 76.7 million as of 2021. The account ownership has been surged from 27% in 2011 to 51.4% in 2021. The proportion of women with active account holdings grew from 34% in 2011 to 47% in 2021 whereas for male this stood at 19% in 2011 to 56% in 2021. Access to formal credit appears among the major hindrances that financial institutions suffer from. However, securing a lending mandate of 8% towards the MSME has changed the landscape drastically so far as access to formal credit is concerned (Hannig & Jansen, 2010).

Myanmar

The status of financial inclusion has not been promising across Myanmar. Out of a total of 40.5 million of population, 21.2 million has been found to be unbanked (representing 52%). The account ownings especially the usage of digital platforms for performing a transaction has been astoundingly reported at 47.8% and 40% respectively in 2021. The growth in account ownership by an individual with 15+ age was nearly twice (22.8% in 2014), whereas performing a digital transaction (both receipts and payments) has grown 10 times (4% in 2014). Female account ownership grew from 17% in 2014 to 46% in 2021 and that for male it stood from 29% in 2014 to 50% in 2021. Following the IMF's country report poor economic conditions, lack of agricultural finance coupled with weak financial regulations, high inflation in association with inadequate reform policies has stuck the drive for financial inclusion (IMF Country Report,

2021). However, from 2017 onwards, Myanmar experienced a double digit growth in account ownership (a 22-percentage point increase) with a meager gender gap so far account ownership is concerned. Yet the income gap in account ownership is still standing in the double digit figures (ranging from 45%–66%). The change in the financial landscape in Myanmar is however driven by the participation of the young people as compared to the older ones. Younger people exhibits account ownership as much as 11% point higher than compared to the older counterparts.

Turkey

Following the 2021 survey, 16.6 million people in Turkey are still unbanked out of a population of 64.1 million. Moreover, 74.1% have access to an active transaction account and 67.6% uses digital means to transact money. The proposition of account holdings has certain improvements from 2011 to 2021 across the male population though insignificant (from 82 to 85%). Among the unbanked population yet females holds a significant share (71%). However, the female participation in account ownership (from 33 to 63%) has significantly improved. So far the performance of digital transactions is concerned a hike of 18 percentage points is observed between 2014 and 2021. Insurgence of Participation Banks (PB's) has been the driving force behind the achievements. These banks acts as an enabler of finance where the conventional banks fail to operate especially due to religious sensitizations (Aysan et al., 2013).

Pakistan

However, Pakistan's experiences in attaining financial inclusion are not entirely gratifying. As a member of the AFI, it introduced the NSFI in 2015. While achieving the country's target of obtaining universal financial access, Pakistan's primary NFIS goal is to extend formalized financial services to a minimum of 50% of individuals, especially women, and youth, particularly to boost the share of SME loans in lending by banks to 15% by 2020. However, significant features emerge from the IMF's financial access figures. In accordance with Global Findex 2021, only twenty one percent of Pakistanis operate a formal account, and women in Pakistan make up more than half of the total unbanked population. Though the major of the respondents (especially females) cite that in their family a member already have and operate such kind of an account. Penetration of digital financial products is also considered to be quite low. In Pakistan, just 8% of individuals possess a debit card, while only 2% use the mobile phone or the internet to check their accounts statement. In terms of credit allocation, just 7% of total available lending funds are channeled to Micro, Small, and Medium Enterprises (MSME). However, there is some respite in digital banking, where 14% of male in Pakistan have a mobile account as compared to the South Asian average of less than 17%.

Vietnam

The financial position of Vietnam as a country is largely volatile due to political instability, geographical heterogeneity and cultural disparities (Tsang et al., 2004). Following the survey of 2021, out of a 74.8 million population 32.9 million are still unbanked. Adults with a transaction account have merely shown a growth from 21% in 2011 to 56% in 2021. Growth for the female has been from 19% in 2011 to 53% in 2021, whereas for the male it shows a growth from 24% in 2011 to 60% in 2021. Further, the liberalization of controls that previously had been rested upon the government has made the country promoted to the category of lower-middle income group from the low-income group. As a result, the proportion of population owning a debit card rose from 15% in 2011 to 49% in 2021. For the same age group, the usage of debit card has escalated from 4% in 2014 to 19% in 2021. The in-store card purchases stands at 77% in 2021 and individuals transacting electronically has doubled to 46% in 2021 from 23% in 2017. In 2021, nearly 96% of the population above the age of 15 owns a mobile phone of which 80% has an active internet connection. Additionally, 48% of the active internet users in 2021 have made online payments while making purchases online which in 2017 were only at 15%. Capitalizing on the telecommunication infrastructure, players like Viettel, VNPT, and Mobifone have changed the payment landscape of Vietnam. As of May 2024, around 8.8 million users are onboard of which 6.3 millions are from geographically constrained areas (mostly from mountainous and remote areas) which represents 72% of the total user base (Vnn, 2024). Consequently, the number of transactions (including deposits, withdrawals, transfers, and payments) has exceeded 119 million valuing 4.462 trillion VND (approximately 186.5 million USD) (Vnn, 2024).

South Africa

Although having more than an average population above the upper-middle income group the proportion of unbanked stands at 14.6% (i.e. 6.2 million out of 42.2 million) as of 2021. Following the 2021 survey, the account owing stands at 85.4% whereas the performance of digital transactions remains with 80.8% of the population. The female account owing (86.2%) is relatively more than the male counterpart (85%). However, the factors leading to such results are multiple. Stable economy coupled with sustainable financial inclusion policies, cross border banking is some of the prominent strategies that has led to the present progressive stage (Beck et al., 2014).

Kenya

The financial inclusion attainments are relatively stable across Kenya serving 33 million population of which merely 6.9 million are unbanked in 2021. The introduction of M-Pesa back in 2007 has drastically changed the landscape of bank-based transactions covering 82% of the population accessing directly a bank account (Hannig & Jansen, 2010). This was furthered by initiatives that were promoting small entrepreneurs (Kalunda, 2014). Transactions taking place electronically has also gone up to 78% as of 2021. Account ownership by men has also grown from 46% in 2011 to 83% in 2021, whereas for women this has shown a growth from 39 to 75% during 2011–21.

Nigeria

Nigeria is among the promising nations of Africa so far as the achievements in financial inclusion are concerned. Following the 2021 survey, 63.7 million of Nigerians are still unbanked from among a total population of 116.5 million. The number of active account holders is at 40.5%. However, the growth in electronic transactions has been slowed down to 34% in 2021 from 37% in 2014. Female participation has also slightly improved with active account holdings growing from 26% in 2011 to 35% in 2021. Although growth in the male account holdings has been fair enough from 33% in 2011 to 55% in 2021. The advancement in financial inclusion is however backed by sufficient financial awareness campaigns and promotion of cashless transactions through proper incentives (Bayero, 2015). Although the impact of spillovers due to oil price shocks has been a major economic drawback that indirectly penalizes the growth of financial inclusion (Adeola & Evans, 2017).

Tanzania

The economic status of Tanzania is mainly represented by the most populous low-income group which further exhibits 16.2 million of unbanked population out of a total of 33.7 million nationwide population. The percentage of transaction account grew from 17% in 2011 to 52% in 2021. Almost 50% of the population in 2021 has participated in performing transactions electronically. Gigantic growth has been observed in account owing for both male (from 21% in 2011 to 59% in 2021) and female (from 14% in 2011 to 46% in 2021). The major driving force behind such inclusive financial environment has been the insurgence of mobile banking (De Koker & Jentzsch, 2013).

Morocco

This country on an average has 44% of its adults accessing a bank account of which 30% uses digital means to transact money as of 2021. The share of men and women accessing a bank account stands at 56% and 33% of their respective populations. Besides the promising initiatives adequate regulatory policies, vigilant supervision, reduction of interest rate caps on microloans along with infusion of technology to the conventional banking system has transformed the sculpture of the financial system (Pearce, 2011).

Egypt

The number of active account holders has risen from 12.18% to 24% during 2014–21. Similar sort of growth has been marked in the number of electronic transactions performed which had risen from 8 to 20% during 2014–21. However, the country exhibits a higher proportion of population that is still unbanked in relation to its total population i.e. 49.08 million out of 67.6 million as of 2021. The account holdings for the male and female separately shows a growth of approximately thrice from 13 to 39% and from 7 to 27% during 2011 to 2017 respectively. However, there has been a fall in the account owning for both male and female during 2017–21. This growth has however been attributed to the attitude of the government, the existing regulatory framework and the orientation towards digitization with especial reference to the MENA (The Middle-East and North Africa region) countries (Pearce, 2011).

Ethiopia

Following the Findex, 2021, 46% of the adults on an average own a transaction account and merely a 20% share of population performs transactions electronically. The account ownership across the genders varies largely. 55% of the male own a transaction account whereas this share across the female category stands only at 39%. A large share (almost 54%) of population (i.e. 37.3 million out of 69 million) are still unbanked. Therefore, the desired outcome is yet to receive, so far FI achievement scores are concerned. Introspection to the facts exhibits the non-availability of suitable financial products, conservative lending policy and limited coverage restricts the intended expansion (Zwedu, 2014).

Mozambique

Following the 2021 survey, the number of active account holders stood at 42.6% whereas 42% of the population in the country is using any of the electronic gateways to perform a transaction. The male female account ownership for the same period kept at 61% and 39% respectively. Introspection to the facts shows that Information and Communication Technology (ICT) has been the major driver behind the inclusive environment especially across Mozambique (Kpodar & Andrianaivo, 2011).

Cote d'Ivoire

Geographically being from the sub-Saharan Africa as of 2021, the country has a substantial degree of population (almost 49%) out of the purview of formal financial system. The country serves almost a population of 15.4 million of which 7.58 million have been found to be unbanked. Following the Findex 2021, on an average 51% of the adults holds a transaction account. However, the holding of a transaction account significantly deviates across the genders with 64% of male on an average hold a transaction account whereas for the female category it stands at 37%. The performance of electronic transactions on an average for the adult population is at 48%. Even with a comprehensive coverage by the micro finance institutions (Lafourcade et al., 2005), the dependency on the informal finance has been much higher for both at the individual as well as institutional levels. Among the several factors identified the structure of the economy and society, the level of understandability and awareness as well as the management of financial institutions and the regulatory environment has been cited as the prominent reasons behind such a standing (Demirguc-Kunt & Klapper, 2013).

Rwanda

Following the data from 2017, the total number of unbanked populations in Rwanda is more than 50% of its total population (i.e. 4.1 million out of 7 million). However, 50% of them have access to a transaction account and nearly 39% of them perform transactions electronically. The progress in the account ownership for female stood at 45% in 2017 from 28% in 2011 whereas for male it remained at 56% in 2017 from 38% in 2011. Although the country suffers from severe poverty, hunger and mortality rates which reduces its economic capacity to a great extent and put restrictions on the expansions of formal financial system (Alter, 2015).

Zambia

Financial services especially for the economically disadvantaged or the low-income group through mobile be it access to a savings account or banking services as a whole has changed the financial landscape of this country dramatically (Ouma et al., 2017). From 21% in 2011, the account owning has gone to 49% in 2021. This growth has been corroborated by the rise in transactions performed digitally, which has grown from 29% in 2014 to 46% in 2021. The male proposition of account owing as of 2021 has been still on the higher side (i.e. 52%) than compared to the female counterpart (i.e. 45%). However, the number of unbanked populations has still been a major concern (5.25 million from among a total population of 10.3 million).

Congo-Kinshasa

Being the second largest African country situated in the Central Africa this nation has an unbanked population of 35.44 million from a total population of 48.55 million as of 2021. The holding of a formal financial account on an average almost rises as much as seven times from 4% in 2011 to 27% in 2021. The average usage of a digital platform to perform transactions is at 26% as of 2021 (rising from 15% as of 2014). Whereas the position of the holdings of formal financial accounts for male grew from 5% in 2011 to 34% in 2021 and for female it moved from 3% in 2011 to 20% in 2021 (World Bank Group, 2024).

Brazil

Brazilian experience demonstrates an amazing development of the national correspondent banking network, microfinance institutions, and cooperative growth, which has raised FI ratings over the last decade or so. Government transfer programs, such as *Bolsa Familia*, are among the significant measures that raise income, particularly across the surface levels of the economic pyramid, resulting in account penetration levels of 56% (the highest among developing nations) which rose further to 84% by 2021. However, the extent of electronic transactions performed has not shown substantial growth during the 2014–21 sessions. It had only an increment from 59 to 77%.

Mexico

The unserved population so far as banking services is concerned stands at 48.8 million out of 95.6 million representing 51% of exclusion as of 2021. Out of this 49% of financially inclusive population, 46% has been actively using the transaction accounts and 44% have used any of the electronic means to perform a transaction. The gender disparity so far as growth in the account holdings are concerned between male and female shows a relatively better position for the female (growing from 22% in 2011 to 42% in 2021, whereas this for the male is found at 33% in 2011 growing to 56% in 2021). Availability of different delivery channels, financial literacy among the population along with tremendous growth in the private sector has been among the common citations from previous literature (Chibba, 2009).

Peru

Following the 2021 survey, 52% of the adult population in Peru has an active account of which 49% of people are actively making transactions electronically. The unbanked population stood at 10.7 million out of 24.8 million. The growth in the account holdings for females during 2011–21 was from 18 to 53%, whereas for male it remained from 23 to 62%. Penetration of financial services for Peru has been on the relatively lower side due to geographical extremities, socio-economic structures and many other factors that has undermine the inherited potential (Izquierdo & Tuesta, 2015).

Colombia

Following the World Bank Data 2021, the total number of unbanked population (15.96 million) to total population (39.6 million) is considerably higher for Colombia. However, this present situation shows doubling the growth in holdings of transaction accounts from 30% (in 2011) to 60% (in 2021). This evidence is further substantiated by the electronically performed transactions from 33% in 2014 to 52% in 2021. The proportion of female account owning has grown from 25% in 2011 to 56% in 2021. This for the male counterpart remained at 64% in 2021 from 36% in 2011.

This growth in penetration of financial services was mainly driven by reducing the restrictive collateral requirements, reduction in the cost of finance, promotion of the banking and digitization of payment facilities along with the conditional cash transfer programs (Karpowicz, 2014; Maldonado & Tejerina, 2010).

Along with each country's central bank, the IMF, G-20, International Finance Corporation (IFC), Alliance for Financial Inclusion (AFI) and Consultative Group to Assist the Poor (CGAP) are some of the international groups that have taken on the task of collecting micro-level data so that national achievements in financial inclusion can be evaluated and compared. However, comparisons across nations or governments within a geographical region are problematic due to the lack of agreement on how to measure the successes of financial inclusion.

2.3 Challenges in Financial Inclusion

Infrastructural limitations pose a significant barrier to financial inclusion, particularly in rural and underserved areas. The reason behind the poor performance in outreach gets vibrant especially across the rural and low-income areas. This evidence emancipates from the fact that the majority of the individuals are devoid of accessibility to basic banking services due to the physical absence of a proper banking outlet. About 1.7 billion adults worldwide have not yet been banked, and a significant part of this population lives in areas without adequate banking facilities (World Bank, 2021). This situation is further worsened by the digital divide where people are not able to make use of online and mobile banking and payment service, mainly due to less availability of internet connectivity, less access to technology devices, and other similar infrastructures. According to a McKinsey & Company's 2020 report, developing digital infrastructure can significantly enable greater access to financial services among all, especially in poor nations. Even in a developed area, inadequate transport cannot facilitate people to reach to a bank; therefore, it cannot be possible for someone to access formal financial sector.

Besides the above, cultural aspects also impact financial inclusion. To a large extent, in many communities, previous experiences by financial institutions have resulted in mistrust, and although most people know of banking systems, these are hardly fully interacted with (Khan et al., 2018). Social norms and cultural practices influence perceptions and even the management of money, which in turn affects behavior in terms of saving or borrowing. For instance, in some cultures, people prefer cash as they feel that cash transactions are safer than bank transactions (Buchmann & Hannum, 2001). Language is also an entry barrier to accessing financial services because products are mainly marketed in dominant languages, leaving others feeling excluded and unsure of their options. The IFC has argued that culturally appropriate financial products should be developed to engage more marginalized groups (IFC, 2019).

Gaps in financial literacy are also another major challenge to financial inclusion. Most people do not have a basic understanding of financial concepts such as interest rates, savings, and investment opportunities, which prevent them from effectively using the services available. According to OECD (2017), financial literacy rates vary widely across countries, with many lacking the skills needed to manage their finances effectively. This gap is usually perpetuated by educational systems that do not include financial education in their curriculum, leaving young people unprepared to navigate financial landscapes. Furthermore, misinformation about financial products leads to poor decision-making and pushes people further into poverty cycles (Klapper et al., 2015). Without targeted financial literacy programs, many individuals will continue to struggle with understanding and accessing financial services, perpetuating exclusion from essential economic opportunities.

References

Abu-Bader, S., & Abu-Qarn, A. (2008). Financial development and economic growth: The Egyptian experience. *Journal of Policy Modeling, 30*(5), 887–898. https://doi.org/10.1016/j.jpolmod.2007.02.001

Adeola, O., & Evans, O. (2017). Financial inclusion, financial development, and economic diversification in Nigeria. *The Journal of Developing Areas, 51*(3), 1–15.

Alter, A. (2015). *Financial inclusion and development in the CEMAC*. International Monetary Fund.

Aysan, A. F., Dolgun, M. H., & Turhan, M. I. (2013). Assessment of the participation banks and their role in financial inclusion in Turkey. *Emerging Markets Finance and Trade, 49*(5), 99–111.

Bayero, M. A. (2015). Effects of cashless economy policy on financial inclusion in Nigeria: An exploratory study. *Procedia-Social and Behavioral Sciences, 172*(8), 49–56.

Beck, T., Senbet, L., & Simbanegavi, W. (2014). Financial inclusion and innovation in Africa: An overview. *Journal of African Economies, 24*(1), i3–i11.

Bist, J. P. (2018). Financial development and economic growth: Evidence from a panel of 16 African and non-African low-income countries. *Cogent Economics & Finance, 6*(1), 1–17. https://doi.org/10.1080/23322039.2018.1449780

Bittencourt, M. (2012). Financial development and economic growth in Latin America: Is Schumpeter right? *Journal of Policy Modelling, 34*(3), 341–355. https://doi.org/10.1016/j.jpolmod.2012.01.012

Buchmann, C., & Hannum, E. (2001). Education and stratification in developing countries: A review of theories and research. *Annual Review of Sociology, 27*(1), 77–102.

Chibba, M. (2009). Financial inclusion, poverty reduction and the millennium development goals. *The European Journal of Development Research, 21*(2), 213–230. https://doi.org/10.1057/ejdr.2008.17

Datta, S., & Sahu, T. N. (2023). *Financial inclusion and livelihood transformation: Perspective from microfinance institutions in rural India*. Springer Nature.

De Koker, L., & Jentzsch, N. (2013). Financial inclusion and financial integrity: Aligned incentives? *World Development, 44*, 267–280.

Demirguc-Kunt, A. (2006). *Finance and economic development: Policy choices for developing countries*. The World Bank. https://doi.org/10.1596/1813-9450-3955

Ghosh, S., Sahu, T. N. (2022). Financial Inclusion: Measurement and Comparative Analysis of Countries from Europe and Asia. In: Mukherjee, P. (eds) Revisiting the Indian Financial Sector. India Studies in Business and Economics. Springer, Singapore. https://doi.org/10.1007/978-981-16-7668-0_3

Demirguc-Kunt, A., & Klapper, L. (2013). Measuring financial inclusion: Explaining variation in use of financial services across and within countries (The Global Findex Database). *Brookings Papers on Economic Activity*. Washington, DC: Brookings Institution.

Hannig, A., & Jansen, S. (2010). *Financial inclusion and financial stability: Current policy issues* (Working Paper No. 259). Asian Development Bank.

IFC. (2019). *Financial inclusion in Africa: The role of women*. International Finance Corporation.

IMF Country Report No. 17/153 browsed on 23.10.2021 at www.imf.org/-/media/Files/Publications/CR/2017/cr1715.

Izquierdo, N. C., & Tuesta, D. (2015). Factors that matter for financial inclusion: Evidence from Peru. *Aestimatio: The IEB International Journal of Finance, 10*, 10–31.

Kalunda, E. (2014). Financial inclusion impact on small-scale tea farmers in Nyeri county, Kenya. *World Journal of Social Sciences, 4*(1), 130–139.

Karpowicz, M. I. (2014). *Financial inclusion, growth and inequality: A model application to Colombia* (Working Paper No. 14/166). International Monetary Fund.

Khan, S., Akter, S., & Hossain, M. (2018). Financial inclusion and the role of microfinance in Bangladesh: A comparative study. *Journal of Financial Services Marketing, 23*(2), 1–14. https://doi.org/10.1057/s41264-018-0026-1

Klapper, L. F., Lusardi, A., & Panos, G. A. (2015). Financial literacy and the financial crisis. *Journal of Banking & Finance, 60*, 1–13.

Kpodar, K., & Andrianaivo, M. (2011). *ICT, financial inclusion, and growth evidence from African Countries* (Working Paper No. 11/73). International Monetary Fund.

Lafourcade, A. L., Isern, J., Mwangi, P., & Brown, M. (2005). *Overview of the outreach and financial performance of microfinance institutions in Africa*. Microfinance Information eXchange (MIX), Washington, DC. http://www.mixmarket.org/medialibrary/mixmarket/Africa_Data_Study.pdf. Accessed on September 16, 2022.

Levine, R. (1997). Financial development and economic growth: Views and agenda. *Journal of Economic Literature, 35*(2), 688–726. http://www.jstor.org/stable/2729790

Leyshon, A., & Thrift, N. (1995). Geographies of financial exclusion: Financial abandonment in Britain and the United States. *Transactions of the Institute of British Geographers, New Series, 20*(3), 312–341.

Maity, S. (2018). *Role of public sector banks and private sector banks in financial inclusion in India* (Doctoral dissertation, Vidyasagar University). http://hdl.handle.net/10603/291690

Maldonado, H. J., & Tejerina, L. (2010). *Investing in large scale financial inclusion: The case of Colombia*. Inter-American Development Bank.

McKinsey & Company. (2020). *Global Payments 2020: Fast Forward to the Future*.

OECD. (2017). *Financial literacy for youth: The role of schools*. Organization for Economic Cooperation and Development.

Ouma, S. A., Odongo, T. M., & Were, M. (2017). Mobile financial services and financial inclusion: Is it a boon for savings mobilization? *Review of Development Finance, 7*(1), 29–35.

Pal, R., & Vaidya, R. R. (2009). Outreach of banking services across Indian states, 1981–2007. *India Development Report, 2011*, 141.

Pearce, D. (2011). *Financial inclusion in the Middle East and North Africa: Analysis and roadmap recommendations* (Working Paper No. 5610). The World Bank.

Rosengard, J. K., & Prasetyantoko, A. (2011). If the banks are doing so well, why can't i get a loan? Regulatory constraints to financial inclusion in Indonesia. *Asian Economic Policy Review, 6*(2), 273–296. https://doi.org/10.1111/j.1748-3131.2011.01205.x

Sarma, M. (2008). Index of Financial Inclusion. Working Paper No. 215, New Delhi: Indian Council for Research on International Economic Relations. https://www.econstor.eu/bitstream/10419/176233/1/icrier-wp-215.pdf

Tsang, E. W., Nguyen, D. T., & Erramilli, M. K. (2004). Knowledge acqui-sition and performance of international joint ventures in the transition economy of Vietnam. *Journal of International Marketing, 12*(2), 82–103.

Vnn. (2024, July 12). Vietnam sees surge in mobile money adoption with 8.8 million users. *Copyright (C) by Bộ Thông Tin Và Truyền Thông*. https://english.mic.gov.vn/vietnam-sees-surge-in-mobile-money-adoption-with-88-million-users-197240712153444122.htm

World Bank. (2021). *Global Findex Database 2021: Financial Inclusion. How People Use Financial Services*.

World Bank Group. (2024). Data. In *World Bank*. https://www.worldbank.org/en/publication/globalfindex/Data#sec3

Yang, Y., & Yi, M. (2008). Does financial development cause economic growth? Implication for policy in Korea. *Journal of Policy Modeling, 30*(5), 827–840. https://doi.org/10.1016/j.jpolmod.2007.09.006

Zwedu, G. A. (2014). *Financial inclusion, regulation and inclusive growth in Ethiopia* (Working Paper No. 408). Overseas Development Institution.

CHAPTER 3

Sustainable Development: Achievements and Strategic Roadmap

3.1 Introduction

The Sustainable Development Goals (SDGs), a universal set of 17 targets taken by the member countries of the United Nations in 2015 aiming to achieve a global agenda by the end of 2030. The SDGs exhibit a framework for achieving sustainable development by resolving the pressing issues on poverty, inequality, environmental degradation, climate change, and social injustice (Fund, 2015). These goals appear as an outcome of the historic summit held at New York among the 193 member countries on 25th September 2015 and are designed to be interconnected and comprehensive, recognizing the complex and interrelated nature of these challenges (Stafford-Smith et al., 2017). They are also based on the principle of leaving no one behind and are meant to be inclusive and equitable, ensuring that the needs of all people, including marginalized communities, are prioritized (Osborn et al., 2015; Weber, 2017). It is commonly understood that the global success of the 2030 agenda would be heavily reliant on India's progress on the SDGs (United Nations, 2024a). It is due not just to the sheer magnitude of the Indian population, but also to the strength and endurance of the Indian economy. Furthermore, worldwide India takes the lead in international climate action programs. The philosophical background behind these targeted goals also forms part of the India's longstanding tradition and heritage of "*Vasudhaiva Kutumbakam*" indicating the world to be an undivided

S. Ghosh et al., *Financial Inclusion*, Sustainable Development Goals Series, https://doi.org/10.1007/978-981-95-3385-5_3

family and has been long practiced even before its official inception globally. In this regard on September 2015, the speech of Hon'ble Prime minister of India, Shri Narendra Damodardas Modi at the UN's Sustainable Development Summit is of great significance. To quote him,

> Much of India's development agenda is mirrored in the Sustainable Development Goals. Our national plans are ambitious and purposeful; Sustainable development of one-sixth of humanity will be of great consequence to the world and our beautiful planet.

3.2 Genesis

The Millennium Development Goals (MDG's) were among the first steps focusing to rescue from the key universal challenges (8 targets) faced by the nations globally within a period of 15 years (2000–15). This further got replaced by The Sustainable Development Goals to carry forward the momentum and concentrate on the issues with a greater detail from 2012 onwards (Sachs, 2012). The inception of these 17 targets was primarily drawn from the outcome of discussions in the UN Conference on Sustainable Development at Rio de Jenairo, Brazil on the occasion of 20th Anniversary of Earth Summit titling "*The Future we Want*". A unique and elementary structure of involvement was initiated to promote a comprehensive discourse on the formation of SDGs. As a consequence, a 30-member Open Working Group (OWG) was formulated to draft the proposal under the aegis of the UN General Assembly (UNGA). India was also a participant in the OWG. At the 69th session during September 2014, the report of OWG got a formal acceptance by the UNGA after thirteen meetings held during March 2013 to July 2014. Global leaders formally supported the resolution titled "*Transforming our World: The 2030 Agenda for Sustainable Development*" at the special UN Summit held in New York from September 25 to 27, 2015, which enunciates 17 SDGs and 169 linked targets.

The developmental goals were mainly screened under a proper evaluation and surveillance to achieve the desired targets. The resolution at the United Nations had promoted research from the elementary level (i.e. regional levels) to macro level (i.e. both the national and at global levels). To measure SDG progress, a Global Indicator Framework comprising 243 indicators has been designed (Indicators, 2016; Kim, 2023). On a worldwide scale, the High-Level Political Forum (HLPF) on the Sustainable

Development Goals was established. It is crucial in the global follow-up and assessment of the SDGs, as well as in offering further guidance. As a part of the standard follow-up and review mechanism countries are encouraged to conduct periodic assessments of achievements across the regional as well as the national levels (Eustachio et al., 2019). These evaluations will be conducted on a voluntary, state-led basis by nations that are both developed and developing. As a result, they are known as Voluntary National Reviews (VNRs). These were part of the execution of the 2030 agenda through sharing experiences, those include the accomplishments, challenges and the lessons learned in that connection.

3.3 Sustainable Development Goals—India's Journey

The SDG's primarily aims to emphasize on the elimination of poverty, ensuring access to equitable education, safe drinking water with appropriate sanitization facilities, providing access to alternative sources of conventional energy that ultimately leads to ecologically sound communities and building up of sustainable cities. These goals were so designed to have certain specific targets and pre-defined indicators that measure the progress of initiatives taken. The design of the SDGs is in a way that accomplishment in one global target may lead to advancement in another. In reality, one set of SDGs serves primarily as a vehicle for accomplishing other SDGs. One could certainly consider SDG 8 (i.e. Decent employment and economic development) paving the way to effectively achieve SDG 1 (i.e. No Poverty). As a result, success on the SDGs mentioned in the document additionally suggests progress on several other SDGs.

The Prime Minister of India's approval on the agenda for Sustainable Development at the United Nations Summit in New York in 2015 have demonstrated India's development plan "*Sabka Saath, Sabka Vikas*" as a pathfinder for other nations. This reflects the fact that, prior to the formulation and adoption of the SDGs, India had launched a number of developmental activities. India, being an affiliate to the Open Working Group (OWG) vigorously highlighted the problems that majority of the impoverished nations are suffering from. It emphasized the need of rapid and inclusive economic growth in lifting a significant number of poor people out of poverty. It proposed a reasonable argument that, despite significant advances in infrastructure development, delivering advancement to all areas of society remains a massive job. All emerging countries throughout the globe are still facing tremendous difficulties to materialize. According to India, the SDGs must serve as an improvement agenda.

The structural composition of the sustainable development can broadly be classified under the three pillars that emphasizes on the social targets (Goals 1–6), economic goals (Goals 7–11), and goals aimed at achieving environmental sustainability (Goals 12–15). A supplementary aim (Goal - 16) for peaceful communities and functional institutions may serve as a facilitator to such developments. In this connection, the stand-alone objective (Goal - 17) seeking collaborations across the globe provide for requisite financial, technical, and structural help that developing nations need to grow.

While poverty eradication and equitable economic growth remain top priorities for development, more resources and capacity-building activities are required to address these vast obligations. The acceleration of development has been throttled by India in favor of the poor nations through increased Official Development Assistance (ODA) and ensuring the transfer of technology at advantageous terms. Thus, the initial evaluation of the efforts is replicated following the development strategy of India, consequently the SDGs. Primarily the emphasis is on the framework that requires mutual comprehension and coordination leading to uniformity of the monitoring and reporting standards especially under the backdrop of the federal structure that Indian government operates through.

Various bodies at the federal and state levels have been charged with leading the work of carrying out ministries, departments, and agencies. The Prime Minister will head the National Institution for Transforming India (NITI Aayog), which would offer overarching coordination and leadership (Jain & Mishra, 2019). The Indian statistics system and agencies collect data to report on certain criteria in a variety of industries. Regional SDG indicators are now being developed to better reflect India's context and requirements. In addition to producing national SDG indicators, the statistics system is intended to assess progress at the sub-national level.

The Ministry of Statistics and Programme Implementation (MoSPI) has reached the completion of a project to produce national indicators based on SDG indicators from across the world. The evaluation strategy for the SDG indicators was motivated by several talks in the form of organized workshops. MoSPI has prepared a draft of national indicators based on lengthy talks and has made it available for broader comment. It is encouraging to see that the MoSPI has received a plethora of feedback from a wide range of stakeholders around the nation. Eventually this will lead to the selection of indicators that as a consequence will help to measure the retrospective progress in this regard.

Under the same strategic plan, NITI Aayog has decomposed the route into an agenda of achieving a subset of Three-Year plan covering 2017–18 to 2019–20. The nation's action plan tackles the country's particular issues and provides methods to accelerate the national development plan. Concurrently, the aimed vision of 15- year and a strategy for the next 7-years is about to complete. These reports are being developed in cooperation with the state authorities following the conventional federal structure of the government.

Implementation of the targeted goals (i.e. SDG's) across India and its provinces has been phenomenal from its inception back in 2015. Some of the major policy interventions in this connection are detailed below.

1. National Action Plan for Climate Change—To materialize the pressing issues on climate change in 2008, the Government of India formulated its first National Action Plan for Climate Change (NAPCC) in 2008. Further, this plan classifies and aims to address six specific areas (Preserving the water resources, Sustainable agriculture, and forests, Preservation of the natural eco-systems along with the coastal zones, and emphasizing on issues related to health, energy, and infrastructure).
2. Swachh Bharat Abhiyan—This initiative was launched back in 2014 with a mission to keep India hygienic and devoid of open defecation. It evolved as a nationwide cleanliness program across the several provinces that India is made up of.
3. Digital India—The mission to digitize, primarily the services rendered by the government to extend hassle free services to every citizen of India was launched back in 2015. This initiative is committed to deliver digital infrastructure, inexpensive access to the internet, and digital proficiency to every Indian resident.
4. Skill India—This initiative aims to address the pressing issues of unemployment especially for a populous country like India. It was launched in 2015, as a nationwide effort that provides adequate skills through training in order to improve their employability.

5. Make in India—This initiative was launched back in 2014 and aims to transform the structural composition of the Indian economy. It is an effort to convert the agrarian based economy into an economy which is driven by capital intensive manufacturing sector.
6. Jan Dhan Yojana—This is an effort to provide the basic banking services especially to them who have difficulties to be a part of the formal financial system back in 2014. By giving access to a no-cost savings bank account, this initiative aims to encourage the savings behavior of the citizens and build confidence in the formal financial system. Further, this also aims to strengthen the economy through improving the rate of capital accumulation.
7. Pradhan Mantri Awas Yojana—This is an effort launched in 2015 to provide affordable homes to poor people (both rural and urban) especially those who are unable to afford a living. It basically extends a loan at a subsidized rate of interest and that helps people to have their own residence.
8. National Health Mission—The National Health Mission was initiated with the aim to improve the health conditions of the citizens of India back in 2013. This scheme had particularly aimed to address the health issues of women and children, by strengthening the traditional health systems and services.
9. Beti Bachao, Beti Padhao—This scheme evolved as a nationwide program, launched in 2015, with the aim to promote education specially for the girl child of the household. It specifically resolves the problem of the conventional gender-biased tradition of the Indian culture that had in some cases restricted girls for taking formal education specially pursuing their higher education.
10. National Rural Livelihood Mission—In order to combat poverty and provide a healthy living to the citizens of India, the National Rural Livelihood Mission came into existence in 2011. This mission particularly seeks to eliminate poverty in rural regions by encouraging self-employment and offering sustainable livelihood options.

Among other things, these projects illustrate India's commitment to reaching the SDGs and creating a sustainable future for all of its residents.

The SDGs serves as an organized blueprint that at a macro-level government, the corporations and the civil society organizations follow to arrive at a future that is relatively sustainable. It lays a mechanism of cooperation between the citizens and the organized bodies that creates a society that is fair, egalitarian, and ecologically sustainable for current and future generations.

3.3.1 Achievements Towards the SDGs—Indian Scenario

A faster rate of development (SDG 8) is a particularly potent instrument that helps a nation to overcome poverty (Baldacci et al., 2004). In one side, it creates certain opportunities through creation of adequately compensated employment, which gives rise to the purchasing power of the household helping them to afford their basic needs (e.g. Food, Clothes, Housing, Education, and Access to Health Care services). Alternatively, it provides the government with ever-increasing income to pay social expenditures (Fosu, 2010). India has pursued its economic reform strategy in order to attain sustained fast development. Among the several reforms that have taken place to accelerate the pace of development notable initiatives are the consolidation of finances, targeting inflation, increased development of infrastructure (SDG 9), corruption reduction (SDG 16), and altogether a better governance. In this connection, the Aadhaar Act (to provide a unique identity to each citizen), the Insolvency and Bankruptcy Code and the Goods & Services Tax have been instrumental to shape the operating environment. Further liberalization of the conditions in relation to foreign direct investments to overcome capital constraints while treatment in the form of organized liquidation of the sick units to release the blockage of government funds are noteworthy. These efforts have positioned India as the world's fastest-expanding major economy. The growth rates for the economy shows an escalation from 7.5% in 2014–15 to 8% in 2015–16 (Das, 2011). There is now strong evidence that India's tremendous development since the start of economic reforms in 1991 has resulted in considerable poverty reduction. India expanded at an average yearly rate of 6.2% between 1993–94 and 2003–04, and 8.3% between 2004–05 and 2011–12. Both eras had considerable declines in poverty, although the later period's fall was substantially steeper due to higher development. However, data suggests that poverty has fallen in all social, economic, and cultural categories worldwide and in every state since the reforms.[1]

[1] https://sustainabledevelopment.un.org/content/documents/16693India.pdf.

Growth has boosted the revenues, allowing the government to maintain an elevated amount of social expenditure and target poverty directly. Public–private partnerships have created opportunity for fresh employments thus, leading to anti-poverty programs. These partnerships have helped to enhance infrastructure for farming, manufacturing facilities, and entrepreneurship-based opportunities for livelihood. The Mahatma Gandhi National Rural Employment Guarantee Act (MGNREGA) solely as a government scheme ensures at least a 100 day of work and thus earning per year especially in the rural areas through unskilled workers. During the previous year, the project created almost two billion man-days worth of jobs (SDG 8). Women and economically disadvantaged individuals have been the program's major recipients (SDG 5 and SDG 10) generating 56% and 39% of man-days of employment created respectively last year. Further, as an extensive policy approach from governments point of view is the Pradhan Mantri Jan Dhan Yojana (PMJDY), launched back in 2014 with a focus to serve initially a no-cost savings account and subsequently invite for participation in all types of financial services, particularly banking, insurance and pension to the citizens of India is praiseworthy. Among others, the Deendayal Antyodaya Yojana (often termed as the National Livelihoods Mission) aims to develop skill-based jobs for the disadvantaged sections. The Mission's purpose is to eventually enroll one female participant into Self-Help Groups from each of a huge number of poor rural households. As a broader perspective, the 'Skill India' initiative by equipping the workforce with the skills necessary for a rapidly changing economy focuses on vocational training and skill development. It directly addresses SDG 4, which aims to ensure inclusive and equitable quality education and promote lifelong learning opportunities for all. The initiative enhances employability and fosters economic growth (SDG 8), thereby helping to reduce poverty (SDG 1) and inequality (SDG 10). Additionally, Skill India promotes gender equality (SDG 5) by encouraging women's participation in various sectors through targeted training programs. Through building a skilled workforce, the initiative supports innovation and sustainable industrialization (SDG 9) while contributing to the overall social and economic development of the country.

Further, in its momentum to achieve sustainable development the 'Make in India' initiative plays a pivotal role by promoting manufacturing, innovation, and sustainable practices across various sectors. By fostering an ecosystem that encourages local production and entrepreneurship, the initiative aims to create jobs, enhance skill development, and reduce poverty (SDG 1). Furthermore, it supports industry modernization and the adoption of clean technologies, contributing to responsible consumption and production (SDG 12) and climate action (SDG 13). By strengthening infrastructure and encouraging foreign investment, Make in India also aids in building resilient infrastructure (SDG 9) and promoting inclusive and sustainable economic growth (SDG 8). Pensions to the aged, widows, and those with disabilities are offered as a measure of social security through the National Social Assistance Programme. In addition, three schemes have been provided for to insure the life of a citizen (PMJJBY), insure the incidence of a personal accident (PMSBY), and pensions for unorganized sector employees (APY). These insurance products have covered a total of approximately 130 million citizens.

A crucial component of an effective poverty-reduction strategy is access to high-quality basic education and low-cost healthcare (Januja & Kamal, 2014). A wide range of activities aimed at universalizing primary healthcare are being undertaken as part of the National Health Mission. Dedicated programs like the Integrated Child Development Services (ICDS) focus to monitor and rehabilitate the nutritional status of both the mother and child in a systematic way. Furthermore, the Janani Suraksha Yojana (JSY) promotes institutional births among the women especially living in the distant locations through conditional cash transfers. There is a particular focus on filling human resource shortfalls in healthcare at all phases of service delivery. In a similar vein, the National Education Mission that focuses to extend universal accessibility to school, addressing gender gaps, and increasing student learning results. The Right to Education Act established a successful legal structure that rights every kid (6–14 years old) to free and obligatory schooling that adheres to the tenets of fairness and non-discrimination.

'Housing for All by 2022' is yet another grandiose scheme. This program provides help to impoverished families for home construction (for example, in the way of credit-linked interest subsidies). Approximately 5 million dwellings for the impoverished in rural regions will be erected this year. Food insecurity is a direct result of poverty. The National Food Security Act in this connection has successfully catered nearly 75% of the rural dwellers and approximately 50% of the urban dwellers through providing a basic level of food grains at subsidized pricing based on the strategically positioned Public Distribution System (Pradhan, 2018). This distribution ensures an entitlement of 5 kg of wheat or rice per participant per month. A parallel scheme offers an entitlement of 35 kg of subsidized wheat or rice to the poorest families each month based on the economic stratification within which an individual/ household fall into. The usage of traditional fuels for cooking in India has observed a significant decline over the years, primarily due to efforts to promote cleaner cooking alternatives. In 2011, approximately 67% of Indian households relied on traditional biomass fuels, such as wood and dung, for cooking (National Sample Survey Office, 2011). By 2019, this figure had decreased to about 38%, reflecting the impact of government initiatives like the Pradhan Mantri Ujjwala Yojana (Launched in 2016), which aimed to provide subsidized liquefied petroleum gas (LPG) connections to low-income families. As of 2021, around 29% of households still used traditional fuels, indicating a continued shift towards cleaner energy sources (Ministry of Petroleum & Natural Gas, 2021). This transition is crucial for improving health outcomes, reducing indoor air pollution, and advancing environmental sustainability in the country. Furthermore, there is a focus on providing enough clean water for drinking in addition to sanitation for the whole community. Under the National Rural Drinking Water Programme, that aims to ensure accessibility to safe drinking water has covered 77.5% of the total households across India having access to 40 liters of daily water per person. So far, an additional 18.9% of homes were completely or partially covered. The Swachh Bharat Mission (Clean India Mission) is a government flagship effort that intends to eliminate open defecation by 2019 in India. Over the last two and a half years, our Mission has installed over 39.5 million residential toilets. Furthermore, 193,000 villages with 531 cities were successful in eliminating open defecation (Kumar et al., 2021). Local governments are being graded on cleanliness standards in order to promote such kinds of changes. So far, the Prime Minister's Rural Roads Programme has connected 70% of the intended rural dwellings with all-weather roads. All-weather roads are expected to link all habitations with a population of 500 or more by March 2019. Green technology is increasingly being used in the development of rural roadways.

There has been significant progress in addressing security of nutrition and food. Nonetheless, difficulties persist. Between 2005–06 and 2015–16, for example, there was a significant drop in stunted and underweight children.[2] However, the worldwide proportion of stunted and

[2] https://sustainabledevelopment.un.org/content/documents/16693India.pdf.

underweight children continues to be substantially high (Ssentongo et al., 2021).

The outreach and effectiveness of the food security programs operative in India is the largest across the world, which persistently caters nearly 800 million people through providing inexpensive access to grains. Conventional research exhibits that women are more concerned about the food security of a household. Thus, government schemes regulating the distribution of food grains among the households has enforced a compulsion to cater food only to those households having a ration card entitlement against the name of the household's most aged female member. Food governance has considerably improved when it comes to responsiveness, openness, and accountability. Almost 232 million people are benefitted through digitizing the traditional ration card that entitled individuals to food and other necessities. Furthermore, 77% of ration cards were connected to a unique identification number, allowing financial transfers to promote nutritional variety. Almost 20 of India's 29 states have automated distribution networks and fair-price shops (i.e. the delivery outlets). A state-wide electronic dispute redressal system has also been established. There are certain projects that capitalize on the nutritional security of specific groups classified on the basis of demography. ICDS, for example, is entrusted to ensure the dietary needs of nearly eighty-three million young children and nineteen million pregnant and breastfeeding women in the nation. The Mid-Day Meal Programme provides healthy prepared lunches to 100 million primary school pupils.

India is the world's greatest producer of milk, pulses, and spices. It additionally comprises the most rice and cotton farming land. Rain-fed croplands account for around 55% of India's croplands, and extensive coastal areas are cultivated. As a result, the country's food production is sensitive to climate change. The National Mission on Sustainable Agriculture (NMSA), in partnership with different organizations under the National Action Plan on Climate Change, is working to reduce the effect of climate change while still maintaining agricultural production. Soil Health Cards are supplied to farmers under the NMSA to provide crop-specific nutrient management advice and to help them increase soil fertility and crop yield. So far, almost sixty-two million cards have already been distributed. Over the previous decade, the amount of land dedicated to organic farming has more than tripled.

Efforts have also been undertaken to improve crop insurance assistance in order to reduce farmer losses and provide single-window extensive risk coverage for a variety of crops. In comparison to an objective of covering 50% of total planted land in the nation over 2016–19, 30% has been successfully met. Farmers' access to new knowledge and skills is also being improved. For example, a total number of 652 management agencies focusing on Agricultural technology have been established around the country to communicate the most recent technologies to farmers. These Agencies interacted with approximately 1.2 million farmers in 2016–17, nearly half of them were women.

Marginal and small-scale farmers account for roughly 80% of all farmers in India. Over ninety percent of them practice rain-fed agriculture. It is consequently critical to increase farmers' revenue in various ways. Several Indian states have made headway in modifying the land leasing Acts in accordance with a Model Act created by NITI Aayog. The Model Act is intended to preserve the tenant's rights while also guaranteeing that the owner of the land does not lose ownership of the property to the renter. Micro-irrigation was used on almost 1.3 million hectares between 2015 and 2017. Fruit and vegetable yield has grown by 42.3% and 52.8%, respectively, as a result of crop spacing, careful use of water, and other inputs. Approximately 38,000 hectares of rain-fed land were covered by the integrated farming system in 2016–17. For increased production, this program connects the agriculture system with activities such as horticulture, animal husbandry, and fishing. It also protects farmers' revenue from natural disasters. The computerized National Agricultural Marketing Platform has been connected with around 250 wholesale agricultural marketplaces in ten states. Other agricultural initiatives being explored in collaboration with state legislatures include direct purchases of agricultural commodities from farmers, encouragement of e-trading, and assuring the legitimacy of a state-wide trade permit. These achievements that have contributed to the advancement of the SDGs in India has been briefly summarized below.

1. Strong political commitment to the SDGs

India has demonstrated strong political commitment to achieving the SDGs by setting up a high-level mechanism for monitoring and assessing SDG progress at the national level.

2. Reduction in poverty

Over the last several decades, India has made great progress in decreasing poverty. From 2005 to 2015, the poverty rate in India has reduced from 52 to 30%.

3. Access to education

India has made tremendous progress in providing access to education, especially at the primary level. In 2014, 96% of children aged 6–14 were enrolled in schools.

4. Improved access to healthcare

India has made significant progress in improving access to healthcare services, particularly in rural areas. Several initiatives have been launched to improve primary health services and reduce maternal and child mortality rates.

5. Renewable energy

India has made significant progress towards advancing renewable energy and lowering greenhouse gas emissions. India ranked fifth in worldwide rankings in terms of installed capacity for renewable energy in 2018.

3.4 Sustainable Development Goals—Global Perspective

Many stakeholders throughout the globe utilize the SDGs to guide their growth. Since 2015, 169 objectives have been established for the 17 goals, nearly 3898 events have occurred, 1348 articles were published, and 7835 actions have been carried out (United Nations, 2024b).

Prospects for global sustainable development continue to differ. In the middle of the COVID-19 epidemic two years ago, the Inter-agency Task Force cautioned of a worldwide divergence that might result in an entire decade being lost for development (United Nations, 2024c). By 2022, these concerns had shown themselves: a large financial disparity had translated into a growing divide in terms of development. The conflict in Ukraine, dramatic rises in food and energy costs, and quickly tightening financial circumstances have worsened many nations' issues, exacerbated poverty, and hunger while stalled momentum on the goals for sustainable development (SDGs).

Notwithstanding some encouraging signs, the macroeconomic future globally seems in turmoil, especially leaving behind the majority of the world's vulnerable countries facing increased debt payment costs and severe budgetary constraints. Recent bank collapses throughout the USA and Switzerland highlight vulnerabilities of financial legislative and supervisory institutions once again. Financing and equitable growth possibilities are separating even more starkly in today's highly tough global financial backdrop.

In the midst of a tremendously tough climate, the outlook for the global economy remains shaky. Though certain of the clouds lurking over the world's economic situation seem to be lifting, the overall prognosis remains unclear. Task Force members expect global growth to decline in 2023, albeit with a larger prediction range than in the past. Task Force members anticipated worldwide growth of 1.9% to 2.4% in 2023, based on market exchange rates, after growth of 3.0% in 2022. Higher-than-expected price increases, which could result in to a spiralling wage gap, a quick and disorderly deterioration of global financial circumstances, and a rise in geopolitical tensions are among the downside possibilities.

Availability of finance is crucial to restrain the sustainable development divide among the nations. SDG financial requirements are increasing, however, funding for development is gaining adequate momentum. There is an immediate need for expanded international assistance for vulnerable nations, particularly numerous LDCs, African countries, including the developing states in small islands (SIDS). At the same time, inadequate investment, especially in many developing nations, is worsening the development gap. Delaying financing for sustainable transitions will make the Agenda 2030 Sustainable Development and Climate Goals hard to achieve, while also exacerbating future financial issues.

Today's sustainable and profitable investments may modify and diversify economies while also strengthening resistance to future shocks, particularly inflationary supply-side crises. Such investments, as highlighted by the Financing for Sustainable Development Report (2022), allow nations to deploy resources over time and sustainably repay debt.

Weaker-than-expected prices have fuelled predictions for a slower pace of tightening of monetary policy in recent months, contributing to the relaxation of global financial conditions. Nevertheless, inflation as a whole is anticipated to continue high in many countries, raising fears that inflation expectations may become unanchored. In this environment, central banks are anticipated to maintain tight monetary policy.

Countries are forced to make challenging fiscal and monetary policy choices. Despite varied recovery rates from the COVID-19 outbreak, rising prices has forced central banks throughout the world to take an aggressive stance on tightening monetary policies. The fast stiffening of the world's financial markets has also raised worries about debt sustainability in a number of emerging nations.

High borrowing rates will be especially detrimental for nations with existing heavy debt service obligations and foreign currency debt. The rise in food and energy costs has put a burden on the public budgets of nations that depend significantly on commodity imports. As budgetary consolidation demands mount, severe delays or reductions in spending on sustainable development, including climate action, are possible. Furthermore, economic austerity often results in cuts to social spending, which adversely affect the most disadvantaged people, notably women and children.

These shocks have the potential to impede progress towards the SDGs, particularly for the least fortunate and most vulnerable nations. Years of progress in numerous aspects of sustainable development, such as poverty, healthcare, and education, have been undone as a result of the Ukraine conflict and epidemic. The recent convergence of crises has disproportionately impacted the world's most vulnerable nations, notably less developed nations (LDCs) and developing nations with small islands (SIDS). Majority of the nations are exposed to catastrophic outcome induced by the rise in the frequency as well as extremity due to the climate-related shocks. The sharp increase in prices has also caused negative growth in real wages in numerous countries, reducing families' buying power and disproportionately harming low-income groups. Food and gasoline price increases have pushed tens of millions into abject poverty and serious food shortages. Such a detrimental macroeconomic scenario has triggered the investor sentiments and has created an adverse momentum especially for long-term investments, in ultimate has stalled the governments' effort towards a sustainable and fair structural transformation. As policy flexibility shrinks, many developing nations risk being trapped in a vicious cycle of slow development, unsustainable debt, and austerity.

The global production growth rate has dropped from 5.8% in 2021 to a rate of 3.0% in 2022. The Russia-Ukraine War has added to the surge in the cost-of-living expenses, especially when the majority of the nations are still combating the devastating effects of the pandemic. Acute supply interruptions pushed prices for both food and energy to record highs, disproportionately affecting the most vulnerable populations. With global inflation expected to reach its highest level in two decades of 9.0 percentage points by 2022, many central banks worldwide squeezed monetary policy to offset demand-side rising prices. The United States Federal Reserve's swift rate rises had significant spill-over effects on developing nations, with several facing significant capital outflows and currency devaluation throughout the year. The unclear GDP forecast, continuing inflation, and ongoing geopolitical concerns all hampered investor risk appetite. The significant stiffening of financial markets worldwide has exacerbated balance-of-payment concerns and vulnerability to debt in a number of emerging nations. Almost 60% of the world's countries with low incomes were on the verge of financial disaster by the time of 2022.

Following the predictions of the United Nations World Economic Situation and Prospects 2023, global growth will fall to 1.9% for 2023, the lowest level in the last ten years. As many of the present global challenges fade, global production growth is predicted to accelerate to 2.7% in 2024. Most developing nations' investment prospects remain bleak, heightening the danger of deeper and longer-lasting damage. Since the COVID-19 outbreak, several developed countries have proposed massive budgetary packages, involving escalates for public investment, to help their economies recover. In contrast, developing nations have faced significant restraints, with many of the poorest forced to reduce infrastructure and education spending. A protracted period of low investment is worsening poor nations' already significant climate and SDG investment disparities.

Growth in emerging nations is predicted to remain around 3.9% in 2023, although growth expectations differ greatly between regions and countries. Consistently increasing food and gasoline costs will have diverse effects on household spending in all areas. Energy price hikes are likely to help the economies of energy exporters, particularly those in Western Asia, recover. Tighter worldwide economic conditions and inflationary orientations will have the greatest impact on nations that are already susceptible to macroeconomic shocks. Rising global commodity prices will continue to have an effect on net commodity importers, notably numerous African and South Asian countries.

Following the COVID-19 epidemic, worldwide food prices increased, due to increased global demand, higher fertilizer and oil prices, more expensive transportation, as well as frequent disruptions in the supply chain. The crisis in Ukraine hampered global food supply and delivery, causing global food prices to hit an all-time high in March 2022. Based on the UN Food and Agriculture Organization's Food Price Index, food costs would be 50% higher in 2022 than they had been in 2019. More than 90% of developing nations saw food price inflation of more than 5%, with a number of nations in Africa, Latin America, the Caribbean, and South Asia observing a double-digit rise. Since the pandemic started, the number of people facing severe food insecurity has nearly tripled, rising from 135 million in 2019 to a predicted 345 million in 2023. Despite a drop in global food prices in the second portion of 2022, they remained elevated in comparison to pre-pandemic levels. Food inflation has the greatest impact on low-income people since food accounts for as much as fifty percent of their family budget. Aside from the catastrophic effect on the lives of people, the food crisis has significant economic repercussions.

The deteriorating trend for per capita income is nowhere different. Growth globally has nosedived from 5.0% in 2021 to 2.1% in 2022, with a further decline to 1.0% in 2023 projected. South Asia has suffered the greatest income per capita loss relative to pre-pandemic forecasts. Soaring food and energy prices, along with rising interest rates, have triggered recessions in a few countries, causing people to experience extreme poverty and starvation for the first time. This is due to insufficient policy buffers and massive external imbalances. However, in Western Asia, per capita income has returned to pre-pandemic levels due to the surge in global oil prices that have boosted the economy of the fuel exporters. In 2022, rich nations lost less productivity per capita than poor countries, owing in part to increased financial assistance efforts to mitigate the consequences of a global pandemic and cost-of-living difficulties. Tighter monetary policies and higher energy costs, on the other hand, are likely to have a greater influence on growth in developed nations in 2023, resulting in bigger losses in production relative to developing countries. Additionally, such losses are projected to continue to be severe for developing nations that were already trailing. Because of their dependence on commodity imports, numerous LDCs, LLDCs, and SIDS have been adversely affected by the large rise in global commodity prices.

The epidemic and the conflict in Ukraine have undone nearly thirty years of poverty reduction progress. According to a recent World Bank study, the worldwide severe poverty rate jumped from 8.4% to 9.3% in 2020 alone, as the epidemic pushed an extra 70 million people into extreme poverty (World Bank, 2022). Poverty alleviation faced a worse setback in 2022 as global GDP slowed and prices rose. In 2022, an extra 70 to 89 million individuals lived in severe poverty, relative to pre-pandemic forecasts. Current trends predict that 574 million people, or over seven percent of the global population, will be living in severe poverty by 2030.

The measure of inequality has not been an exception to this and has been worsened to touch the extreme limits. By 2020, the earnings losses of the world's poorest 40% were double those of the richest 20%.

Current crises have caused more and longer-lasting harm to already vulnerable categories of workers, such as women, youth, and low-skilled workers. Many high-skilled occupations (which include executives, specialists, and technicians) were back in business by 2022, whereas many low- and medium-skilled ones were yet to return (United Nations, 2023). Simultaneously, informal employment, in which employees sometimes lack social security, has been rebounding rapidly, especially in low- and low-middle-income nations. This might jeopardise the formalization trend seen over the previous 15 years (ILO, 2023). While women had a greater employment rebound after the outbreak, this recovery was mostly driven by informal labor. By 2022, around four out of every five jobs created for women will be informal (ILO, 2023). In the meantime, young adults continue to confront major barriers to finding legitimate employment. Youth employment suffered the most during the epidemic, and its recovery has lagged significantly behind the progress of adults, leaving more than one in every five youths, not in school, work, or training (ILO, 2023).

Governments are going to require a careful policy balance in order to keep inflation under control without impeding development. They must also solve the current crisis while pursuing productivity over the long term and the SDGs. Such investments have the ability to create jobs in the near term while also fuelling a virtuous cycle of rising GDP and tax revenues while also stabilizing long-term debt ratios. To address the plethora of issues that emerging nations confront, substantial and effective global policy actions are required. This includes attempts to more effectively manage spillovers from developed-country policies, address rising financial distress risks, increase SDG investment, expedite climate action, and assist people impacted by crises and famine.

References

Baldacci, E., De Mello, L., & Inchauste, G. (2004). Financial crises, poverty, and income distribution. In *Macroeconomic policies and poverty* (pp. 269–307). Routledge.

Das, D. K. (2011). Indian economy: Growth performance and prospects of transitioning the growth trajectory. *Journal of Asia-Pacific Business, 12*(2), 171–193.

Eustachio, J. H. P. P., Caldana, A. C. F., Liboni, L. B., & Martinelli, D. P. (2019). Systemic indicator of sustainable development: Proposal and application of a framework. *Journal of Cleaner Production, 241*, Article 118383.

Fosu, A. K. (2010). Inequality, income, and poverty: Comparative global evidence. *Social Science Quarterly, 91*(5), 1432–1446.

Fund, S. (2015). *Sustainable development goals*. https://www.un.org/sustainabledevelopment/inequality

ILO. (2023). *World employment and social outlook: Trends 2023*. ILO.

Indicators, G. S. (2016). *Tier classification for global SDG indicators*.

Jain, A. K., & Mishra, S. N. (2019). Role of NITI Aayog in the implementation of the 2030 agenda. *Agenda and India: Moving from Quantity to Quality: Exploring Convergence and Transcendence, 2030*, 239–254.

Janjua, P. Z., & Kamal, U. A. (2014). The role of education and health in poverty alleviation a cross country analysis. *British Journal of Economics, Management & Trade, 4*(6), 896–924.

Kim, R. E. (2023). Augment the SDG indicator framework. *Environmental Science & Policy, 142*, 62–67.

Kumar, P., Kumar, K., Mathur, S., & Singh, M. (2021). Aligning developmental programs to implement SDGs in India: challenges and the way forward. *Journal of Contemporary Issues in Business & Government*, *27*(6).

Ministry of Petroleum and Natural Gas. (2021). *Annual report 2020–21*. Government of India. https://www.petroleum.nic.in

National Sample Survey Office. (2011). *Key indicators of household consumer expenditure in India 2009–10* (NSS 66th round). Ministry of Statistics and Programme Implementation. https://www.mospi.gov.in

Osborn, D., Cutter, A., & Ullah, F. (2015). Universal sustainable development goals. *Understanding the Transformational Challenge for Developed Countries, 2*(1), 1–25.

Pradhan, M. (2018). *Understanding the demand side of social protection programmes: The case of public distribution of food in India* (Doctoral dissertation, University of East Anglia).

Sachs, J. D. (2012). From millennium development goals to sustainable development goals. *The Lancet, 379*(9832), 2206–2211.

Ssentongo, P., Ssentongo, A. E., Ba, D. M., Ericson, J. E., Na, M., Gao, X., Fronterre, C., Chinchilli, V. M., & Schiff, S. J. (2021). Global, regional and national epidemiology and prevalence of child stunting, wasting and underweight in low-and middle-income countries, 2006–2018. *Scientific reports*, *11*(1), 5204.

Stafford-Smith, M., Griggs, D., Gaffney, O., Ullah, F., Reyers, B., Kanie, N., Stigson, B., Shrivastava, P., Leach, M., & O'Connell, D. (2017). Integration: The key to implementing the Sustainable Development Goals. *Sustainability science*, *12*, 911–919.

United Nations. (2023). *Financing for sustainable development report 2023: Financing sustainable transformations*. New York: United Nations https://developmentfinance.un.org/fsdr2023)

United Nations. (2024a). *Our work on the sustainable development goals in India*. UN India Digital Library. https://india.un.org/en/sdgs

United Nations. (2024b). *Sustainable development*. Department of Economic and Social Affairs. https://sdgs.un.org/goals

United Nations. (2024c). *COVID—19 could lead to a lost decade for development*. Department of Economic and Social Affairs. https://www.un.org/en/desa/un-covid-19-could-lead-lost-decade-development

Weber, H. (2017). Politics of 'leaving no one behind': Contesting the 2030 Sustainable Development Goals agenda. *Globalizations, 14*(3), 399–414.

World Bank. (2022). *Poverty and shared prosperity 2022: Correcting course*. World Bank.

CHAPTER 4

Assessing Financial Inclusion for Inclusive Growth and Development

4.1 Introduction

This section is dedicated to materialize the first two objectives of this study. It attempts to estimate the significant factors and measure the extent of their contribution towards building up of financial inclusion index. Additionally, it aims to measure financial inclusion across the different samples that we have considered in this study. Section 4.3 focuses to estimate the achievements in financial inclusion across the 28 states and 7 union territories of India for 2003–2022 applying the Max–min. approach for normalization of the three variables and the inverse of the Euclidean distance to consolidate the different dimensions. Section 4.4 aims to estimate the financial inclusion achievements across the 26 Asian countries for 2013–2022 using Factor Analysis along with the weighted arithmetic mean in the 1st stage and the inverse of Euclidian distance solely in the second stage. Factor analysis has been used to find the common structure and derive the relative weight of the theoretically relevant variables. Section 4.5 considers measuring financial inclusion across the 12 Asian and 12 European nations using Max–min. approach for normalization and applying the inverse of Euclidean distance for consolidation. This sample uses the principal component analysis as weighing strategies to allocate relevant weights to the 20 variables those are found to be theoretically relevant. Further, Sect. 4.6 provides a global perspective through measuring the financial inclusion achievement scores

S. Ghosh et al., *Financial Inclusion*, Sustainable Development Goals Series, https://doi.org/10.1007/978-981-95-3385-5_4

across the 46 countries taken out of 5 continents using 14 variables for 2012–21. This sample uses the Max–min. approach for normalization along with the weighted arithmetic mean for consolidation of the dimensions.

4.2 Characterizing Financial Inclusion

Inclusion of previously excluded groups into the regulated financial system offers an extensive selection of low-cost (Leyshon & Thrift, 1995), fair, and safe formal financial services (Carbo et al., 2005) in the form of credit, deposits, insurance, and payments (Conroy, 2005; Mohan, 2006) whenever needed while minimizing the expansion of unregulated sources of funding (e.g., moneylenders), that usually are prone to appear predatory (Rangarajan Committee; 2008). Kempson and Whyley (2001) describe financial exclusion as those persons who have been denied the opportunity to utilize conventional financial products and services (Kempson & Whyley, 1998). Sinclair (2001) agrees with the above opinions and characterizes financial marginalization as being unable to get required financial services suitably. In this connection, the survey from the World Bank highlights the probable reasons as to why people are outside the purview of formal finance (see Fig. 4.1).

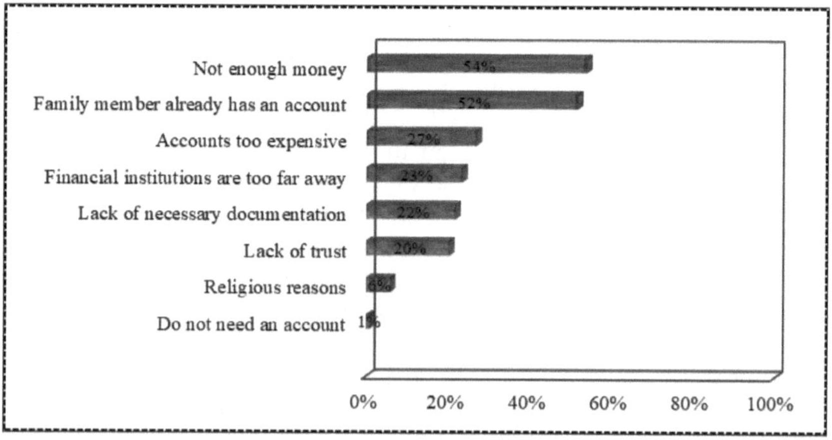

Fig. 4.1 Percentage without a financial institution account, age 15+ *Source* Presented by authors

Sarma (2008) as a pioneering work followed by Kumar and Mishra (2011), Sarma (2012), Gupte et al. (2012) and Chakravarty and Pal (2013) in Indian context, Rahman (2012) in Malaysian context and Amidžić et al. (2014) considering a substantial cross-section of countries are research those are worth noting in terms of measuring the degree of financial inclusion attained.

Normalizing the data following the United Nations Development Program's (UNDP's) computation of the Human Development Index, or HDI for short and consolidating the same in to a single value applying the inverse of Euclidean distance, Sarma (2008) calculates the composite measure of financial inclusion (IFI). As a result, the objection of complete substitutability is avoided. The study's findings demonstrate that OECD nations dominate in both categories (the 55-country group with three dimensions available and the 100-country group with just two dimensions available). Malaysia and Singapore are large exceptions in the three-dimensional category, with higher IFI values than Norway (an OECD member). Nevertheless, Indian accomplishments in each category falls short. Arora (2010) in a cross-country setup; Rahman (2012) in the Malaysian context; Chattopadhyay (2011), Laha and Kuri (2014) and Kaleeswaran and Meera (2017) across the Indian provinces are examples of academic works that apply similar approach to Sarma (2008) in a different setting. Following the key performance indicators on financial inclusion (KPI's) as proposed by the Central Bank of Malaysia, Rahman (2012) using four important dimensions of Convenient Accessibility, Take-Up Rate, Responsible Usage and Satisfaction Level measures financial inclusion across the provinces of Malaysia. Results show that the FI score of Malaysia stood high at 0.77. This was further evaluated for different income classes to observe whether the income-specific policy interventions on financial inclusion are required or not. It was observed that nearly 40% of the population are categorized under the low-income group and hence are devoid of benefits that are extended for people at large. Considering 27 states of India for 2010, Laha and Kuri (2014) observed significant differences across the states while measuring financial inclusion from the two complementary aspects of demand for and supply of financial services. Results show that states mostly from the western and southern region takes the lead with a significant rural–urban divide between the attainment scores. A special emphasis on the state of West Bengal was given by Chattopadhyay (2011), to observe the district-wise extent of financial inclusion attained. Using three dimensions (i.e.

Penetration, Availability and Usage), the study finds Kolkata as a leading district with financial inclusion scores crossing the 0.5 mark. However, a large proportion of the districts remained at the lower end of the FI scores attained (i.e. 0–0.3). A micro-level analysis of the MGNREGA scheme that helped to increase financial inclusion was evaluated for the Virudhunagar district in Tamil Nadu by Kaleeswaran and Meera (2017). Considering the number of bank branches, number of bank accounts and the amount of wage disbursals associated with the MGNREGA scheme this study aimed to observe the FI achievements across the 11 blocks and 450 panchayats under the jurisdiction of the district authorities. Results show that the Tiruchuli block holds the top position so far as the holding of bank accounts and wage disbursements connected to the MGNREGA scheme is concerned, whereas the Sivakasi block is found to have a greater number of bank branches than compared to other blocks of the district. The FI score shows Tiruchuli (0.543) to take the lead, whereas the majority of the blocks (4 out of 11) having a score of 0–0.3. Sarma (2012) also introduces an idea of placing non-parametric weights to indicators and determines the numeric average of both the Euclidean distance and its inverse to highlight the importance of statistical integrity when assessing financial inclusion achievement. The literature's conclusions are consistent in comparison to the previous version, in which OECD nations were discovered standing at the peak during the entire research period (i.e. 2004–2010). Furthermore, it is not unexpected that Overseas Financial Centres (OFCs) like Cyprus, Malta, and Singapore have taken a significant place in terms of the IFI calculated throughout the years. A clear picture of India's situation demonstrates that it is regularly on the borderline of the lower-middle IFI category.

Chotia and Rao (2014) is among the researches that is based on the application of methods used in Sarma (2012), though largely different apart from the variables involved in such estimate and the contextual background. Rather evaluating a cross-country setup, the study emphasizes on the Indian states relying on a different set of factors discovering that the northern area ranks first, while the north-east region ranks worst. Furthermore, states with a greater degree of IFI have a substantial GDP per capita and a decent Human Development Index.

Similarly, Kumar and Mishra (2011) proposed changes to earlier research by way of alteration in the composition of variables along with the technique that has been applied in to. Rather utilizing the maximum-minimum methodology, giving biased results owing to the inclusion of extreme values, the research employs the distance from the average method that creates both independent indexes from distinct yet interlinked perspectives of demand and supply for financial services. An examination of the supplying side reveals a substantial discrepancy across states, which exacerbates, when the rural–urban division is included. The findings from the demand side shows that roughly one-third is set aside for savings, a fraction of a fifth for credit, and one-eighth for insurance, yielding a combined number of 0.208 so far financial inclusion achievements are concerned.

Gupte et al. (2012) investigated another version of approach in which variables under a category were combined applying the arithmetic mean and the categories were aggregated to arrive at a final index applying the geometric mean. Considering a composition of variables in four distinct groups during the year in question (i.e. 2008–2009) reveals that India's ranking has improved, but not at the desired rate. Similarly, Chakravarty and Pal (2013) proposes a 'Marginal Rate of Substitution' (MRS) across the variables assuming a value greater than zero but less than one as a conceptual revision to Sarma (2008). The analysis utilizing particular axiomatic criteria on data of the supply-side indicators reveals that setting MRS to 1 implicitly presupposes that variables are entirely substitutable (Ghosh & Sahu, 2020). The results show that financial inclusion achievements have undergone significant changes between pre-social banking (1972–1976) and social banking period (1977–1990). However, supply of regulated financial services shows enhanced concentration over the chosen segments of society in post-social banking period as compared to the social banking period.

Furthermore, Amidžić et al. (2014) normalize variable values using an adjusted geometric mean and the distance to a point of reference to minimalize the condition of perfect substitution between the variables of interest alongside avoiding the influence of extreme values. Applying factor analysis, the research statistically characterizes the variables under each dimension and suitable weights are derived using the squared values of factor loadings. According to the research, nations with higher and upper-middle incomes are at the higher side of the inclusivity hierarchy, with little regional variation, whereas those with lower and lower-middle incomes are at the lower end. Considering the role of cooperatives, Saravanabhavan (2018) investigates the financial inclusion scores across all states of India with the same technique. However, Saravanabhavan (2018) presents a significant rise in the financial inclusion achievements across Indian states from 1984 to 2016 through employing the max.-min. technique instead of considering the reference point as employed by Amidžić et al. (2014).

4.3 Measuring the Index of Financial Inclusion

Financial participation is a broad concept to explore. Presentation of the concept by a single number involves consolidation of a number of variables. However, the variables that build up the index or are significant contributors are numerous as well as diverse in nature. By considering the maximum number of possible variables to retain the diversity of variables ends up with a clearer picture of the status of financial inclusion. We form a group of variables sharing high theoretical affinity among them and further using the parametric assignment of weights (both CFA and PCA) verifies that the theoretical affinity conforms to the data-driven statistical evidence. The whole set of variables available are classified into groups based on homogeneity and for each of the variables selected for the study we intend to normalize the values of attainments using the max.-min. technique following either Eqs. 1 or 2. The primary equation (Eq. 4.1) serves as the basis for variables that are having a positive effect on the final results, while the following formula (Eq. 4.2) is intended for variables with an adverse effect.

$$d_j = w_i \frac{A_i - m_i}{M_i - m_i} \tag{4.1}$$

$$d_j = w_i \frac{M_i - A_i}{M_i - m_i} \tag{4.2}$$

where d_j = the dimensional achievements of financial inclusion, where 'j' represents a positive discrete number ranging between 1 to n, in a way, that $1 \leq j \leq$ n with n having a finite value. w_i = the associated weights of the ith variable, where, $0 \leq w_i \leq 1$. A_i = the achievements of the concerned variable in the jth category. m_i = the observed lowest value of the concerned variable belonging to the jth dimension. M_i = the observed highest value of the concerned variable in the jth dimension.

Using the max.-min. approach as suggested in the widely accepted index of human development (as constructed by the UNDP) as a measure of normalization helps to recover from the problem of scale invariance.

We consider measuring the achievement scores of financial inclusions across the four samples as elaborated in Table 4.1.

After computing the achievement of a particular variable, we use the two-stage principal component analysis and factor analysis (differs from case to case[1]) for assigning parametric weights to the variables forming a group. This assignment of weight highlights the contribution of a particular variable towards the group (i.e. the dimension). Further using the squared factor loadings as weight helps to clarify the theoretically established relation with statistical significance. A two-stage PCA has been selected for the study (in Sample 3) wherein its first stage weights are

Table 4.1 Sample profile of the study

Samples	*Context*	*Study period*	*Normalization technique*	*Consolidation technique*
Sample 1	28 states and 7 union territories of India	2003–22	Max–min approach	Inverse of Euclidean Distance
Sample 2	26 Asian countries	2013–22	Max–min approach	Factor Analysis & Weighted Arithmetic Mean (1st stage) Inverse of Euclidean Distance (2nd stage)
Sample 3	12 Asian and 12 European countries	2013–22	Max–min approach	Principal Component Analysis & Inverse of Euclidean Distance (at both stages)
Sample 4	46 countries from the continent of Africa (16), Asia (10), Europe (11), North (3) and South (6) America	2012–21	Max–min approach	Factor Analysis & Weighted Arithmetic Mean (at both stages)

Source Authors Consideration

[1] Selection between PCA and Factor Analysis depends on the characteristics of the data to be dealt with. When variables share high and significant correlation with each other, Factor Analysis appears to be a preferred selection over PCA.

found for the variables in order to consolidate into a dimension and at the second stage, weights are used to subsume the categories into a final achievement score.

Let $\overrightarrow{X}$ be the vector of intended variables (say v_i) observed across the Tables 5.3–5.5, $\overrightarrow{F}$ be vector of m_i random unobservable variables referred as common factors of $\overrightarrow{X}$ and $\overrightarrow{\epsilon}$ be the vector of specific factors of $\overrightarrow{X}$ where, by construction $\overrightarrow{F} \leq \overrightarrow{X}$ or, ($m_i \leq v_i$). Therefore, the m -factor model would appear in Eq. (4.3).

$$\vec{Y} = L\vec{F} + \vec{\in} \tag{4.3}$$

where, $\overrightarrow{Y} = \overrightarrow{X} - \overrightarrow{\mu}$ (the centered values) and Cov ($\overrightarrow{X}$) = $\sum$, L = (l_{ij}) $(1 \leq i \leq v)$ & $(1 \leq j \leq m)$ is the factor loadings matrix whereas, l_{ij}, being the loading factor of the ith variable Y_i on the jth common factor F_j.

The conventional assumptions of the model being[2]:

- *Assumptions on Mean*:

E (ε_i) = 0 and E (l_i) = 0 (where, $i = 1, 2, 3, \ldots, j$) □ E ($\overrightarrow{X}$) = $\overrightarrow{\mu}$

- *Assumptions on Variance*:

σ (ε_i) = ψ_i (i.e. the specific variance) and σ (l_i) = 1 (where, $i = 1, 2, 3, \ldots, j$).

- *Assumptions on Association*:

Cov (l_i, l_j) = 0 (for $i \neq j$); Cov (ε_i, ε_j) = 0 (for $i \neq j$) and Cov (ε_i, l_j) = 0 (for $i \neq j$);

2 8 The model inherently assumes that the variables ('i's) are linear function of the common factors ('j's). Since the common factors are unobservable, we cannot check for linearity. The variance-covariance matrix forms a symmetric matrix. That means the variance between variable i and common factor j is same between j and i.

The variance-covariance matrix is going to have i (i + 1)/2 unique elements of $\sum$ (in Eq. 4.4) approximated by (i × j) factor loadings in matrix L and I specific variances. Implying [i (j + 1)] < < [i (i + 1)/2].

These assumptions are necessary to uniquely estimate the parameters of the model among an infinite number of equally well-fitted models.

These assumptions lead to the following equations:

$$\sum = LL' + \psi \tag{4.4}$$

$$L = \text{Cov}\,(\vec{Y}, \vec{F}) \tag{4.5}$$

$$\sigma^2_{Y_i} = \sum_{j=1}^{m} l^2_{ij} + \psi_i \tag{4.6}$$

where, $l^2{}_{ij}$ is the ith commonality (i.e. the proportion of Y_i explained by the common factor j) and ψ_i is the specific variance of that ith commonality.

This could be also viewed as a regression specification.

Where,

$V_1 = \mu_1 + l_{11}m_1 + l_{12}m_2 + \cdots\cdots + l_{1j}m_j + \varepsilon_1$ (a)

$V_2 = \mu_2 + l_{21}m_1 + l_{22}m_2 + \cdots\cdots + l_{2j}m_j + \varepsilon_2$ (b)

$V_3 = \mu_3 + l_{31}m_1 + l_{32}m_2 + \cdots\cdots + l_{3j}m_j + \varepsilon_3$ (c)

Similarly,

$V_i = \mu_i + l_{i1}m_1 + l_{i2}m_2 + \cdots\cdots + l_{ij}m_j + \varepsilon_i$ (d)

where, $V_1, V_2, V_3, \ldots, V_i$ denotes the data for the observed variables in the study, $m_1, m_2, m_3, \ldots, m_j$ denotes the unobservable common factors to be extracted, ε_1 to ε_j shows up the specific variance of each variable v, and μ_1 to μ_i are the intercepts of the specified regression equations.

The l_{ij}'s (partial slopes or the elasticities) for all the multiple regressions are referred as factor loadings and represents loading of the ith variable on the jth factor.

This in principal component analysis (PCA) is represented as a linear programming problem with:

Objective function: Maximize $\mu_1\, \mu_1{}^T$ S; Subject to constraint: $\mu_1\, \mu_1{}^T = 1$.

Incorporating the constraint and first order derivative in respect of μ_1 gives, $S\, \mu_1 = \lambda\, \mu_1$ or, $\mu_1\, \mu_1{}^T\, S = \lambda$, where, λ represents the eigenvalues from the covariance matrix S. Therefore, such a maximization problem can be solved by optimizing the λ (i.e. the Lagrange Multiplier).

Further, where the number of variables is not much high (i.e. in case of sample 1) we consider transposing the values from Euclidean Distance for consolidation following Eqs. 4.7 or 4.8.

$$\text{Variables sharing a positive affinity}: \text{IFI}_t = 1 - \frac{\sqrt{\sum_1^n \left[1 - \left\{\sum_1^n w_i \left(\frac{A_i - m_i}{M_i - m_i}\right)\right\}\right]^2}}{\sqrt{n}} \tag{4.7}$$

$$\text{Variables sharing a negative affinity}: \text{IFI}_t = 1 - \frac{\sqrt{\sum_1^n \left[1 - \left\{\sum_1^n w_i \left(\frac{M_i - A_i}{M_i - m_i}\right)\right\}\right]^2}}{\sqrt{n}} \tag{4.8}$$

Before directly going for PCA, we have also gone through a number of checklists, which validates the usage of PCA for our study. We have tested the sampling adequacy using the Kaiser–Meyer–Olkin's test (KMO), along with Bartlett's test of sphericity in order to fix whether the variables considered for the study are so related that enables us to detect a latent structure behind them. Similarly, cases where Factor analysis has been used, we consider studying whether the covariance matrix is diagonal, spherical and compound symmetric. While running PCA we have used two types of rotations based on the values found in the component correlation matrix. Rotation of axes is performed in PCA to amplify the variation contained in the dataset which in turn helps to present a clearer picture of the actual state. Considering the value of 0.32 or above to be significant as a proxy of association between factors identified by the PCA, we use the Direct Oblimin as a method of rotating the axes. Similarly, where the values fall below 0.32, we prefer to go with the orthogonal rotation of axes (i.e. the Varimax as a tool to rotate the axes). Such rotation of axes gives us the opportunity to capture the maximum of the variation in the dataset, which leads to having with us a clearer picture of the state of affairs. In order to consolidate the attainments of variables into dimensions and further into the final index, we have analyzed the changes relating to the variables over the study period. By plotting the changes for each variable over time, we are able to fix the method that seems appropriate for the purpose of consolidation. The movement of a particular variable over time may show a linear growth or fall or a non-linear growth or fall. Where a non-linear change may stand convex or concave or L-shaped change to its origin. A number of studies have used the arithmetic mean as a measure to consolidate variables into dimensions or dimensions into the final index or both Arora (2010), and Rahman (2012).

Using arithmetic mean presupposes that the variables are linearly growing or falling and retains the concept of perfect substitutability within variables. Some of the studies have opted for the geometric mean for consolidation assuming that the variables are growing or falling with a non-linear (concave to origin) curve Gupte et al. (2012), Amidžić et al. (2014), and Saravanabhavan (2018). A very few studies have assumed the growth or falling trend to be an L-shaped (also known as "The Leontief Curve") curve to its origin. Studies like Chakravarty and Pal (2013) have assumed the growth or fall to be in the form of convex to origin curve where the convexity is assumed to be within a range of $0 < r < 1$. This range of convexity addresses the demerit of perfect substitutability within the variables. Among others, a majority of studies uses the transpose of the Euclidean distance method to combine the values of variables into dimensions and further into the index. The concept evolves its root from the study dated back to 1974 by Zeleny. In our study, after observing the changes we assign the appropriate techniques for consolidating the variables which remove any sort of assumption that might lead to divergence in between theoretical and statistical (data-driven) evidence.

4.4 Investigation of Indian States and UTs

The union territory of Chandigarh drives ahead in regards to average IFI scores, utilizing the data collected from India's 28 states and 7 union territories for 2003–2022, obtaining an average score of 0.715, then come the individual states of Goa (0.702) and Kerala (0.539). Furthermore, the North-East states [Manipur (0.041) and Nagaland (0.026)] occupy the bottom of the ranking (Chotia & Rao, 2014). Additionally, when the variance in average IFI values is considered, the observation shows that, among every state and union territory, Andaman and Nicobar (0.088) seem to be the steadiest, while West Bengal (0.94) appears to be the most variable.

These scores are however based on three indicators that represents the dimension of service outreach (i.e. per capita number of bank branches, savings, and credit accounts) so far, the measure of financial inclusion is concerned.

Results for the top three contenders (i.e. Chandigarh, Goa, and Kerala) in financial inclusion achievements are due to the fact that these states/ UT's have a relatively high standard of living backed by relatively high per capita income and educational facilities (Kumar & Mishra, 2011). Further the states like Manipur and Nagaland have relatively poor performance due to their geographical position, unemployment and low-level of education. However, due to the unavailability of disintegrated data based on rural–urban classifications, emphasis with regard to social groups, wealth and other relevant aspects, this study could not emphasize on the pressing issues like the inequality in the provisioning of the financial services especially the allocation of credit and its respective utilization. This is very likely that industrialized states and UT's might exhibit better relative positions than compared to their alternatives in financial inclusion standing due to fact that industries solely can absorb credit to a large extent compared to individuals. This may lead to an upward bias which the present study is supposed to suffer from. Further, the data that we are proceeding with is limited to the organized sector. Putting it in other way, the substantial amount of finance from the unorganized sources is omitted from quantification of the effects. Thus, collection of disintegrated data as part of a primary survey might improve the insights to a greater detail (Table 4.2).

4.5 Assessment Across the Asian Continent

Using an average of the degree of inclusion attained through the investigation (12th column of Table 4.3) across the 26 Asian nations, the study discovers that South Korea leads with a score of 0.641, ahead of countries from the whole continent's eastern and northernmost regions. Further, the South Korea is acclaimed to have the highest number of bank account ownership (Bae, 2022). This sample takes into account 14 distinct metrics that include critical characteristics such as Stability; Service Potential; Accessibility, Availability, and Efficiency, which characterize financial inclusion comprehensively. In parallel, a review of the country experiences across the time based on the reported variations, we observe South Korea has maintained a steadier presence than its rivals in the overall order of the IFI rankings. However, when it comes to IFI accomplishments, Myanmar's performance is quite weak. It displays substantial changes over time, with an average accomplishment value of 0.170 contrasted with South Korea (the leader). India, for example, holds

Table 4.2 Overview of the financial inclusion index

UT's	*Average*	*SD*	*Coefficient of Variation*[b]	*Position*[a]
Andaman & Nicobar Islands	0.205	0.017	0.088	1
Puducherry	0.476	0.047	0.104	2
Delhi	0.517	0.069	0.14	3
Lakshadweep	0.31	0.046	0.155	4
Daman & Diu	0.213	0.036	0.177	5
Chandigarh	0.715	0.18	0.264	6
Dadra & Nagar Haveli	0.193	0.077	0.418	7
States	*Average*	*SD*	*Coefficient of Variation*[b]	*Position*[a]
Uttar Pradesh	0.125	0.013	0.102	1
Himachal Pradesh	0.38	0.038	0.103	2
Kerala	0.539	0.056	0.107	3
Karnataka	0.424	0.047	0.118	4
Goa	0.702	0.08	0.12	5
Punjab	0.407	0.048	0.126	6
Uttarakhand	0.336	0.042	0.131	7
Gujarat	0.216	0.029	0.143	8
Maharashtra	0.329	0.056	0.176	9
Odisha	0.215	0.038	0.186	10
Rajasthan	0.148	0.027	0.19	11
Tamil Nadu	0.49	0.095	0.202	12
Haryana	0.299	0.062	0.218	13
Jharkhand	0.141	0.029	0.222	14
Arunachal Pradesh	0.145	0.033	0.239	15
Jammu & Kashmir	0.245	0.056	0.239	16
Sikkim	0.289	0.068	0.247	17
Madhya Pradesh	0.137	0.033	0.251	18
Andhra Pradesh	0.328	0.085	0.272	19
Mizoram	0.229	0.08	0.368	20
Tripura	0.257	0.091	0.374	21
Bihar	0.072	0.026	0.383	22
Meghalaya	0.175	0.066	0.398	23
Assam	0.121	0.048	0.423	24
Chhattisgarh	0.117	0.05	0.457	25
Nagaland	0.026	0.019	0.765	26
Manipur	0.041	0.037	0.939	27
West Bengal	0.216	0.193	0.94	28

[a] Positions are based on the values of [b] Coefficient of Variation (CV) obtained
Source Authors Calculation

the third position in overall mean IFI values obtained for the study period, and holds the top position in the South Asian Association for Regional Cooperation (SAARC) category to pass the 0.307 (2nd quartile) thresholds. The study found that the coefficient of variation, or CV, for the collection of countries according to average IFI scores from 2013 to 2022 shows that a large number of nations in the Southeast as well as the East are ahead. As far as mean IFI values during 2013–2022 are concerned, Japan (ranked first), Malaysia (ranked second), and South Korea (ranked third) are among the most stable economies. Nevertheless, the inclusive atmosphere in countries such as Turkey (23rd), Azerbaijan (24th), Jordan (25th), and Mongolia (26th) is determined to be particularly volatile (see Table 4.4). Year on year financial inclusion achievements along with the dimensional contributions throughout the study period (i.e. 2013–2022) can be studied from Table 4.3.1–4.3.10 provided in the appendix of the study. So far as the achievement in the dimension of Service Potential is concerned, Japan takes a consistent lead across the Asian group except in 2014, where Thailand took the top position. Service potential mainly stresses on the capacity of the service providers to render financial services on a continuous basis to the clients at affordable prices. The effectiveness of the service provider is an essential element that determines the attainment of financial inclusion. More efficient service providers are able to render services for a longer period of time. Among the 26 Asian countries Maldives holds the top position consistently across 2013–2016, whereas during 2017–2022, Malaysia took the first position in terms of the dimension of efficiency is concerned. Comparative view across the dimension of stability among the service providers show that Cambodia to be placed at the top consistently for 2013–2022, except in 2017 where Brunei Darussalam takes the lead. This sample uses the strength of legal right index as a sole indicator that represents the stability of the service provider. It basically measurers the degree of protection that the law of the land offers to the borrower as well as the lender to protect their right and facilitate free flow of funds between the deficit and surplus sectors. So far, the dimension of outreach is concerned basic financial services are extensively available across the Maldives holding the top position during 2013–2016 and 2021–2022 whereas South Korea remaining at the top during 2017–2020. In the dimension of accessibility across 2013–2022 South Korea remains at the top, except in 2015 where Japan takes the lead. Review of India's position shows that across all the five dimensions (except the dimension of availability) it shows a consistent growth from 2016 onwards. However, the dimension of availability shows a substantial fall during 2016–2017. These results were part of consequences of the aggressive policy initiatives taken by the government of India to promote accessibility to financial services more comprehensively especially across the rural economy which forms the background of the Indian economy. Availability of financial services is however related to building up of adequate infrastructure where it may take some time to reflect the progress.

So far, the outreach of financial services is concerned the position of Afghanistan and partially for Mongolia as compared to the group leaders seems to be fragile. Although with relatively lesser availability, the accessibility to financial services for Pakistan is reasonably high.

Table 4.4 shows a strong positive relationship between income and mean IFI values across countries categorized on GNP per head (PPP terms on constant international $). As demonstrated by the statistical conclusion in the next section of the research, shifting between a low-income to a high-income nation boosts the mean IFI score.

Among the stated income groups, South Korea, Georgia, India, and Nepal had the strongest IFI average scores for the full research period. When average IFI scores across the research period are taken into account, it is clear that only South Korea (3rd), Georgia (15th), India (20th), and Nepal (5th) have an acceptable degree of consistency.

While elaborating on its performance, the Deputy Governor, Bank of Korea, states that the economy is sufficiently stable and secure to manage its financial transactions (Bae, 2022). The account ownership remains as high as 95% of its population and is barely achieved by any of the advanced economies. Further, the penetration of digitization allows a real-time transfer over a 24 × 7 window from 2001 onwards, when Korea became the first country to introduce a retail based fast payment system under the aegis of 'Electronic Banking System'. With a high smartphone penetration rate (i.e. 95%) offering a steady growth in the number of internet and mobile banking users, usage have shown a leapfrog in the numbers especially after the COVID-19 pandemic. Faster authentication (e.g. passwords, fingerprints etc.), pre-stored information in the mobile application and payment gateways at the terminals are some of the initiatives those are accelerating the numbers to deliver hassle free payment and transfers. The whole system was part of the strategic move of the Bank of

Table 4.3 Index of financial inclusion scores for the study period

Country Name	*IFI_ 2013*	*IFI_ 2014*	*IFI_ 2015*	*IFI_ 2016*	*IFI_ 2017*	*IFI_ 2018*	*IFI_ 2019*	*IFI_ 2020*	*IFI_ 2021*	*IFI_ 2022*	*Average*	*S.D*	*C.V*	*Rank*
Japan	0.605	0.585	0.637	0.569	0.613	0.605	0.609	0.613	0.619	0.625	0.608	0.019	0.032	1
Malaysia	0.458	0.436	0.440	0.428	0.480	0.460	0.463	0.467	0.471	0.474	0.458	0.017	0.038	2
Korea, Rep	0.610	0.595	0.618	0.600	0.657	0.646	0.656	0.666	0.676	0.686	0.641	0.033	0.051	3
Brunei Darussalam	0.414	0.408	0.382	0.390	0.457	0.430	0.437	0.444	0.450	0.457	0.427	0.027	0.064	4
Nepal	0.307	0.299	0.319	0.331	0.314	0.316	0.307	0.295	0.279	0.259	0.303	0.021	0.069	5
Cambodia	0.374	0.381	0.382	0.402	0.412	0.419	0.429	0.439	0.449	0.458	0.414	0.030	0.071	6
Myanmar	0.179	0.197	0.164	0.149	0.187	0.167	0.166	0.165	0.164	0.163	0.170	0.014	0.081	7
Saudi Arabia	0.270	0.273	0.268	0.261	0.312	0.298	0.305	0.313	0.320	0.327	0.295	0.024	0.083	8
Afghanistan	0.260	0.250	0.218	0.234	0.224	0.217	0.213	0.210	0.207	0.205	0.224	0.019	0.083	9
United Arab Emirates	0.333	0.333	0.361	0.352	0.376	0.382	0.393	0.403	0.414	0.424	0.377	0.032	0.086	10
Thailand	0.433	0.425	0.432	0.424	0.499	0.482	0.495	0.508	0.521	0.534	0.475	0.043	0.090	11
Bhutan	0.334	0.331	0.333	0.341	0.345	0.358	0.372	0.389	0.408	0.431	0.364	0.035	0.096	12
Vietnam	0.378	0.365	0.389	0.389	0.447	0.442	0.458	0.475	0.491	0.507	0.434	0.051	0.117	13
Maldives	0.487	0.499	0.444	0.448	0.431	0.413	0.396	0.380	0.363	0.347	0.421	0.051	0.120	14
Georgia	0.508	0.507	0.483	0.488	0.432	0.432	0.415	0.398	0.381	0.363	0.441	0.053	0.120	15
Lebanon	0.450	0.427	0.484	0.468	0.456	0.446	0.425	0.397	0.361	0.318	0.423	0.051	0.121	16
Indonesia	0.399	0.407	0.390	0.407	0.383	0.373	0.355	0.333	0.307	0.276	0.363	0.045	0.124	17
Pakistan	0.252	0.263	0.245	0.264	0.316	0.306	0.319	0.332	0.345	0.358	0.300	0.041	0.137	18
Bangladesh	0.377	0.374	0.344	0.327	0.319	0.299	0.283	0.267	0.250	0.234	0.307	0.050	0.161	19
India	0.416	0.413	0.413	0.399	0.436	0.455	0.491	0.536	0.590	0.653	0.480	0.087	0.181	20
Philippines	0.228	0.223	0.229	0.223	0.307	0.289	0.305	0.321	0.337	0.353	0.281	0.051	0.182	21

Country Name	*IFI_ 2013*	*IFI_ 2014*	*IFI_ 2015*	*IFI_ 2016*	*IFI_ 2017*	*IFI_ 2018*	*IFI_ 2019*	*IFI_ 2020*	*IFI_ 2021*	*IFI_ 2022*	*Average*	*S.D*	*C.V*	*Rank*
Armenia	0.408	0.394	0.372	0.406	0.400	0.430	0.466	0.512	0.568	0.635	0.459	0.086	0.188	22
Turkey	0.376	0.374	0.358	0.359	0.385	0.407	0.443	0.490	0.547	0.615	0.435	0.088	0.203	23
Azerbaijan	0.254	0.316	0.253	0.231	0.301	0.302	0.332	0.369	0.415	0.469	0.324	0.075	0.232	24
Jordan	0.182	0.154	0.239	0.212	0.258	0.272	0.294	0.315	0.336	0.357	0.262	0.066	0.252	25
Mongolia	0.370	0.357	0.360	0.336	0.383	0.409	0.456	0.516	0.589	0.676	0.445	0.114	0.256	26
Mean	**0.372**	**0.369**	**0.368**	**0.363**	**0.390**	**0.387**	**0.395**	**0.406**	**0.418**	**0.431**				

Source: Authors Calculation

Table 4.4 Financial inclusion index over the study period (based on income groups)

Country Name	*IFI 2013*	*IFI 2014*	*IFI 2015*	*IFI 2016*	*IFI 2017*	*IFI 2018*	*IFI 2019*	*IFI 2020*	*IFI 2021*	*IFI 2022*	*Average IFI*
Korea, Rep	0.610	0.595	0.618	0.600	0.657	0.646	0.656	0.666	0.676	0.686	0.641
Japan	0.605	0.585	0.637	0.569	0.613	0.605	0.609	0.613	0.619	0.625	0.608
Brunei Darussalam	0.414	0.408	0.382	0.390	0.457	0.430	0.437	0.444	0.450	0.457	0.427
United Arab Emirates	0.333	0.333	0.361	0.352	0.376	0.382	0.393	0.403	0.414	0.424	0.377
Saudi Arabia	0.270	0.273	0.268	0.261	0.312	0.298	0.305	0.313	0.320	0.327	0.295
High	0.446	0.439	0.453	0.434	0.483	0.472	0.480	0.488	0.496	0.504	0.470
Georgia	0.508	0.507	0.483	0.488	0.432	0.432	0.415	0.398	0.381	0.363	0.441
Maldives	0.487	0.499	0.444	0.448	0.431	0.413	0.396	0.380	0.363	0.347	0.421
Lebanon	0.450	0.427	0.484	0.468	0.456	0.446	0.425	0.397	0.361	0.318	0.423
Malaysia	0.458	0.436	0.440	0.428	0.480	0.460	0.463	0.467	0.471	0.474	0.458
Thailand	0.433	0.425	0.432	0.424	0.499	0.482	0.495	0.508	0.521	0.534	0.475
Armenia	0.408	0.394	0.372	0.406	0.400	0.430	0.466	0.512	0.568	0.635	0.459
Turkey	0.376	0.374	0.358	0.359	0.385	0.407	0.443	0.490	0.547	0.615	0.435
Azerbaijan	0.254	0.316	0.253	0.231	0.301	0.302	0.332	0.369	0.415	0.469	0.324
Jordan	0.182	0.154	0.239	0.212	0.258	0.272	0.294	0.315	0.336	0.357	0.262
Upper-Middle	0.395	0.392	0.389	0.385	0.405	0.405	0.414	0.426	0.440	0.457	0.411
India	0.416	0.413	0.413	0.399	0.436	0.455	0.491	0.536	0.590	0.653	0.480
Indonesia	0.399	0.407	0.390	0.407	0.383	0.373	0.355	0.333	0.307	0.276	0.363
Cambodia	0.374	0.381	0.382	0.402	0.412	0.419	0.429	0.439	0.449	0.458	0.414
Vietnam	0.378	0.365	0.389	0.389	0.447	0.442	0.458	0.475	0.491	0.507	0.434
Mongolia	0.370	0.357	0.360	0.336	0.383	0.409	0.456	0.516	0.589	0.676	0.445
Bangladesh	0.377	0.374	0.344	0.327	0.319	0.299	0.283	0.267	0.250	0.234	0.307

Country Name	*IFI 2013*	*IFI 2014*	*IFI 2015*	*IFI 2016*	*IFI 2017*	*IFI 2018*	*IFI 2019*	*IFI 2020*	*IFI 2021*	*IFI 2022*	*Average IFI*
Bhutan	0.334	0.331	0.333	0.341	0.345	0.358	0.372	0.389	0.408	0.431	0.364
Pakistan	0.252	0.263	0.245	0.264	0.316	0.306	0.319	0.332	0.345	0.358	0.300
Philippines	0.228	0.223	0.229	0.223	0.307	0.289	0.305	0.321	0.337	0.353	0.281
Myanmar	0.179	0.197	0.164	0.149	0.187	0.167	0.166	0.165	0.164	0.163	0.170
Lower-Middle	0.331	0.331	0.325	0.324	0.354	0.352	0.363	0.377	0.393	0.411	0.356
Nepal	0.307	0.299	0.319	0.331	0.314	0.316	0.307	0.295	0.279	0.259	0.303
Afghanistan	0.260	0.250	0.218	0.234	0.224	0.217	0.213	0.210	0.207	0.205	0.224
Lower	0.284	0.275	0.269	0.282	0.269	0.266	0.260	0.253	0.243	0.232	0.263

Source Authors Calculation

Korea and the government in action back in 2019 that allows the financial institutions to compete in open market and customize their services that fits the consumer needs and preferences. However, concerns are still there on the fairness of credit provisioning. Individuals with no or less collateral are devoid of getting loans from institutional arrangements and are exposed to predatory money lenders charging high rate of interest. The reliance on credit card seems to be exceptionally high (almost 53.8%) due to the promotion of credit cards following the 1997's financial crisis. Certain policies that promote to increase consumption and allow refunds of income tax based on credit card purchases somehow infuses instability in the individual's asset portfolio due to enhanced debt burdens.

Further while discussing the scenario of second rank holder (i.e. Japan) in terms of mean achievement score in financial inclusion and the top contender in terms of stability, we find a prominent positioning of the postal savings system that drives the financial inclusion initiatives across the Japanese population. In fact, in Japan the postal savings system is one of the largest savings institutions that are operational across the world with a widespread network of post offices through which it carries on the operations (Yoshino et al., 2018). The postal savings products were so attractive to the Japanese citizens that the accumulated savings had reached an amount of ¥262 trillion as of 1999 which was almost 50% of the country's GDP. This attraction emancipates from the regulatory liberty that the postal savings institutions enjoy of issuing a ten-year time deposit that other players cannot offer. While identifying the driving forces behind the intensity to save, Horioka et al. (2007) find three major factors. Firstly, growth in income or the economy per se is followed by the savings behavior of an individual. That means when an individual observes his/her income is growing they start to save more. Secondly, the amount and intensity to save is driven by the number of children a particular household contains. People tend to save more when they have more children. Finally, the condition of the social welfare system determines the individual's drive to save for his/her future or retirement per se.

However, the growth in provisioning finance for all across Japan is mainly driven by the dominance of government banks infused by the Financial Investment Loan Program (FILP) that the Ministry of Finance has taken charge of. Savings collected is channelized into loans through government banks or directly put into loans through by post offices. Government banks assume the leading role while extending credit and often join hands with the private players. However, the direction of the loans is guided by the government owned banks (often referred as the 'Cow bell effect'. Almost 50% of the corpus was allocated for improving the standard of living, involving the housing facilities, promotion of social welfare and education following the FILP in 1999–2000. This however declined to 29% in 2010 and fell to 19% in 2016. However, upgradation of the sectors exhibiting lower levels of productivity was made by improving the allocation through doubling the amount by 2010.

The attraction of the postal savings schemes among the Japanese citizens were not only due to extensive network coverage (almost 24,167 branches as of 2017) which had better accessibility but also due to the fact that they offer a better safety as an option especially under the backdrop of the burst of the bubble in 1989 when the majority of the banks were faced with bankruptcy.

Further review of the countries lying down the list (e.g. Myanmar) with high fluctuations in the achievement scores, we find a lot of reasons for such weak performance especially following the research of Tambunlertchai (2018). This article is based on the primary survey of 5100 respondents across the country and has attempted to outline the demand-side barriers that act as hindrances for the financial sector to grow. Following the FinScope Myanmar, a nation-wide survey of individuals in 2013 shows that only 30% of the adults have access to formal financial services whereas 39% of the adults have been excluded from the ambit of formal finance. So far, the usage of informal finance is concerned the proportion comes to 31% leading to rising costs of financing (Chamberlain et al., 2014). The financial landscape of Myanmar is underdeveloped and dominated by banks [mainly state-owned banks (GIZ, 2013)] holding nearly 99% of the formal financial sectors' assets (Turnell, 2014). However, the concentration of ownership is on transition from the state ownership to private hands, expansion of the branches to remotely located areas while the available infrastructure is yet limited (Duflos et al., 2013).

So far economic transition is concerned the growth rates for the year 2015 and 2016 stands at 7% and 5.9% respectively with forecasted growth at 8% for 2017 and 2018 fiscal years[3] compared to the average growth for the SAARC group at 5% (ADB, 2017). Despite a surge in the digits,

[3] Fiscal year being from 1st April to 31st March.

Myanmar is still one of the poorest nations having 61 million of population with an average income of USD 3.5 per day and half of which can hardly surpass USD 2 per day threshold. The agrarian based economy has an attainment of higher education of just 7% with employment rate of 3.4% of the total available labor force and with strikingly low internet connectivity and mobile phone penetration rates of just 1.2% (including 0.2% having fixed broadband connectivity) and 12.8% respectively. These numbers are far below the SAARC group average and even below than the attainments of the lower-income groups (World Bank, 2015). Thus, the review shows that the economy along with the development of the financial sector is still in its nascent stage and justifies its standing as part of our empirical findings.

4.6 Investigation of Asia and Europe

Considering simply the mean of the inclusion rates achieved during the study's entire period (12th column of Tables 4.5 and 4.6), the investigation finds that the Netherlands dominates having a score of 0.968, next to Italy (0.962) then Japan (0.959) from the Asian continent. Furthermore, despite being near the bottom, the Pakistan has a value that is higher than the average (i.e. 0.527). India ranks 11th considering all the nations taken together and 6th against the other Asian countries with an inclusion index score of 0.787.

Evaluating the components of stability and viability in Table 4.5.1–4.5.10 in the appendix, it was discovered that a favorable atmosphere continually increased financial inclusion in countries such as the Maldives, Turkey, and the Netherlands during the study period. Except in 2016, when India outperforms Pakistan, data reveal that both Pakistan and Bhutan do better among India's neighbors. Furthermore, assessing the scale of efficiency indicates mix evidence where countries like Saudi Arabia, Spain from Europe and Japan, Pakistan and Maldives from Asia have been among the most consistent countries across the period of study, while Bhutan and Greece were among the least consistent ones. Additionally, with regard to depth, Switzerland and Japan have constantly been at the top during the period of study except in 2022 where Malaysia takes the lead, whereas Pakistan has continually been at the bottom.

Likewise, rankings in the banking and financial services outreach component remain stable, having Japan and Switzerland continuing to lead and Pakistan with Latvia (in 2014) at the bottom.

Table 4.5 Financial inclusion index over the study period (based on continents)

Countries	*IFI (2013)*	*IFI (2014)*	*IFI (2015)*	*IFI (2016)*	*IFI (2017)*	*IFI (2018)*	*IFI (2019)*	*IFI (2020)*	*IFI (2021)*	*IFI (2022)*	*Average IFI*
Latvia	0.693	0.655	0.659	0.633	0.628	0.621	0.615	0.610	0.605	0.601	0.632
Italy	0.965	0.920	0.973	0.957	0.961	0.963	0.966	0.969	0.972	0.975	0.962
Estonia	0.795	0.758	0.704	0.757	0.718	0.710	0.704	0.698	0.693	0.688	0.723
Croatia	0.898	0.862	0.884	0.858	0.858	0.854	0.850	0.847	0.845	0.842	0.860
Czech Republic	0.769	0.668	0.786	0.660	0.638	0.579	0.508	0.425	0.329	0.220	0.558
Greece	0.914	0.885	0.872	0.840	0.819	0.796	0.772	0.749	0.726	0.702	0.807
Netherlands	0.940	0.916	0.981	0.962	0.971	0.975	0.979	0.983	0.986	0.988	0.968
Spain	0.978	0.915	0.952	0.932	0.923	0.918	0.914	0.911	0.908	0.905	0.925
Switzerland	0.911	0.873	0.940	0.923	0.937	0.947	0.958	0.968	0.978	0.988	0.942
Bosnia and Herzegovina	0.761	0.695	0.753	0.702	0.705	0.700	0.696	0.692	0.689	0.686	0.708
Macedonia, FYR	0.746	0.732	0.760	0.669	0.675	0.655	0.637	0.618	0.601	0.583	0.668
Georgia	0.659	0.686	0.656	0.679	0.676	0.678	0.679	0.680	0.681	0.682	0.676
Average IFI (Europe)	0.836	0.797	0.827	0.798	0.792	0.783	0.773	0.762	0.751	0.738	0.786
Saudi Arabia	0.720	0.751	0.770	0.757	0.775	0.781	0.785	0.789	0.793	0.796	0.772
Israel	0.820	0.779	0.833	0.860	0.866	0.883	0.901	0.918	0.935	0.953	0.875
Japan	0.958	0.911	0.954	0.975	0.960	0.963	0.965	0.966	0.968	0.969	0.959
Malaysia	0.800	0.768	0.875	0.822	0.834	0.822	0.799	0.766	0.724	0.671	0.788
Maldives	0.836	0.780	0.804	0.821	0.800	0.798	0.796	0.795	0.793	0.792	0.802
Turkey	0.807	0.830	0.806	0.795	0.752	0.696	0.623	0.532	0.426	0.302	0.657
Armenia	0.698	0.773	0.720	0.697	0.708	0.702	0.697	0.691	0.686	0.680	0.705
Pakistan	0.586	0.635	0.597	0.529	0.533	0.514	0.495	0.477	0.460	0.443	0.527
Indonesia	0.689	0.753	0.683	0.644	0.641	0.622	0.604	0.586	0.569	0.552	0.634

(continued)

Table 4.5 (continued)

Countries	*IFI (2013)*	*IFI (2014)*	*IFI (2015)*	*IFI (2016)*	*IFI (2017)*	*IFI (2018)*	*IFI (2019)*	*IFI (2020)*	*IFI (2021)*	*IFI (2022)*	*Average IFI*
Philippines	0.484	0.568	0.685	0.588	0.667	0.686	0.702	0.716	0.729	0.740	0.656
India	0.741	0.774	0.785	0.773	0.790	0.795	0.799	0.802	0.806	0.808	0.787
Bhutan	0.650	0.750	0.656	0.614	0.618	0.600	0.582	0.565	0.548	0.532	0.611
Average IFI (Asia)	0.732	0.756	0.768	0.740	0.745	0.738	0.729	0.717	0.703	0.687	0.731
Grand Average of IFI	0.784	0.776	0.797	0.769	0.769	0.761	0.751	0.740	0.727	0.712	

Source Authors Calculation

Table 4.6 Financial Inclusion Index Over the Study Period (Based on Income Groups)

Countries	*IFI (2013)*	*IFI (2014)*	*IFI (2015)*	*IFI (2016)*	*IFI (2017)*	*IFI (2018)*	*IFI (2019)*	*IFI (2020)*	*IFI (2021)*	*IFI (2022)*	*Average IFI*
Latvia	0.693	0.655	0.659	0.633	0.628	0.621	0.615	0.610	0.605	0.601	0.632
Italy	0.965	0.920	0.973	0.957	0.961	0.963	0.966	0.969	0.972	0.975	0.962
Estonia	0.795	0.758	0.704	0.757	0.718	0.710	0.704	0.698	0.693	0.688	0.723
Croatia	0.898	0.862	0.884	0.858	0.858	0.854	0.850	0.847	0.845	0.842	0.860
Czech Republic	0.769	0.668	0.786	0.660	0.638	0.579	0.508	0.425	0.329	0.220	0.558
Greece	0.914	0.885	0.872	0.840	0.819	0.796	0.772	0.749	0.726	0.702	0.807
Netherlands	0.940	0.916	0.981	0.962	0.971	0.975	0.979	0.983	0.986	0.988	0.968
Spain	0.978	0.915	0.952	0.932	0.923	0.918	0.914	0.911	0.908	0.905	0.925
Switzerland	0.911	0.873	0.940	0.923	0.937	0.947	0.958	0.968	0.978	0.988	0.942
Saudi Arabia	0.720	0.751	0.770	0.757	0.775	0.781	0.785	0.789	0.793	0.796	0.772
Israel	0.820	0.779	0.833	0.860	0.866	0.883	0.901	0.918	0.935	0.953	0.875
Japan	0.958	0.911	0.954	0.975	0.960	0.963	0.965	0.966	0.968	0.969	0.959
Average IFI (High Income)	0.863	0.824	0.859	0.843	0.838	0.833	0.826	0.819	0.811	0.802	0.832
Bosnia and Herzegovina	0.761	0.695	0.753	0.702	0.705	0.700	0.696	0.692	0.689	0.686	0.708
Macedonia, FYR	0.746	0.732	0.760	0.669	0.675	0.655	0.637	0.618	0.601	0.583	0.668
Georgia	0.659	0.686	0.656	0.679	0.676	0.678	0.679	0.680	0.681	0.682	0.676
Malaysia	0.800	0.768	0.875	0.822	0.834	0.822	0.799	0.766	0.724	0.671	0.788
Maldives	0.836	0.780	0.804	0.821	0.800	0.798	0.796	0.795	0.793	0.792	0.802
Turkey	0.807	0.830	0.806	0.795	0.752	0.696	0.623	0.532	0.426	0.302	0.657
Armenia	0.698	0.773	0.720	0.697	0.708	0.702	0.697	0.691	0.686	0.680	0.705
Pakistan	0.586	0.635	0.597	0.529	0.533	0.514	0.495	0.477	0.460	0.443	0.527
Indonesia	0.689	0.753	0.683	0.644	0.641	0.622	0.604	0.586	0.569	0.552	0.634

(continued)

Table 4.6 (continued)

Countries	*IFI (2013)*	*IFI (2014)*	*IFI (2015)*	*IFI (2016)*	*IFI (2017)*	*IFI (2018)*	*IFI (2019)*	*IFI (2020)*	*IFI (2021)*	*IFI (2022)*	*Average IFI*
Philippines	0.484	0.568	0.685	0.588	0.667	0.686	0.702	0.716	0.729	0.740	0.656
India	0.741	0.774	0.785	0.773	0.790	0.795	0.799	0.802	0.806	0.808	0.787
Bhutan	0.650	0.750	0.656	0.614	0.618	0.600	0.582	0.565	0.548	0.532	0.611
Average IFI (Low Income)	0.705	0.729	0.732	0.694	0.700	0.689	0.676	0.660	0.642	0.623	0.685
Grand Average	0.784	0.776	0.795	0.769	0.769	0.761	0.751	0.740	0.727	0.712	0.758

Source Authors Calculation

The verification of India's position across the sample nations reveals a reasonable success in terms of stability and viability. Furthermore, considering accomplishments in the dimension of efficiency suggests that India is in a considerably better situation. However, the situation with respect to depth and service extension is not very encouraging, since accomplishments are significantly below the sample average. Review of Table 4.5.11 in the appendix shows the factor loadings of the variables applying PCA (1st stage). This gives us a primitive idea on the contribution of these variables towards building up of the index.

Among the high-income OECD countries 83% of the adult's report to have an account in a financial institution of which 95% as of 2014 agrees that they use their Debit-cum ATM card for making payments directly (Demirgüç-Kunt et al., 2015). Netherlands being one of the member countries of the OECD high-income group exhibits to have 99% of its adults to hold an account with a formal financial intermediary. Further across the group of adults belonging to the poorest 40%, the households reported to have accessibility to formal financial accounts stood at 99% (World Bank, 2015). Among the Asian countries Japan shows an account penetration rate of 97% of which 95% belongs to the poorest 40% of households. While going down at the bottom, Bhutan is reported to have an account penetration rate of 34% where the marginalized 40% in terms of their income shows to have account ownership at the rate of 25% only. Therefore, the findings of our study are in consonance with the results as provided by the World Bank (2015).

Among the latest buzz going around the adoption of revised Payment Services Directive (PSD2) and Open Banking (OB) across the European Union, very ample evidence shows up the implications on the underbanked or even the unbanked population (Preziuso et al., 2023). Although across the EU this proportion of unbanked is merely close to 4%, but the far-flung consequences of adopting PSD2 and OB are yet to emphasize. The conventional financial intermediaries have become incompetent to innovate especially after the financial crisis of 2008 (Arner et al., 2016). The digital wave as rippled by the introduction of the FinTechs have captivated the customers confidence by offering tailored services to their clients making the transactions more agile, transparent and affordable (Gomber et al., 2017). Through offering personalized and data-driven solutions the FinTechs are able to conquer the 1.4 billion of unbanked worldwide (Findex, 2022). Through democratizing the services FinTechs aim to ease out the economic, social and geographical barriers (Philippon, 2019). By improving transparency, the reliance on intermediation can be mitigated resulting into crowdfunding to collect funds in somewhat unconventional way though being under the regulations (Nguyen et al., 2021) and by aborting blockchain platforms (Scott et al., 2017). This could indeed help the vulnerable especially the disadvantaged groups (e.g. lower-income groups, women and immigrants) to survive and eventually earn a living (Dorfleitnet et al., 2021). The disruptions in the conventional banking habits are already in place through rising the income and consumption levels (Lashitew et al., 2019) alongside the enhanced usage of the mobile phones to transact money (Ouma et al., 2017). As a downside this can also contribute to increase indebtedness and infuse instability which might shake the overall stability of an economy (Bateman et al., 2019). Therefore, the need of a suitable balance becomes an obvious trade-off between the conventional banking and the FinTechs to have consumer protection as well as financial stability associated to innovative products and healthy competition (Van Loo, 2018). In this regard, the open banking (OB) regulations can be instrumental to make the system flexible to adaptations to the changing needs of the clients (Zachariadis & Ozcan, 2017). Adoption of OB can increase competition offering affordable rates of financial products while maintaining confidentiality of consumer rights simultaneously (European Commission, 2015). Penetration of technology can improve productivity and profitability while offering customized solutions to clients at competitive rates (Berger, 2003; Romanova et al., 2018). Thus, joining hands between FinTechs and the conventional banks could offer benefits and safely avoid the disruptive consequence that interventions of technology usually have (Schumpeter, 1942).

4.7 Global Perspective

Considering the Sample 4 consisting of 46 countries from the 5 continents over 2012–2021 it is observed from Table 4.7 that Namibia, Bangladesh, Turkey, Costa Rica and Ecuador seems to be the most stable nations among the African, Asian, European, North and South American continents so far as the values of coefficient of variation in financial inclusion scores are concerned. Continental average of achievements in financial inclusion during the study period across the sample nations show Europe (0.417) to take the lead followed by Asia (0.407), South America (0.360), North America (0.339) and Africa (0.263). Further,

Singapore (0.579), Mauritius (0.528), Italy (0.473), Brazil (0.463) and Costa Rica (0.393) from the continents of Asia, Africa, Europe, South America and North America respectively appears as the top contenders in regards to the average IFI scores throughout the study period. Testing the assumptions of the covariance matrix whether it is diagonal, spherical and compound symmetric of the variables and subsequently of the dimensions are provided in Table 4.7.1 and 4.7.2 respectively in the appendix. Results show that almost all the variables and dimensions during the period of study satisfies the stipulated assumptions of using factor analysis by rejecting the null hypothesis at an acceptable level of significance (mostly at 1 percent). The dimensional achievements across the study period are shown in Table 4.7.3 along with factor loadings of the variables and dimensions in Table 4.7.4 and 4.7.5 respectively in the appendix.

Review of the dimensional achievements show that on an average throughout the period of study (i.e. 2012–21) considering the dimension of Capital Formation, China (0.760) takes the lead followed by Brazil (0.567), Mauritius (0.521) and Singapore (0.504). This dimension particularly aims to emphasize on the basic finance function and thus represents the intrinsic capability of an economy to create surplus out of the intermediation between the surplus and deficit units (Beck et al., 2007). However, review of the positions shows Burundi (0.163) to be at the bottom. Further, review of the dimension of Efficiency shows China (0.656) at the top followed by Malaysia (0.636), Singapore (0.627) and Malta (0.605), whereas Cameroon (0.202) to be placed at the bottom. This dimension particularly focuses to represent the effectiveness and efficiency of service providers to extend financial services to their intended clients. Finally, across the dimension of Outreach this study finds Seychelles (0.696) at the top followed by Croatia (0.615), Singapore (0.596) and Estonia (0.543). Across the dimension of Outreach, Burundi (0.014) finds place at the bottom. This dimension particularly gives an overview of the attainments that the service providers have achieved while extending the contours of the formal financial system within an economy.

Results show that China seems to be the top contender in terms of Capital Formation and the dimension of Efficiency. This result is substantiated by the fact that the savings rate in China is extremely high (Yang, 2012). Further, reviewing the status of account ownership, the proportion of adults with access to a formal financial account in China remains at 66%, which is way higher than its peers in the BRICS group. Among other members the proportion of account ownership stood at 55% for

Table 4.7 Index of financial inclusion

Country Name	*FI 2012*	*FI 2013*	*FI 2014*	*FI 2015*	*FI 2016*	*FI 2017*	*FI 2018*	*FI 2019*	*FI 2020*	*FI 2021*	*Average IFI*	*Standard Deviation*
Namibia	0.3384	0.3529	0.3592	0.3573	0.3582	0.385	0.359	0.3447	0.3454	0.3498	0.3550	0.0127
Botswana	0.2973	0.3081	0.3018	0.2915	0.2674	0.299	0.2898	0.2836	0.2821	0.2969	0.2918	0.0117
Djibouti	0.2281	0.2333	0.2168	0.2075	0.2176	0.2383	0.2326	0.2367	0.2169	0.2116	0.2239	0.0111
Mauritius	0.5207	0.5757	0.5544	0.5542	0.5348	0.5451	0.5028	0.4926	0.4946	0.501	0.5276	0.0294
Ghana	0.2359	0.2526	0.2478	0.2478	0.2335	0.2849	0.2797	0.2776	0.2873	0.3091	0.2656	0.0254
Seychelles	0.3543	0.4107	0.4097	0.4364	0.4856	0.491	0.47	0.4513	0.4794	0.4391	0.4428	0.0426
Eswatini	0.2608	0.2514	0.252	0.2424	0.2223	0.2546	0.2358	0.2294	0.2664	0.3163	0.2531	0.0262
Zambia	0.2013	0.2077	0.2146	0.2143	0.2018	0.2551	0.2203	0.2262	0.2329	0.2996	0.2274	0.0300
Lesotho	0.1978	0.2138	0.1994	0.1858	0.184	0.2783	0.2601	0.2659	0.2299	0.2476	0.2263	0.0350
Rwanda	0.2251	0.2094	0.2082	0.2082	0.2053	0.2523	0.238	0.2677	0.2937	0.3316	0.2440	0.0427
Nigeria	0.1689	0.1731	0.1687	0.2208	0.2254	0.2656	0.2605	0.2573	0.2591	0.2529	0.2252	0.0407
Burundi	0.1452	0.1563	0.1483	0.1378	0.1444	0.1804	0.1878	0.2331	0.2374	0.2489	0.1820	0.0431
Cameroon	0.1322	0.1344	0.1182	0.1236	0.1324	0.1543	0.1781	0.1936	0.2037	0.23	0.1601	0.0388
Uganda	0.1539	0.1575	0.1331	0.1314	0.186	0.2582	0.2389	0.2633	0.2515	0.2882	0.2062	0.0599
Guinea	0.1197	0.126	0.1232	0.1239	0.1466	0.1946	0.2231	0.2483	0.2313	0.2507	0.1787	0.0562
Madagascar	0.1416	0.1347	0.1312	0.1396	0.1434	0.1963	0.25	0.2558	0.253	0.2934	0.1939	0.0633
African Continents' Scores											0.2627	0.1000
Bangladesh	0.2211	0.2358	0.2354	0.2324	0.2343	0.2322	0.2473	0.2447	0.259	0.244	0.2386	0.0104
Malaysia	0.5062	0.5249	0.5133	0.488	0.4645	0.4536	0.4755	0.4714	0.471	0.4802	0.4849	0.0230
United Arab Emirates	0.3907	0.4057	0.418	0.4499	0.4505	0.4445	0.4421	0.4271	0.4352	0.4495	0.4313	0.0207
Philippines	0.2755	0.3052	0.3014	0.2852	0.2791	0.2918	0.2847	0.2969	0.3176	0.324	0.2961	0.0161
Thailand	0.5152	0.54	0.5509	0.5429	0.5281	0.5257	0.5015	0.4655	0.4819	0.4866	0.5138	0.0289
China	0.5164	0.5238	0.5429	0.538	0.528	0.5355	0.574	0.579	0.592	0.6207	0.5550	0.0344
Singapore	0.567	0.6253	0.6268	0.611	0.5965	0.5846	0.5714	0.5457	0.5304	0.5259	0.5785	0.0369

Country Name	*FI 2012*	*FI 2013*	*FI 2014*	*FI 2015*	*FI 2016*	*FI 2017*	*FI 2018*	*FI 2019*	*FI 2020*	*FI 2021*	*Average IFI*	*Standard Deviation*
Israel	0.4831	0.4733	0.4722	0.4514	0.4448	0.4559	0.4183	0.3716	0.3963	0.4069	0.4374	0.0373
Pakistan	0.2285	0.226	0.1872	0.1884	0.1806	0.231	0.2053	0.2125	0.2341	0.2353	0.2129	0.0212
Uzbekistan	0.2848	0.3018	0.3017	0.2614	0.2725	0.357	0.3319	0.3608	0.3981	0.3708	0.3241	0.0463
Asian Continents' Scores											0.4073	0.1314
Turkey	0.3769	0.4014	0.4095	0.4293	0.4283	0.4542	0.4344	0.405	0.4208	0.4106	0.4170	0.0213
Georgia	0.3794	0.4077	0.4215	0.4235	0.4172	0.4547	0.4407	0.3977	0.401	0.4183	0.4162	0.0215
North Macedonia	0.3883	0.4021	0.4089	0.4073	0.3934	0.4077	0.3752	0.3549	0.3533	0.3402	0.3831	0.0257
Poland	0.3947	0.4118	0.4249	0.4215	0.4127	0.4038	0.3961	0.3631	0.3481	0.3431	0.3920	0.0300
Moldova	0.3037	0.3582	0.3425	0.3574	0.3484	0.3628	0.3255	0.3123	0.2968	0.2994	0.3307	0.0262
Malta	0.4776	0.4871	0.4832	0.5238	0.5235	0.5011	0.4409	0.4156	0.4377	0.4217	0.4712	0.0400
Italy	0.5365	0.5151	0.5214	0.5146	0.4953	0.4835	0.44	0.3985	0.4157	0.4062	0.4727	0.0526
Croatia	0.512	0.5095	0.5099	0.4983	0.5006	0.4918	0.4704	0.4042	0.3929	0.3738	0.4663	0.0543
Latvia	0.4539	0.45	0.4323	0.4165	0.3911	0.4267	0.3625	0.3205	0.3212	0.3182	0.3893	0.0548
Estonia	0.5011	0.5255	0.5214	0.5055	0.4904	0.4676	0.436	0.3769	0.3833	0.3546	0.4562	0.0642
Ukraine	0.5024	0.4729	0.4561	0.4089	0.3812	0.3889	0.3594	0.3659	0.3298	0.2863	0.3952	0.0666
European Continents' Scores											0.4173	0.0453
Costa Rica	0.3882	0.4116	0.4282	0.3986	0.4005	0.4282	0.3993	0.3528	0.3568	0.3643	0.3929	0.0273
Dominican Republic	0.2943	0.3033	0.3077	0.2956	0.2674	0.3147	0.2745	0.2498	0.262	0.2727	0.2842	0.0218
El Salvador	0.3234	0.3425	0.3968	0.3595	0.3507	0.3542	0.3174	0.3077	0.3176	0.3168	0.3387	0.0274
North American Continents' Scores											0.3386	0.0032
Ecuador	0.288	0.2998	0.299	0.2873	0.2682	0.3078	0.2843	0.2847	0.2855	0.3055	0.2910	0.0120
Paraguay	0.3018	0.3154	0.3211	0.2912	0.282	0.3096	0.3147	0.3097	0.317	0.3223	0.3085	0.0132
Brazil	0.5233	0.4967	0.4915	0.4617	0.4503	0.4671	0.4342	0.4158	0.4376	0.4477	0.4626	0.0328

(continued)

Table 4.7 (continued)

Country Name	*FI 2012*	*FI 2013*	*FI 2014*	*FI 2015*	*FI 2016*	*FI 2017*	*FI 2018*	*FI 2019*	*FI 2020*	*FI 2021*	*Average IFI*	*Standard Deviation*
Uruguay	0.3216	0.3376	0.3409	0.3554	0.3326	0.4061	0.378	0.3842	0.3824	0.3578	0.3597	0.0272
Colombia	0.3386	0.3644	0.3791	0.3926	0.3871	0.4259	0.4029	0.3836	0.3495	0.3244	0.3748	0.0308
Peru	0.3178	0.3357	0.3519	0.4154	0.3938	0.4196	0.3824	0.3519	0.3486	0.3401	0.3657	0.0349
South American Continents' Scores											0.3604	0.0101
Average IFI	0.3399	0.3523	0.3518	0.3486	0.3431	0.3678	0.3510	0.3404	0.3443	0.3486		
Standard Deviation	0.1294	0.1327	0.1364	0.1343	0.1289	0.1114	0.1037	0.0903	0.0923	0.0874		

Source Authors Calculation

Brazil, 44% in Russia, 37% in India, and 57% in South Africa. In fact, the proportion of population having a bank account in China is way above the average across the world (Demirgüc-Kunt & Klapper, 2013). These cross-sectional deviations can be explained by the country's economic development status, which is replicated by the per capita output (i.e. GDP per capita) (Demirgüc-Kunt & Klapper, 2013). However, per capita GDP for China is not the highest among the BRICS members showing that the usage of formal financial accounts is certainly higher than compared to the other members in the group (Fungáčová & Weill, 2015). Following the survey results of Fungáčová and Weill (2015), 82% of the respondents who saved in the past 12 months reported to save their earnings in a financial institution, which is far better than the BRICS results lying within the range of 50–72% and well above the world average of 36% and is corroborated further by the high aggregate deposits to GDP ratio for China among the G20 nations (Iorgova & Lu, 2013). The well positioning in terms of account openings with financial intermediaries those are having a formal recognition is highly motivated by the extremity of high savings rate across the country (Zhang et al., 2012). Further, it is observed that this high savings rate leads to higher propensity to park savings with formal financial intermediary rather going into the traps of informal channels unlike Russia where a majority of the depositors fall in the trap of shadow banking. The leading factors behind the high savings rate being the lesser availability of the social as well as private insurance alongside the gender imbalance with which Chinese households are associated with (Wei & Zhang, 2011).

However, following the survey of Fungáčová and Weill (2015), the extent of formal credit accessibility for China seems to be very low at 7% for the past year. This is far below the average of the high-income countries (i.e. 14%) as well as compared to BRICS average. The status of other BRICS members show that use of formal credits remained at 7% for Brazil, 9% for Russia, 8% for India and 11% for South Africa. Notwithstanding the Banks' private credit to GDP ratio remained at 121% following the Global Financial Development Database which classifies China in the high-income category (Čihak et al., 2012). The reason behind the divergence is the high concentration of credit channelized towards the large state-owned firms while individual loan applicants being completely neglected (Hale & Long, 2010).

Further in the dimension of Capital Formation and Outreach, Burundi is found to be at the bottom from among the sample countries considered in the study. Burundi, as classified under the category of Least Developed Countries (LDC's) is landlocked from its geographical positioning and is characterized with low and volatile growth rates (Nkurunziza et al., 2012). Faced with political and economic turmoil the country is constrained with flexibility in utilizing the public and private resources. Thus, it survives with regard to drag investment, create entrepreneurship and accelerate the growth rates in the economy. High rates of poverty and unemployment forces the stagnation of the income and thus put severe constraints on the mechanism of capital formation as well as restricts people to access the available financial services (Nkurunziza et al., 2012).

4.8 Conclusions

The term financial inclusion can have a multi-faceted view. Since, it captures a broad category of aspects, characterizing the concept primarily becomes an essential prerequisite for representation. Further, issues concerning the measurements also requires a detailed characterization to portray the essence of the achievements. Measurements of financial inclusion achievements are based on a diverse set of methodologies those aim to capture the achievements in a fair way. The present study reviews the status of financial inclusion achievements through classifying the available data into four samples covering Indian states and UTs, the Asian continent, the Asian and European continent and a holistic view through considering all of the five continents taken together. Results show that Chandigarh to take the lead among the Indian states and UTs, so far mean achievement scores are concerned throughout the study period. Similarly, across the countries from Asia, South Korea takes the lead and across the sample of European and Asian countries Netherlands was found to take the lead. While considering a broader cross-sectional view (reviewing the five continents together), results exhibit a common observation that demographics with better economy tend to place themselves in better financial inclusion achievement scores. This evidence was quite in line with the proposition of Robinson (i.e. Where enterprise leads finance follows) dated back in 1952.

Further, the observations across the Indian states and UTs indicates geographical positioning, the status of unemployment and low-level of education to significantly influence the performance of financial inclusion initiatives. Similar evidences was also a part of observation when introspecting the countries from the Asian continent (especially the

Afghanistan and Mongolia) where geographical positioning was one among the prominent reasons behind poor achievement scores.

References

ADB. (2017). *Asian Development Outlook 2017 Update*. Asian Development Bank, Manila, Philippines.

Amidžić, G., Massara, M. A., & Mialou, A. (2014). *Assessing Countries' Financial Inclusion Standing—A New Composite Index. IMF Working Paper* WP/14/36. https://doi.org/10.5089/9781475569681.001

Arner, D. W., Barberis, J., & Buckley, R. P. (2015). The evolution of Fintech: A new post-crisis paradigm. *Geo. J. Int'l l., 47*, 1271.

Arora, R. U. (2010). Measuring Financial Access. *Griffith Business School*, Discussion Papers No. 2010–07.

Bae, J. (2022). The Bank of Korea's CBDC research: current status and key considerations. *BIS Papers*, 107.

Bateman, M., Duvendack, M., & Loubere, N. (2019). Is fin-tech the new panacea for poverty alleviation and local development? Contesting Suri and Jack's M-Pesa findings published in Science. *Review of African Political Economy, 46*(161), 480–495. https://doi.org/10.1080/03056244.2019.1614552

Beck, T., Demirguc-Kunt, A., & Levine, R. (2007). Finance, Inequality, and the Poor. *Journal of Economic Growth, 12*(1), 27–49. https://doi.org/10.1007/s10887-007-9010-6

Berger, A. N. (2003). The economic effects of technological progress: Evidence from the banking industry. *Journal of Money, credit and Banking*, 141–176. https://www.jstor.org/stable/3649852

Carbo, S., Gardener, E. P., & Molyneux, P. (2005). *Financial Exclusion*. Palgrave MacMillan.

Chakravarty, S. R., & Pal, R. (2013). Financial inclusion in India: An axiomatic approach. *Journal of Policy Modeling, 35*, 813–837. https://doi.org/10.1016/j.jpolmod.2012.12.007

Chamberlain, D., Bester, H., Smit, H., Loots, C., Mburu, S., Dermish, A., Gidvani, L., & Saunders, D. (2014). *Myanmar: Demand, Supply, Policy and Regulation*. Centre for Financial Regulation and Inclusion (CENFRI), Bankable Frontier Associates (BFA).

Chattopadhyay, S. K. (2011). Financial Inclusion in India: A Case-Study of West Bengal. Working Paper Series (DEPR): 8/2011). *Department of Economic and Policy Research*. RBI Working Paper Series.

Čihák, M., Demirgüç-Kunt, A., Feyen, E., & Levine, R. (2012). Benchmarking financial systems around the world. *World Bank policy research working paper* (6175). https://ssrn.com/abstract=2152254

Chotia, V., & Rao, N. M. (2014). Financial Inclusion and Human Development: A State- wise Analysis from India. *International Journal of Economics, Commerce and Management, II, 5*, 1–24.

Conroy, J. (2005). APEC and Financial Exclusion: Missed Opportunities for Collective Action? *Asia- Pacific Development Journal, 1*, 53–80. https://doi.org/10.18356/339ee7fc-en

Demirgüç-Kunt, A. & Klapper, L. (2013). Measuring Financial Inclusion: Explaining Variation in Use of Financial Services across and within Countries (The Global Findex Database). *Brookings Papers on Economic Activity.*

Demirgüç-Kunt, A., Klapper, L. F., Singer, D., & Van Oudheusden, P. (2015). The global findex database 2014: Measuring financial inclusion around the world. *World Bank Policy Research Working Paper* (7255).

Dorfleitner, G., Oswald, E. M., & Zhang, R. (2021). From credit risk to social impact: On the funding determinants in interest-free peer-to-peer lending. *Journal of Business Ethics, 170*, 375–400. https://doi.org/10.1007/s10551-019-04311-8

Duflos, E., Luchtenburg, P., Ren, L., & Chen, L. Y. (2013). *Microfinance in Myanmar sector assessment*. CGAP, IFC.

European Commission. (2015). Payment services. *The European Commission.* https://ec.europa.eu/info/law/paymentservices-psd-2-directive-eu-2015-2366_en

Findex. (2022). *The global findex database 2021.* World Bank Group

Fungáčová, Z., & Weill, L. (2015). Understanding financial inclusion in China. *China Economic Review, 34*, 196–206. https://doi.org/10.1016/j.chieco.2014.12.004

Ghosh, S., & Sahu, T. N. (2020). How financially inclusive the Asian's are—An empirical analysis. *Rajagiri Management Journal, 14*(1), 1. https://doi.org/10.1108/RAMJ-01-2020-0003

GIZ. (2013). *Myanmar's Financial sector: A challenging environment for banks.* Deutsche Gesellschaft für Internationale Zusammenarbeit (GIZ) GmbH.

Gomber, P., Koch, J. A., & Siering, M. (2017). Digital Finance and FinTech: Current research and future research directions. *Journal of Business Economics, 87*, 537–580. https://doi.org/10.1007/s11573-017-0852-x

Gupte, R., Venkataramani, B., & Gupta, D. (2012). Computation of Financial Inclusion Index for India. *Procedia—Social and Behavioral Sciences, 37*, 133–149. https://doi.org/10.1016/j.sbspro.2012.03.281

Hale, G., & Long, C. (2011). What are the sources of financing for Chinese firms?. In *The evolving role of Asia in global finance* (Vol. 9, pp. 313–339). Emerald Group Publishing Limited. https://doi.org/10.1108/S1574-8715(2011)0000009018

Horioka, C. Y., Suzuki, W., & Hatta, T. (2007). Aging, Savings, and Public Pensions in Japan. *Asian Economic Policy Review, 2*(2), 303–319.

Iorgova, S. & Liu, Y. (2013). Structure of the Banking Sector and Implications for Financial Stability. In U. Das, J. Fiechter, T. Sun (Eds.). *China's Road to Greater Financial Stability: Some Policy Perspectives* (pp. 123–139). IMF.

Kaleeswaran, P., & Meera, M. (2017). Financial Inclusion Index through MGNREGS in Virudhunagar District. *Asian Journal of Managerial Science, 6*(1), 5–9.

Kempson, E. & Whyley, C. (1998). *Access to Current Accounts*. British Bankers' Association.

Kempson, E., & Whyley, C. (2001). *Payment of Pension and Benefits*. Department for Work and Pension.

Kumar, C. & Mishra, S. (2011). Banking Outreach and Household Level Access: Analyzing Financial Inclusion in India. In *13th Annual Conference on Money and Finance in the Indian Economy* (pp. 25–26).

Laha, A., & Kuri, P. K. (2014). Demand for and Supply of Financial Inclusion in India: An Inter-State Analysis. *Elk Asia Pacific Journal of Finance & Risk Management, 5*(1), 103–116.

Lashitew, A. A., van Tulder, R., & Liasse, Y. (2019). Mobile phones for financial inclusion: What explains the diffusion of mobile money innovations? *Research Policy, 48*(5), 1201–1215. https://doi.org/10.1016/j.respol.2018.12.010

Leyshon, A., & Thrift, N. (1995). Geographies of Financial Exclusion: Financial Abandonment in Britain and the United States. *Transactions of the Institute of British Geographers, New Series, 20*(3), 312–341.

Mohan, R. (2006). Economic Growth, Financial Deepening and Financial Inclusion. *Address at the Annual Bankers' Conference 2006*, Hyderabad on November 3, 2006. http://rbidocs.rbi.org.in/rdocs/Speeches/PDFs/73697.pdf). Accessed in June 2023

Nguyen, L. T., Hoang, T. G., Do, L. H., Ngo, X. T., Nguyen, P. H., Nguyen, G. D., & Nguyen, G. N. (2021). The role of blockchain technology-based social crowdfunding in advancing social value creation. *Technological Forecasting and Social Change, 170*, Article 120898.

Nkurunziza, J. D., Ndikumana, L., & Nyamoya, P. (2012). The financial sector in Burundi (No. w18289). *National Bureau of Economic Research*. https://doi.org/10.3386/w18289

Ouma, S. A., Odongo, T. M., & Were, M. (2017). Mobile Financial Services and Financial Inclusion: Is It A Boon for Savings Mobilization? *Review of Development Finance, 7*(1), 29–35.

Philippon, T. (2019). *On fntech and fnancial inclusion (BIS Working Papers 84)*. Bank for International Settlements. https://www.bis.org/events/conf190628/philippon.pdf

Preziuso, M., Koefer, F., & Ehrenhard, M. (2023). Open banking and inclusive finance in the European Union: Perspectives from the Dutch stakeholder

ecosystem. *Financial Innovation, 9*(1), 111. https://doi.org/10.1186/s40854-023-00522-1
Rahman, Z. A. (2012). *Financial Inclusion in Malaysia: Tracking Progress using Index*. IFC Workshop on Financial Inclusion Indicators Co-hosted by Bank Negara Malaysia and Central Bank of Malaysia.
Rangarajan, C. (2008). *Financial Inclusion in India—An Assessment*. Government of India. https://rbidocs.rbi.org.in/rdocs/Speeches/PDFs/MFI101213FS.pdf. Accessed March 9, 2020.
Romanova, I., Grima, S., Spiteri, J., & Kudinska, M. (2018). The payment services Directive II and competitiveness: The perspective of European fintech companies. *European Research Studies, 21*(2), 3–22.
Saravanabhavan, S. (2018). *Understanding the Disparity in Financial Inclusion across Indian States: A Comprehensive Index for the Period 1984–2016*. Working Paper—405. Institute for Social and Economic Change (ISEC).
Sarma, M. (2008). *Index of Financial Inclusion*. Working Paper No. 215. Indian Council for Research on International Economic Relations. https://www.econstor.eu/bitstream/10419/176233/1/icrier-wp-215.pdf
Sarma, M. (2012). *Index of Financial Inclusion—A measure of financial sector inclusiveness*. Springer.
Schumpeter, J. A. (1942). *Capitalism, socialism, and democracy*. Harper & Bros.
Scott, B., Loonam, J., & Kumar, V. (2017). Exploring the rise of blockchain technology: Towards distributed collaborative organizations. *Strategic Change, 26*(5), 423–428. https://doi.org/10.1002/jsc.2142
Sinclair, S. P. (2001). *Financial exclusion: An introductory survey*. Report of Centre for Research in Socially Inclusive Services, Heriot Watt University.
Tambunlertchai, K. (2018). Determinants and barriers to financial inclusion in Myanmar: What determines access to financial services and what hinders it? *The Singapore Economic Review, 63*(01), 9–26.
Turnell, S. (2014). Banking and financial regulation and reform in Myanmar. *Journal of Southeast Asian Economies, 31*, 225–240.
Van Loo, R. (2018). Making innovation more competitive: The case of fintech. *UCLA L. Rev., 65*, 232.
Wei, S. J., & Zhang, X. (2011). The competitive saving motive: Evidence from rising sex ratios and savings rates in China. *Journal of Political Economy, 119*(3), 511–564.
World Bank. (2015). *World Development Indicators*.
Yang, D. T. (2012). Aggregate savings and external imbalances in China. *Journal of Economic Perspectives, 26*(4), 125–146. https://doi.org/10.1257/jep.26.4.125
Yoshino, N., Helble, M., Takeda, A., & Shimazaki, S. (2018). Postal savings and financial inclusion in Japan. *Postal Savings: Reaching Everyone in Asia, 1*, 268–285.

Zachariadis, M., & Ozcan, P. (2017). *The API Economy and Digital Transformation in Financial Services: THe Case of Open Banking*. https://doi.org/10.2139/ssrn.2975199

Zeleny, M. (1974). A concept of compromise solutions and the method of the displaced ideal. *Computers and Operations Research, 1*, 479–496. https://doi.org/10.1016/0305-0548(74)90064-1

Zhang, J., Wang, L., & Wang, S. (2012). Financial development and economic growth: Recent evidence from China. *Journal of Comparative Economics, 40*(3), 393–412. https://doi.org/10.1016/j.jce.2012.01.001

CHAPTER 5

Comparative Analysis of Financial Inclusion Across Regions

5.1 Introduction

Spice AbstractThis section is concerned with fulfilling the third objective of the study. It majorly focuses to appreciate the differences in average achievements in financial inclusion across the continents, time period, as well as the income groups considered in the study. This study makes a comprehensive review of the comparisons across mean FI achievements across four different samples as detailed in the following discussion.

Section 5.2 showcases the performances across the 28 states and 7 union territories of India during 2003–22 and attempts to figure out the significance of change in mean FI scores across the different regions and income groups classified. Following the statistical chronology, we apply both the Analysis of Variance and Independent Samples t- test to figure out significant changes across the regions and income groups respectively. Section 5.3 considers 26 Asian countries across a period of 2013–22 and attempts to outline significant improvements in mean FI scores applying Dependent samples t- test as well as Wilcoxon-Signed Rank test as a non-parametric alternative depending upon the fulfilment of the conditions prescribed. It further classifies the Asian countries into high-income and low-income categories and through application of the Independent samples t- test attempts to observe whether high income countries are significantly better than the lower ones in terms of mean FI achievements.

S. Ghosh et al., *Financial Inclusion*, Sustainable Development Goals Series, https://doi.org/10.1007/978-981-95-3385-5_5

Section 5.4 is in no way different, except that it further outlines the significance of the continental differences between Asia and Europe during the study period of 2013–22 applying the Independent samples t- test. Finally, Section 5.5 focuses to appreciate the differences across mean FI scores during 2012–21 considering an extended cross-sectional band of 46 countries selected from 5 different continents. This sample capitalizes on the magnitude of differences in mean FI scores across the 5 continents applying ANOVA (parametric alternative) as well as the Kruskal-Wallis test (non-parametric alternative).

5.2 Investigation of the Indian States and UTs

Inquiry into data from India's 28 states and 7 union territories (i.e. Sample 1) for 2003–22 reveals a noticeable disparity in the mean IFI scores across regions for both income levels (see Table 5.1 below). Difference is also evident across the average IFI values between both the income groups belonging to a specific region. This study attempts to compare the achievement scores across both the income groups and regions and find whether such mean differences are significant or not. Considering the financial inclusion scores dependent upon the income groups and the regions as categorical independent variables we set up a Two-way/ Factorial ANOVA model.

Results in Table 5.2 shows that both the independent categorical variables (i.e. groups based on income and regions) and their interaction factor significantly influences financial achievement scores. Hence, values of financial inclusion achievement are expected to differ significantly across the income groups and regions. The effect size as measured by the eta-squared shows the dependence to be more vibrant across the regions (0.928). However, robustness of such findings largely depends on the assumptions that underlie behind the functions of the model. Table 5.1.1 in the appendix shows that the data considered is fairly normal (results of Kolmogorov- Smirnov along with Shapiro-Wilk tests) and the variance across the groups are homogeneous (results of the Levene Statistic). The outcomes of the two different Breusch-Pagan and White tests demonstrate that the data is reasonably homoscedastic, implying that this parametric Two-way ANOVA is the best fit under the current circumstances.

According to Table 5.3, the mean variation in financial inclusion ratings between both low- and high-income groups is little (at 5%).

Table 5.1 Region-Wise Attainments

Preliminary Statistics

Financial inclusion index values

NSDP category	*Region*	*N*	*Average*	*Standard deviation (SD)*	*CV*	*3σ confidence interval for mean*		*Ceiling*	*Floor*
						Max.	*Min.*		
High	South	5	0.452	0.036	0.079	0.344	0.559	0.494	0.359
	West	3	0.348	0.032	0.091	0.253	0.442	0.374	0.275
	North	5	0.380	0.036	0.094	0.273	0.487	0.408	0.298
	East	1	0.217	0.036	0.164	0.110	0.324	0.280	0.164
	Central	2	0.198	0.034	0.170	0.097	0.299	0.240	0.125
	North-east	2	0.172	0.046	0.267	0.035	0.310	0.235	0.104
Low	West	4	0.317	0.020	0.064	0.256	0.378	0.348	0.268
	North	1	0.395	0.028	0.070	0.312	0.477	0.440	0.325
	South	1	0.404	0.030	0.073	0.316	0.493	0.466	0.342
	East	4	0.162	0.013	0.082	0.122	0.201	0.185	0.135
	Central	2	0.160	0.014	0.089	0.117	0.202	0.185	0.129
	North-east	5	0.112	0.017	0.154	0.060	0.164	0.142	0.083

Source Authors Calculation

Table 5.2 Result of Factorial ANOVA

Tests of Between-Subjects Effects

Dependent variable: IFI

Source	*Sum of squares (type III)*	*Degree(s) of Freedom*	*Squared mean*	*F-Statistics*	*Level of Significance*	*Partial eta squared*
Corrected model	0.951[a]	11	0.101	111.197	0.002	0.942
Intercept	7.673	1	6.973	8205.758	0.001	0.987
NSDP_CODE	0.054	1	0.042	37.106	0.003	0.316
Region	0.896	6	0.201	228.148	0.000	0.928
NSDP_CODE* region	0.024	6	0.010	3.751	0.008	0.189
Error	0.081	86	0.003			
Total	8.878	98				
Corrected Total	1.215	89				

[a]R squared = 0.967, adjusted R squared = 0.921

Source Authors Calculation

The results are robust to the intrinsic assumption of applying the parametric independent samples *t*- test at an acceptable confidence level (i.e. normality, homogeneity of variance and heteroscedasticity). Therefore, financial inclusion achievements for high-income groups are better prima facie but not significantly in comparison to the low-income groups.

Comparison across the regions in Table 5.4 shows a significant difference in mean FI score achievements after satisfying the intrinsic assumptions (Stoline, 1981). However, identification of the groups where mean values differ significantly depends on the test results shown in Table 5.5.

Tukey's Honestly Significant Difference (HSD) and the Ryan-Einot-Gabriel-Welsch (R-E-G-W-Q) test are used in Table 5.5 to identify particular locations that vary substantially from each other (Shingala & Rajyaguru, 2015). Table 5.5 displays the agreement of each of these post-hoc analyses at the five percent threshold of significance, indicating

Table 5.3 Comparisons Based on Per Capita NSDP

Summary of the Results			*High*	*Low*
Descriptive statistics		Size of the sample	18	17
		Average	0.289	0.253
		SD	0.101	0.111
		CV	0.350	0.437
Predictive/ inferential statistics (*testing the assumptions*)	Tests of normality	Skewness (*z*-value)	−0.076	0.112
		Kurtosis (*z*-value)	−1.735	−1.715
		Kolmogorov-Smirnov[a] test	0.120 (0.079)	0.144 (0.094)
		Shapiro-Wilk test	0.898 (0.087)	0.915 (0.065)
	Homogeneity of variances	Levene's test[b]	1.987 (0.121)	
	Tests of heteroscedasticity	Breusch-pagan test[b]	0.383 (0.536)	
		White test[b]	1.124 (0.289)	
Prescriptive statistics	Level of significance (between two-mean difference)	Independent samples t-test (2-tailed)	1.442 (0.076)	

[a]Lilliefors corrected significance, [b] design: Intercept + NSDP_CODE
p-values are presented in the parenthesis; where, $\alpha = 0.05$

Source Authors Calculation

Table 5.4 Regional Comparisons

Summary of the Statistics			*North*	*North-East*	*East*	*Central*	*West*	*South*
Descriptive statistics		Size of the sample	6	7	5	4	7	6
		Average	0.387	0.142	0.240	0.179	0.332	0.427
		Std. deviation	0.033	0.046	0.043	0.033	0.032	0.042
		Coefficient of variation	0.084	0.326	0.179	0.182	0.095	0.098
Predictive/ inferential statistics (*testing the assumptions*)	Tests of normality	Skewness (ζ-value)	−1.896	1.785	1.378	1.159	−0.869	0.08
		Kurtosis (ζ-value)	1.765	0.038	−0.753	−0.785	−0.963	−1.524
		Kolmogorov-Smirnov[a] test	0.133 (0.200*)	0.191 (0.120)	0.202 (0.081)	0.171 (0.200*)	0.169 (0.200*)	0.151 (0.200*)
		Shapiro-Wilk test	0.903 (0.088)	0.857 (0.067)	0.891 (0.059)	0.907 (0.106)	0.933 (0.268)	0.911 (0.121)
	Testing homogeneity of variances	Levene's test[b]	1.254 (0.421)					
	Tests of heteroscedasticity	Breusch-pagan test[b]	0.785 (0.416)					
		White test[b]	6.574 (0.465)					
Prescriptive statistics	Level of significance (between regions)	ANOVA	154.398 (0.001)					

*Floor level of the true significance, a. Lilliefors corrected significance, b. design: Intercept + region
p-values are presented in parenthesis; where, $\alpha = 0.05$

Source Authors Calculation

that each site seems to be statistically different so far as mean achievements in IFI scores are concerned during the period of study. Figure 5.1 depicts a graphical representation of mean IFI scores across geographies and socioeconomic categories, which adds to the evidence.

Findings of this analysis converge with the results of Laha & Kuri (2014). Such convergence was mainly due to the similarity in the mechanism of constructing the index as well as consideration of both the demand and supply side variables taken in to account. However, the results are somehow different apart from ranking of the areas based on their accomplishments in financial inclusion following Gupte et al. (2012). The source of contradiction results from the non-parametric assignment of weights rather employing a data-driven approach that the present study overcomes from (Ghosh & Sahu, 2020).

Table 5.5 Regional Comparisons (Post-Hoc Tests)

IFI Scores

	Region	*N*	*Subset for alpha = 0.05*					
			1	*2*	*3*	*4*	*5*	*6*
Tukey HSD[a]	North-east	7	0.201					
	Central	4	0.215					
	East	5		0.312				
	West	7			0.452			
	North	6				0.489		
	South	6					0.497	
	Sig.		0.078	1.000	1.000	1.000	1.000	
Ryan-Einot-Gabriel-Welsch range [b]	North-east	7	0.201					
	Central	4		0.215				
	East	5			0.312			
	West	7				0.452		
	North	6					0.489	
	South	6						0.497
	Sig.		1.000	1.000	1.000	1.000	1.000	1.000

Means across the groups in homogeneous subsets are given, [a] assuming the harmonic mean sample size 6.

[b]Non-monotonic critical values are observed across the data. Adequate substitutions are made to keep consistent monotonicity. Therefore, the type I error becomes smaller.

Source Authors Calculation

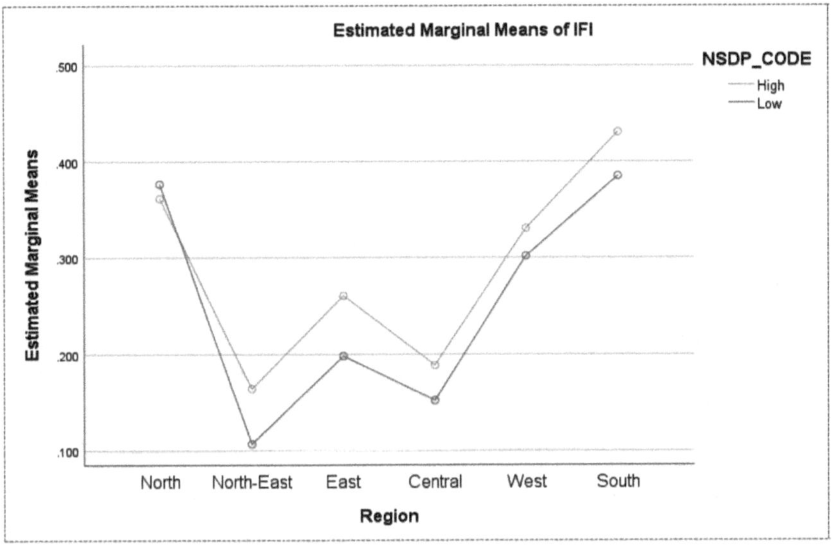

Fig. 5.1 Difference in Mean IFI Scores Across Productivity Groups and Regions *Source* Authors Compilation

Regional disparity in the achievement scores of financial inclusions implies disproportional provisioning of financial assets (Chakravarty & Pal, 2013). A part from analyzing the regional achievements, the North-Eastern part of the country is most vulnerable so far as the dimension of stability is concerned. Due to the difficult terrain and poor connectivity to the rest of the country, the establishment of the banking outlets especially cross the rural and remote areas become more difficult (Bhattacharyya & Dutta, 2018; Gupte et al., 2012). Further, the insufficient mobile network coverage and poor internet penetration hinders the digital drive of financial inclusion (Singh & Choudhury, 2017). Often the ethnic groups in the North-east region of India use their own traditional financial systems (e.g. Chit Funds) due to general lack of trust, which function outside the formal financial system (Bora, 2019). Bhattacharyya & Dutta (2018) show that due to less documentation, familiarity and convenience people in this region prefer informal arrangements even if they are less stable and resilient. Further, the political instability is yet another important factor that often disrupts local economies and functioning of financial institutions (Singh & Choudhury, 2017). Additionally, the gap in the financial literacy among the inhabitants exacerbates the problem of low formal financial participation (Singh, 2020).

The divergence in financial inclusion achievements across the productivity-based groups is statistically weak and further reinforces the states agenda of ensuring finance for all (Cámara & Tuesta, 2014). However, this might have an adverse consequence through increasing the existing levels of non-performing assets that the financial intermediaries may suffer from (Sarma & Pais, 2011). The economic inequality in India is a major element leading to the disparity in financial inclusion. Lower-income demographics have obstacles including restricted income, insufficient savings, and elevated debt levels, which diminish their capacity to obtain or utilize financial products. Individuals in higher-income groups are more capable of affording the expenses related to financial services, including account maintenance fees, loan repayments, and insurance premiums (Kumar & Patel, 2020). Further, Low-income demographics frequently possess insufficient knowledge and comprehension of financial products and services, impeding their capacity to access and utilize them successfully. Financial literacy tends to be diminished among rural populations, women, and individuals from marginalized communities, resulting in reduced adoption rates of formal financial services (Ghosh & Kapoor, 2019).

5.3 Assessment across the Asian Continent

Meeting the requirements of normality across both the dependent and independent samples t-test and variance uniformity within the t-test of samples that are independent while comparing the mean FI scores of 26 Asian nations through 2013 to 2022 (Sample 2) leads to a sound statistical foundation for depending on these parameterized investigations. However, since several constructs violated the normality requirements when a temporal comparison was attempted, the research used a non-parametric alternative (the Wilcoxon signed-rank test) to identify the significance of the difference.

One of the many frequently utilized parametric evaluations for identifying major shifts in IFI ratings over a period of time is the sample-dependent t-test. However, application of the test requires the differences across the paired values to form a normal distribution. Results for the year 2014–15 to 2016–17 and 2018–19 to 2021–22 appears to be fairly normal following the scaled values of skewness and kurtosis (i.e. with ζ-values in a spread of $\pm 1.96\sigma$) (Posten, 1984) and the test results of Shapiro-Wilk (Shapiro & Wilk, 1965). Furthermore, at the 1 and 5 percent thresholds of significance, the distributions for 2013–14 and 2017–18 are non-normal. Table 5.6 summarizes the normality tests, proposing the Wilcoxon signed-rank test as an alternative for the sample-dependent t-test for 2013–14 and 2017–18. The outcome presented in Table 5.6 exhibits that the fluctuation in average IFI scores indicated in Table 4.3 may be adequately ascribed based on the extent of instability present in this information set, having $t\ (25) = 0.201$ as well as $p = 0.923$ during the 2014–2015 period. Additionally, since the variation in average IFI value for 2014–15 was non-significant, we decided not to pursue the investigation of size effects as prescribed by Hedges & Olkin (1985) and Cumming (2012). In contrast, the data for 2016–17, 2020–21, and 2021–22 reveal a significant rise at 1 percent level of threshold significance but with a comparatively lower to moderate effect size of 0.319 to 0.742. Likewise, the growth was significant at 5 percent level during 2018–19 and 2019–20 with moderate effect sizes.

The drive for financial inclusion shows a steady rise from 2016 onwards across the sample comprising Asian countries. However, the trajectory of steady growth is not completely smooth. Therefore, there is mix evidence of growth which is in line with the report of World Bank 2020. Economic slowdown prior to the COVID - 19 pandemic has reduced the pace of efforts targeting to financially include the excluded sections within its periphery. Thus, the erratic move in financial inclusion achievements looks somewhat inconsistent with the approach to attain an all-inclusive financial system, highlighting the inherited instability in the targeting of policy initiatives by countries from the Asian continent. Similar predictions have been made as part of suggestions in the studies of Rahman (2012) in the Malaysian context and of Amidžić et al. (2014) considering a broader cross-sectional band.

When we consider to study the importance of variations across the time, where the spread of variations among pairs becomes irregular (Table 5.7), we see that the change in both 2014 and 2018 across 2013 and 2017, seems to be essential at 10%, suggesting that in conjunction to average IFI scores, Asian countries are going through a statistically minor recession.

Table 5.6 Comparison over time using the sample-dependent t-test

Study Period	*Normality Tests*			*Significance*	
	Skewness (ζ- value)	*Kurtosis (ζ- value)*	*Shapiro-Wilk (Sig.)*	*Paired Samples t-test (2-tailed)*	*Hedges g*
2013–14	−3.973	8.128	0.001	Not applicable	Not applicable
2014–15	−2.081	0.892	0.503	0.201 (0.923)*	Not required
2015–16	1.882	2.998	0.141	0.985 (0.305)	Not required
2016–17	1.013	−0.794	0.189	−3.793 (0.000)	0.319
2017–18	0.875	0.197	0.036	Not applicable	Not applicable
2018–19	0.475	0.785	0.595	1.985 (0.043)	0.435
2019–20	0.156	0.415	0.324	2.156 (0.034)	0.348
2020–21	1.875	0.694	0.745	−3.145 (0.001)	0.543
2021–22	1.548	2.547	0.248	−3.985 (0.001)	0.742

*The values in parenthesis represent the p-values. *Source* Authors Calculation

Table 5.7 Comparison over time using the Wilcoxon signed-rank test

Study Period	*Assumptions*	*Results*	*Level of Significance*	
	Is the Distribution Symmetric?	*Rank Compositions Positive/ (Negative)*	*Wilcoxon- Signed Rank Test (2-tailed)*	*Measure of Effect Size (PS dep)*
2013–14	Yes	9 (19)	−2.059 (0.087)*	Not required
2017–18	Yes	10 (16)	−1.254 (0.078)	Not required

*p-values are provided in the parenthesis. *Source* Authors Calculation

The t-test for independent samples was applied to assess the statistical importance of the income category differences in the average IFI achievements (Table 5.8). With ζ- values for the kurtosis and skewness lying within a spread of ± 1.96 standard deviations from the mean (Schmider et al., 2010) while converging with the acceptance of the Shapiro-Wilk test's null hypothesis (Shapiro & Wilk, 1965), data converges to normality. Further, the variation across the groups appears to be systematic following the Levene's F-test [$F(1, 24) = 1.071$, $p = 0.311$]. This contributes to establish a solid foundation for carrying out the independent samples t-test. The results of the t-test for independent samples additionally demonstrate significant differences ($p < 0.05$) across the mean levels of IFI attained between nations categorized as wealthy and economically disadvantaged. This was further reinforced by an exciting magnitude of the effect of 0.918, which suggests that a country's income has certainly an impact over the level of the financial inclusion accomplished.

Experiences show that accessibility to financial services improves savings behavior (Aportela, 1999; Ashraf et al., 2006), accelerates investment in productive assets (Dupas & Robinson, 2013), and empowers women (Ashraf et al., 2010). Accessibility to financial services promotes growth at a higher pace at the firm level (Beck et al., 2005; Ayyagari et al., 2008) as well as at the aggregate levels (Levine, 2005). Even the poorest of the poor requires accessibility to financial services to earn a living (Collins et al., 2009). Improved accessibility to financial services helps the underprivileged to access growth opportunities that they might not finance with their limited own resources (Demirgüç-Kunt & Klapper 2013). Further, findings of this study are also part of the results in Ayyagari & Beck (2015) (see Fig. 1 at page number 2). Category of countries with higher income (scaled with respect to GDP) is found to have better financial inclusion achievements as compared to the lower ones. Similar evidences were observed in the analysis of Global Findex Database, which had a coverage of 150,000 nationally representative adults above 15 years of age selected at random. This survey covering almost 148 economies by Gallop Inc. revealed a substantial variation in account penetration between the groups classified on the basis of income. Among the countries classified under the high-income group, 90% of the respondents reported to have an account with a formal financial intermediary whereas the percentage of account ownership for the Sub-Saharan Africa reported to be less than 18%. However, the developing Asia has fared better than compared to the Sub-Saharan African region with a score of

Table 5.8 Wealthy vs. Economically Disadvantaged (Applying Independent Samples t- Test)

	Skewness (ζ-value)	*Kurtosis (ζ - value)*	*Shapiro-Wilk (Sig.)*	*Levene's F-test*	*Mean*	*S.D*	*Independent Samples t (2- tailed)*	*Hedges g*
High	0.11	−0.081	0.755	1.071	0.395	0.11	2.334	0.918
Low	−1.022	−0.559	0.289	(0.311)*	0.307	0.073	(0.028)	

*p-values provided in the parenthesis*Source* Authors Calculation

26.7%. Further, the variation within the Asian continent is also substantially high [see Fig. 5 at page number 5 of Ayyagari & Beck (2015)]. In this connection, several Asian countries have organized financial education campaigns to improve financial literacy and thus their earnings especially for the rural populations, women, and low-income groups (Chien & Lin, 2020). Experiences from Philippines and Bangladesh show that through mobile money services like GCash and bKash millions of people got access to financial services those who previously had no bank accounts. This digital drive led to significantly reduce the income disparities and substantial improvements in financial inclusion by lowering the transaction costs hence reaching the underserved population (Li & Zhang, 2021). Further, the inclusive credit system especially through the Microfinance Institutions in this regard has become the enabler to those poor households who otherwise do not qualify for credit under the conventional banking protocols (Rahman & Sultana, 2019). Thus, access to credit serves as a means for individuals to earn their living and beyond that at the macro level reduces inequality within the nations.

5.4 Investigation of Asia and Europe

The introspection of differences in financial inclusion achievements across the sample (Table 5.9) and emphasizing on Europe (Table 5.10) and separately Asia (Table 5.11) has been done applying three distinct sets of dependent samples t-tests on Sample 3.

Examining the shift of the average IFI values over the period from Table 4.5 or Table 4.6 above demonstrates considerable reduction in IFI achievements throughout the study period except in 2014–15, where significant improvement (at 5 percent level) is observed. However, these

Table 5.9 Comparisons Across the Periods (Sample: European and Asian Countries)

	Skewness (ζ-value)	*Kurtosis (ζ- value)*	*Shapiro-Wilk (Sig.)*	*Matched Samples t-test (one-tailed)*	*Hedges g*
2013–14	−1.325	−0.895	0.081	0.708 (0.197)*	–
2014-15	0.206	−0.689	0.796	2.096 (0.046)	−0.189
2015–16	0.901	0.912	0.298	4.144 (0.001)	0.226
2016–17	0.785	0.458	0.365	0.678 (0.787)	–
2017–18	0.985	−0.256	0.045	0.812 (0.184)	–
2018–19	1.254	0.145	0.256	0.912 (0.165)	–
2019–20	1.458	0.364	0.089	1.875 (0.047)	0.298
2020–21	0.458	0.456	0.145	1.945 (0.043)	0.364
2021–22	0.256	0.478	0.785	2.013 (0.035)	0.421

*p-values are presented in the parenthesis *Source* Authors Calculation

reductions in achievement scores were significant during 2019–22 at 5 percent level, whereas extreme reductions are observed during 2015–16 at 1 percent level.[1] Furthermore, when looking at the magnitude of the impact in the shift of the average IFI evaluations throughout the years, it turns out to be quite modest (Hedges & Olkin 1985 and Cumming 2012), giving reasonable evidence that the changes might not be sufficiently large enough.

While focusing on the relevance of differences in IFI throughout the years for the entire Europe with respect to the average IFI attainments, a substantial fall in the mean across is observed across the whole research period, except in 2014–15 where the achievement scores have grown up similar to that of the entire sample space at 5 percent significance.

[1] Data conforms to normality in Table 7.9, with ζ - values for the skewness and kurtosis ranging between within ±1.96σ (Posten, 1984) while rejecting the alternative hypothesis of the Shapiro-Wilk test (Shapiro & Wilk, 1965).

Table 5.10 Comparisons Across the Periods (Sample: European Countries)

	Skenwess (ζ - value)	*Kurtosis (ζ - value)*	*Shapiro-Wilk (Sig.)*	*Matched Samples t-test (one-tailed)*	*Hedges g*
2013–14	−0.099	1.489	0.615	3.487 (0.001)[a]	0.328
2014-15	0.210	−0.010	0.896	2.301 (0.018)	−0.251
2015–16	0.698	1.051	0.215	2.069 (0.019)	0.224
2016–17	0.675	0.478	0.315	0.768 (0.198)*	–
2017-18	0.432	0.789	0.219	0.968 (0.176)*	–
2018-19	0.984	1.312	0.345	0.994 (0.165)*	–
2019-20	1.654	0.984	0.687	1.095 (0.153)*	–
2020-21	0.786	1.089	0.758	1.121 (0.121)*	–
2021-22	0.896	1.048	0.895	1.243 (0.098)*	0.153

*p-values are provided in the parenthesis *Source* Calculated by the Authors

Table 4.5 exhibits that, with the exception of 2015, the mean IFI values show a significant fall during 2013–14, 2015–16 and 2021–22 at 1, 5 and 10 percent significance respectively. Though, revealing a falling tendency throughout the study period, the size of the effect remains at their lower ends. Economic slowdown prior to the COVID - 19 pandemic as highlighted in the policy report by the World Bank 2020 has been among the strongest reasons for exhibition of such falling trend in the financial inclusion achievement scores.

Reviewing the change in average IFI scores for the Asian continent, a considerable decrease in achievements across the countries was recorded in 2015–16 (at 5 percent LOS) with moderate effect size followed by a continuous fall, though not statistically significant during 2017–22. However, a significant growth was observed across 2013–14 (at 5 percent level) and 2014–15 (at 10 percent level), whereas across 2016–17 the growth remained statistically insignificant.

Table 5.11 Comparisons Across the Periods (Sample: Asian Countries)

	Skewness (ζ - value)	*Kurtosis (ζ - value)*	*Shapiro-Wilk (Sig.)*	*Matched Samples t-test (one-tailed)*	*Hedges g*
2013–14	0.398	−1.215	0.315	1.786 (0.043)*	0.358
2014-15	−0.201	−0.896	0.713	1.548 (0.087)	–
2015–16	0.379	0.195	0.876	1.896 (0.037)	0.402
2016–17	0.452	0.475	0.362	1.136 (0.125)	–
2017–18	0.365	0.685	0.156	1.032 (0.136)	–
2018–19	0.785	0.596	0.567	0.789 (0.236)	–
2019–20	0.654	0.381	0.654	1.568 (0.092)	0.212
2020–21	0.875	0.254	0.546	1.478 (0.087)	0.268
2021–22	1.253	0.541	0.715	1.387 (0.079)	0.296

*p-values are provided in the parenthesis *Source* Authors Calculation

As part of the parametric analysis, we intend to utilize the Independent samples t-test to assess the significance of variations in financial inclusion performance across continents and income groups (Table 5.12). This study refrains from using the matched samples t-test due to having no theoretical dependence between the wealth and geographical groups.

The result in Table 5.12 indicates an insignificant ($p>0.05$) variance between Europe and Asia when introspecting their average financial inclusion ratings derived from the countries included for 2013–16. The disparity in mean values found in Table 4.5 for both continents is due to random chance. Apart from analyses, we may conclude that nations throughout Europe and Asia appear to be more or less similar when it comes to evaluating the success in average financial inclusion, at least for the sample we examined.[2] The sample countries considered from the

[2] However, such an analysis requires the data to be distributed approximately normally The ζ values for both the skewness and kurtosis remains within a range of $\pm$ 1.96 σ from the mean (Schmider et al. 2010) and the rejection of the alternative hypothesis of the Shapiro-Wilk contends, the data to be fairly normal (at 5% LOS) (Shapiro & Wilk, 1965).

Table 5.12 Summarized Results of Europe vs. Asia and High vs. Low Income

Continent-wise Comparisons

	Skewness (ζ - value)	Kurtosis (ζ - value)	Shapiro-Wilk (sig.)	Levene's F	Mean	S.D	Independent samples t
Europe	−0.03	−1.543	0.056	1.045 (0.318)*	0.776	0.109	1.432 (0.083)
Asia	0.159	0.198	0.655		0.714	0.103	

Income group-wise comparisons

	Skewness (ζ - value)	Kurtosis (ζ - value)	Shapiro -Wilk (sig.)	Levene's F	Mean	S.D	Independent samples t	Hedges g
High Income	−0.863	−0.917	0.149	1.89 (0.183)	0.808	0.099	3.52 (0.001)	1.387
Low Income	−0.6	−0.555	0.313		0.681	0.076		

*p-values are provided in parenthesis *Source* Calculated by the Authors

continent of Europe are majorly from the middle-income group, whereas the Asian countries taken as sample more or less represent every possible groups. This basically leads to non-significant differences in financial inclusion achievements so far as the achievement scores are concerned.

However, a significant difference is observed between the countries from high-income and low-income group (with $p < 0.01$) during 2013–22. Prior to such parametric comparisons, the data duly satisfies the assumptions of normality and similar pattern of fluctuations across the groups. A substantial impact size of magnitude 1.387 justifies the existing variations in average financial inclusion successes across income groups, as seen in Table 4.6. This certainly indicates that wealthier countries have a substantially better position in financial inclusion achievements than compared to the deprived ones and was part of previous literary evidence given by Demirgüç-Kunt (2006) and Ghosh & Sahu (2020). Additionally, countries with more resources tend to exhibit higher growth rates thus showing greater dependability on increasing the outreach of financial

Furthermore, the insignificance of F- statistics suggests that the independent samples t-test should be used.

services. Thus, the drive of financial inclusion follows the growth in the real sector (Lucas Jr., 1988). The theoretical foundations suggest funding to act as a solvent in the process of economic growth (Levine, 1997) and thus contributing to improve equitable accessibility to finance.

5.5 Global Perspective

Considering as much as continents (i.e. 5 continents) across a large cross-sectional band covering 46 countries over a period of 2012–21[3] in Sample 4, it is observed that the countries within the continents are distributed approximately normally and share a constant variance across the groups. This leads to the application of the parametric One-way ANOVA[4] to find whether the differences across the continents are significant or not. However, a non-parametric estimate (i.e. the application of the Kruskal-Wallis test) is also provided (if non-normality is assumed). The test results of both parametric and non-parametric alternatives suggest a significant difference in mean FI scores for the study period. The differences between Africa (0.263) and Asia (0.407), between Africa (0.263) and Europe (0.417) and between Africa (0.263) and South America (0.3604) are significant. However, after the Bonferroni correction in the p-values only differences between Africa and Asia and between Africa and Europe appears significant. The F-statistic values for equality of the means across the five continents are consistently significant according to Wilks' lamda, Pillai's trace, Lawley-Hotelling trace and Roy's largest root showing the overall model fitness.

Results in Table 5.13 below coincide with Table 5.12 when comparing the continent of Europe with Asia. The continent of Europe even after having a different composition of countries observed over a different study period at prime facie crosses Asia in terms of mean FI scores achieved. This point of convergence shows the consistency of data properties across the different samples that we have used for this study.

While comparing the FI scores across the study period for the countries which are directly observable, we apply the ANOVA (with repeated

[3] Values for the year 2021 are estimates calculated by the authors using the trend of indicators from the preceding years.

[4] Since the FI scores for the continents are not directly observable rather, they are the representation (i.e. mean values) of the countries contained we cannot apply Two-way or the Factorial ANOVA.

Table 5.13 Comparison Across the Continents

Assumption: Normally Distributed FI scores					*Assumption: Non-normally Distributed FI scores*
Comparisons	*Normality*	*Equality of variances*	*F- stat*	*Heteroscedasticity*#	$\chi^2 = 17.199^{***\dagger}$
Overall	4.67*	11.073 (0.026)	$6.02^{***\#\#}$	0.21 (0.6434)	
Africa vs. Asia	Asia 9.12^{**}	Represents the	$15.06^{***\#\#}$	H_0: Constant variance (i.e.	
Africa vs. Europe		chi-square statistic of	$18.24^{***\#\#}$	homoscedasticity)	
Africa vs. North America	Africa 2.27	the Bartlett's test	1.70		
Africa vs. South America			4.87^{**}		
Asia vs. Europe	Europe 0.28		0.06		
Asia vs. North America			1.27		
Asia vs. South America	North America -		0.97		
Europe vs. North America			1.71		
Europe vs. South America	South America -		1.47		
North America vs. South America			0.11		

(continued)

Table 5.13 (continued)

Assumption: Normally Distributed FI scores			*Assumption: Non-normally Distributed FI scores*
	## *Bonferroni adjusted p-values*	# *Breusch-pagan test on fitted values Values in parenthesis represents p-values of the chi-square test*	† Chi-square and its significance applying Kruskal Wallis test

Source Authors Calculation
Note ***, ** and *represents significance at 1, 5 and 10 percent level respectively

measures) and Friedman Repeated test. These tests are commonly used for mean comparisons when the samples are assumed to draw from the same population (i.e. samples are dependent). The study provides both the parametric and non-parametric estimations as the results for testing normality of the specific study periods in Table 5.14 and is further observed to be of parsimonious in nature.

Results in Table 5.15 show that both the F- statistics and Friedman statistics are significant at 5 and 1 percent level respectively, suggesting that given the countries remain constant the change in FI scores over

Table 5.14 Testing of Normality

Year	*Adjusted Chi-square (p-values)*
2012	6.89 (0.0318)
2013	4.20 (0.1224)
2014	4.74 (0.0936)
2015	6.01 (0.0496)
2016	8.46 (0.0146)
2017	5.87 (0.0531)
2018	3.61 (0.1648)
2019	2.84 (0.2416)
2020	2.95 (0.2289)
2021	6.90 (0.0317)

Source Authors Calculation

the years are significant. The difference in the FI achievements across the continent of Europe and Africa and further between Asia and Africa emanates from the fact that countries within these continents are different in terms of their regulatory framework, economic structures, and many other factors those lead to differences in their achievement scores. Most of the countries from the continent of Africa are categorized as the Least Developed Countries (LDC's) based on their economic capacity (World Bank, 2015). Further, countries from the continent of Europe are mostly classified under the high or middle-income group. Thus, the difference in their economic capacity creates a difference in financial inclusion achievement scores which is found significant as a result in this study. Similar analogy is valid while comparing the continent of Asia with Africa. Although the economies across the continent of Asia in terms of their economic capacity are quite symmetric and normally distributed but their average capacity is substantially higher than compared to an average African nation (ADB, 2017).

A significant percentage of the Sub-Saharan African population resides beneath the poverty threshold, with the World Bank projecting that around 41% of individuals in the region subsist on less than $1.90 daily (World Bank, 2020). Elevated poverty rates constrain individuals' capacity to save or invest, hence diminishing their likelihood of interacting with formal financial institutions. This differs from the nations of Europe and certain areas of Asia, where elevated income levels and a more substantial middle class foster a conductive atmosphere for financial inclusion. Moreover, income disparity in Africa intensifies the issue, as money is concentrated within small elite, resulting in the rest of the people being marginalized from financial systems (OECD, 2019). So far the achievements in financial literacy rates are concerned, African nations as compared to their counterparts remain at the bottom which continues to hinder the attainments in the financial inclusion scores. OECD (2020) shows that a substantial segment of the African populace is uninformed about fundamental financial products, including savings accounts, insurance, and credit facilities. This inadequate financial literacy indicates that numerous individuals are oblivious to the advantages and dangers linked to formal financial services. This issue is exacerbated by cultural factors, as informal savings systems, including community-based savings organizations, are frequently favored for their social connections and trust-based characteristics. Although these informal networks are widespread throughout Africa, they exhibit lower efficiency and security compared to official financial institutions (Schreiner, 2001). Conversely, financial literacy rates in numerous regions of Asia and Europe are elevated, bolstered by extensive educational initiatives and governmental programs that advocate for financial understanding and inclusion.

African nations often encounter difficulties concerning the concentration and allocation of financial services, particularly in rural regions. Sub-Saharan Africa exhibits a markedly lower density of bank branches per capita in comparison to places such as East Asia and the Pacific (World Bank, 2021). In numerous regions of Africa, banking services are predominantly located in urban areas, resulting in significant rural populations lacking access to official financial services. Conversely, nations in Asia and Europe possess more comprehensive banking networks and more sophisticated digital payment systems that promote financial inclusion. Moreover, a lack of access to credit and formal financial services is exacerbated by the absence of credit history and reliable data on borrowers, which makes it difficult for individuals to secure loans. The lack of credit bureaus in many African countries further complicates access to credit, whereas in many European and Asian countries, credit systems are more developed and data-driven (World Bank, 2018).

A primary factor hindering financial inclusion in Africa is the lack of robust regulatory frameworks. In numerous African nations, financial rules are either inadequately developed or badly enforced, resulting in a deficiency of trust in formal financial institutions (International Monetary Fund, 2018). The absence of consumer protection legislation and ineffective enforcement procedures further dissuade individuals from interacting with banks and other financial service providers. This scenario is significantly apart from Europe and certain regions of Asia, where financial systems are more stringently regulated and consumer protections are more robust.

Despite the increasing popularity of mobile phones in Africa, access to the internet and smartphones is still restricted in most regions of the continent. Report of the International Telecommunication Union (2020) shows that mobile broadband penetration in Sub-Saharan Africa to remain approximately at 26% as compared to 73% in Europe. This digital divide substantially obstructs the advancement of mobile money services and other digital financial instruments that have transformed financial inclusion in nations such as India, where the extensive use of the Unified Payments Interface (UPI) has drastically broadened access to banking services. Further, political instability and ineffective governance

Table 5.15 Comparison Over Time

Assumption: Normally Distributed FI scores					*Assumption: Non- normally Distributed FI scores*
Source	Partial SS	Df	MS	F- stat	Friedman = 423.1507
Model	5.5814	54	0.1034	80.04***	
Country	5.5540	45	0.1234	95.58***	Kendall = 0.9403
Year	0.0273	9	0.0030	2.35**	
Residual	0.5230	405	0.0013		p-value = 0.0000
Total	6.1044	459	0.0133		

Source Authors Calculation
Note The F- statistic values and its significance can safely be interpreted, since the null hypothesis of the Mauchly's test of sphericity is accepted at all the plausible levels of significance

provide substantial obstacles to financial inclusion in Africa. Corruption, political instability, and fragile institutional frameworks have traditionally deterred both domestic and foreign investment in the banking industry. Consequently, financial services in numerous African nations remain underdeveloped, leading to widespread distrust of the formal financial sector among citizens. The African Development Bank (2020) indicates that inadequate governance and insufficient regulatory control significantly hinder the expansion of financial services in the region. Conversely, nations in Europe and certain regions of Asia enjoy more stable political climates, wherein effective governance and robust institutions foster financial development and inclusivity.

This basically leads to the conclusion that with higher economic capacity countries from the continent of Asia surpasses an average African country in terms of the achievement scores in financial inclusion.

5.6 Conclusions

While considering the primary observations from the previous chapter that the financial inclusion achievement scores might heavily rely on the economic status the region belongs to, this section attempts to evaluate the eventual implications. The review of financial inclusion achievements exhibits disproportionate achievement in the average scores across the regional and income-group classifications across the Indian states and UTs. However, their significance is in question so far, the differences across the classifications are concerned. Review of Indian states and UTs show that differences across the regions are significant. However, income-group based categorizations do not exhibit severe differences. Due to the methodological similarity results converges with the findings of Laha & Kuri (2014) but are slightly different with those reported in the Gupte et al. (2012). The regional disparity in the achievement scores of financial inclusions is also part of findings in Chakravarty & Pal (2013) citing that geographical inconvenience as of the major reasons behind the poor performance of certain specific regions (the North-Eastern region for India). Poor economic environment associated with the instability in the formal financial system is yet another an important factor that supresses the attainments in certain specific regions (Gupte et al., 2012).

Reviewing the changes over the time in financial inclusion achievements shows an unstable growth, especially across the countries from Asia. This instability emanates from the economic recessions prior to the COVID - 19 that majority of the Asian countries were facing with. Contextual evidence for Malaysia were cited in Rahman (2012) and with a broader cross-sectional band observed in Amidžić et al. (2014). However, financial inclusion achievement scores are significantly better for countries those are classified as wealthier as compared to those classified under the disadvantaged group. Research demonstrates that having access to financial services increases the amount of money saved (Aportela, 1999; Ashraf et al., 2006), speeds up the purchase of productive assets (Dupas & Robinson, 2013), and gives women more power (Ashraf et al., 2010). Financial services accessibility encourages faster growth both at the aggregate and firm levels (Beck et al., 2005; Ayyagari et al., 2008) (Levine, 2005). Additionally, the results of this investigation are included in Ayyagari & Beck's (2015) findings (refer to Fig. 1 on page 2). Financial inclusion achievements are reported to be superior in higher income (scaled with regard to GDP) countries than in lower-income ones. Analyses of the Global Findex Database, which included 150,000 randomly selected persons over the age of 15, who were nationally representative, revealed similar findings.

Notwithstanding the Asian evidence, review of the European continent across the period of time exhibits a consistent fall so far attainments in financial inclusion is concerned. However, the difference across the continents is not statistically significant giving an indication that the sample profile across the continents is more or less alike. The income-driven philosophy of financial inclusion attainments is reinforced across the sample nations considered from both the continents of Europe and

Asia which further aligns with previous literary evidence of Demirgüç-Kunt (2006) and Ghosh & Sahu (2020). Following the proposition that the growth in the real sector drives the growth in the financial sector (Lucas Jr., 1988), which eventually shows that countries with more resources tend to exhibit higher growth rates. Thus, showing a greater dependability on increasing the outreach of financial services. This evidence also holds true across a broader cross-sectional band comprising countries from all the five continents. Due to the stable economic positioning Europe and Asia as compared to Africa enjoys significantly better achievements in financial inclusion attainments.

References

ADB (2017). Asian Development Outlook 2017 Update. Asian Development Bank, Manila, Philippines.

African Development Bank. (2020). *Financial inclusion in Africa: Challenges and opportunities*. African Development Bank Group.

Amidžić, G., Massara, M. A., & Mialou, A. (2014). Assessing countries' financial inclusion standing - a new composite index. IMF Working Paper WP/14/36. doi:https://doi.org/10.5089/9781475569681.001.

Aportela, F. (1999). Effects of financial access on savings by low-income people. Mexico City: Banco de Mexico Research Department.

Ashraf, N., Karlan, D., & Yin, W. (2006). Tying Odysseus to the mast: Evidence from a commitment savings product in The Philippines. *The Quarterly Journal of Economics, 121*(2), 635–672. https://www.jstor.org/stable/25098802

Ashraf, N., Karlan, D., & Yin, W. (2010). Female empowerment: Impact of a commitment savings product in The Philippines. *World Development, 38*(3), 333–344. https://doi.org/10.1016/j.worlddev.2009.05.010

Ayyagari, M., & Beck, T. (2015). Financial inclusion in Asia: An overview. *Asian Development Bank Economics Working Paper Series*, (449). https://hdl.handle.net/11540/5142

Ayyagari, M., Demirgüç-Kunt, A., & Maksimovic, V. (2008). How important are financing constraints? The role of finance in the business environment. *The World Bank Economic Review, 22*(3), 483–516. https://www.jstor.org/stable/40282285

Beck, T., Demirgüç-Kunt, A. S. L. I., & Maksimovic, V. (2005). Financial and legal constraints to growth: does firm size matter? *The Journal of Finance, 60*(1), 137–177.

Bhattacharyya, D., & Dutta, S. (2018). Financial inclusion and development in the north-East India: Challenges and prospects. *Journal of Financial Inclusion, 12*(2), 54–66. https://doi.org/10.2139/ssrn.3251243

Bora, A. (2019). Financial inclusion and socio-cultural challenges in north-East India. *Economic and Political Weekly, 54*(32), 40–47.

Cámara, N., & Tuesta, D. (2014). Measuring Financial Inclusion: A Multidimensional Index. *BBVA Research Working Paper* (WP No 14/26).

Chakravarty, S. R., & Pal, R. (2013). Financial inclusion in India: An axiomatic approach. *Journal of Policy Modeling, 35*, 813–837. https://doi.org/10.1016/j.jpolmod.2012.12.007

Chien, C. F., & Lin, K. L. (2020). Financial literacy and its impact on financial inclusion in Asia. *Asian Economic Policy Review, 15*(1), 56–72. https://doi.org/10.1111/aepr.12345

Collins, D., J. Morduch, S. Rutherford, & O. Ruthven (2009). Portfolios of the poor: How the world's poor live on $2 a day. Princeton, NJ: Princeton University Press.

Cumming, G. (2012). The new statistics: Why and how. *Psychological Science, 25*(1), 7–29.

Demirgüç-Kunt, A. (2006). Finance and economic development: policy choices for developing countries. *The World Bank*. https://doi.org/10.1596/1813-9450-3955

Demirgüç-Kunt, A. & Klapper, L. (2013). Measuring financial inclusion: Explaining variation in use of financial services across and within countries (the global Findex database). Brookings papers on economic activity. Washington, DC: Brookings Institution.

Dupas, P., & Robinson, J. (2013). Savings constraints and microenterprise development: Evidence from a field experiment in Kenya. *American Economic Journal: Applied Economics, 5*(1), 163–192.

Ghosh, S., & Kapoor, S. (2019). Financial literacy and inclusion in India: A critical review. *Indian Journal of Economics and Business, 18*(2), 115–130.

Ghosh, S., & Sahu, T. N. (2020). How financially inclusive the Asian's are–an empirical analysis. *Rajagiri Management Journal, 14*(1). https://doi.org/10.1108/RAMJ-01-2020-0003

Gupte, R., Venkataramani, B., & Gupta, D. (2012). Computation of financial inclusion index for India. *Procedia - Social and Behavioral Sciences, 37*, 133–149. https://doi.org/10.1016/j.sbspro.2012.03.281

Hedges, L. V., & Olkin, I. (1985). Statistical methods for meta-analysis. San Diego, CA: Academic Press.

International Monetary Fund. (2018). *Financial inclusion: A comprehensive overview*. International Monetary Fund. https://www.imf.org/en/Topics/Financial-Inclusion

International Telecommunication Union. (2020). *Measuring the information society report 2020*. ITU.

Kumar, S., & Patel, R. (2020). Financial inclusion and income inequality in India: An empirical analysis. *Journal of Financial Inclusion and Development, 12*(3), 89–102. https://doi.org/10.1016/j.jfid.2020.04.001

Laha, A., & Kuri, P. K. (2014). Demand for and supply of financial inclusion in India: An inter-state analysis. *Elk Asia Pacific Journal of Finance & Risk Management, 5*(1), 103–116.

Levine, R. (1997). Financial development and economic growth: Views and agenda. *Journal of Economic Literature, 35*(2), 688–726. http://www.jstor.org/stable/2729790

Levine, R. (2005). Finance and growth: Theory and evidence. In P. Aghion and S.H. Durlauf, eds. Handbook of economic growth Vol. 1. Elsevier: Amsterdam.

Li, Z., & Zhang, X. (2021). The role of digital financial services in enhancing financial inclusion in Asia: Evidence from mobile money. *Asian Development Review, 38*(3), 108–125. https://doi.org/10.1162/asep_a_01432

Lucas, R. E., Jr. (1988). On the mechanics of economic development. *Journal of Monetary Economics, 22*(1), 3–42. https://doi.org/10.1016/0304-3932(88)90168-7

OECD. (2019). *OECD/INFE toolkit for measuring financial literacy and financial inclusion.* Organisation for Economic Co-operation and Development. doi:https://doi.org/10.1787/9789264318859-en.

OECD. (2020). *Financial literacy in Africa: Understanding the barriers.* Organisation for Economic Co-operation and Development.

Posten, H. O. (1984). Robustness of the two-sample t-test. Robustness of statistical methods and nonparametric statistics, Springer, Dordrecht, 92–99.

Rahman, Z. A. (2012). Financial inclusion in Malaysia: Tracking Progress using index. Kuala Lumpur: *IFC Workshop on Financial Inclusion Indicators Co-hosted by Bank Negara Malaysia and Central Bank of Malaysia.*

Rahman, A., & Sultana, S. (2019). Microfinance, income generation, and financial inclusion: Evidence from Bangladesh. *Journal of Development Studies, 55*(2), 200–215. https://doi.org/10.1080/00220388.2018.1458659

Sarma, M., & Pais, J. (2011). Financial inclusion and development. *Journal of International Development, 613-628.* https://doi.org/10.1002/jid.1698

Schmider, E., Ziegler, M., Danay, E., Beyer, L., & Bühner, M. (2010). Is it really robust? Reinvestigating the robustness of ANOVA against violations of the Normal distribution assumption. *Methodology,* 6, 147–151.

Schreiner, M. (2001). A cost-effective way to measure the outreach and sustainability of microfinance institutions. *Microfinance and Development Review, 18*(4), 19–34.

Shapiro, S., & Wilk, M. (1965). An Analysis of Variance Test for Normality (Complete Samples). Biometrika, 52, 591–611.

Shingala, M. C., & Rajyaguru, A. (2015). Comparison of post hoc tests for unequal variance. *International Journal of New Technologies in Science and Engineering, 2*(5), 22–33.

Singh, R. (2020). Financial literacy and its impact on financial inclusion in north-East India. *Asian Journal of Economics and Business, 8*(1), 67–79.

Singh, L., & Choudhury, D. (2017). Political instability and financial inclusion in north-East India. *Journal of Political Economy, 45*(4), 32–46. https://doi.org/10.1080/1234567890

Stoline, M. R. (1981). The status of multiple comparisons: Simultaneous estimation of all pair-wise comparisons in one-way ANOVA designs. *The American Statistician, 35*(3), 134–141.

World Bank (2015). World Development Indicators.

World Bank. (2020). *The global Findex database 2021: Financial inclusion, digital payments, and resilience in the age of COVID-19*. World Bank Group.

World Bank. (2021). *Global financial inclusion and consumer protection*. World Bank Group.

CHAPTER 6

Financial Inclusion and Sustainable Development Goals—Unfolding the Association

6.1 Introduction

The quest of sustainable development has taken front stage on a worldwide agenda in a world getting more linked. With its 17 Sustainable Development Goals (SDGs), the United Nations' 2030 Agenda for Sustainable Development offers a whole road map for reaching a more rich, fair, and ecologically friendly future for all. Even if every SDG tackles a different facet of world problems, their interdependence is indisputable. Financial inclusion is one very important factor supporting development on several SDGs. The complex link between financial inclusion and the Sustainable Development Goals will be discussed in this introduction, with particular attention to how access to and efficient use of financial services might act as a strong accelerator for accomplishing a variety of development objectives.

Broadly defined as the access to and use of reasonably priced, useful, and responsible financial products and services, financial inclusion is essential for social empowerment, economic growth, and poverty reduction (World Bank, 2021). Delivered through official financial institutions like banks, microfinance organizations, and mobile money providers, it spans a range of services including savings, credit, insurance, payments, and remittances. Conversely, the lack of financial inclusion results in financial exclusion, in which people and companies—especially those in underprivileged areas—are deprived of access to or use of these necessary services,

S. Ghosh et al., *Financial Inclusion*, Sustainable Development Goals Series, https://doi.org/10.1007/978-981-95-3385-5_6

so sustaining cycles of poverty and vulnerability (Alliance for Financial Inclusion, 2017).

Poverty is generally characterized by a lack of assets, money, and opportunity. These deprivations are often encountered through offering a varied category of financial services and in this regard, access to financial services becomes the most direct and intuitive way to address poverty. Opening up of savings accounts make individuals resilient to shocks and channelizes investments in productive activities, often allowing acquisition of assets. (Demirgüç-Kunt et al., 2018). Further, microcredit to small businesses unlocks the potential income and job prospects for unemployed youth and population (Morduch & Haley, 2002). Secured payment systems in this regard plays a crucial role through minimizing the potential threats of handling cash with especial reference to receiving remittances and social assistance and further reducing the incidence of corruption (CGAP, 2015). Financial inclusion through effective management of their resources helps the poor to lift out from extreme poverty and emphatically eliminating multidimensional poverty.

Other than poverty, one of the important issues that financial inclusion addresses is to ensure 'Zero Hunger' (i.e. SDG 2). Financial inclusion aims to end hunger and malnutrition through ensuring food security and improving the agricultural output. Following the FAO (2017) report, it is observed that small farmers are often deprived from accessing improved seeds, fertilizers, irrigation and other important agricultural inputs due to lack of funds (Food & Agriculture Organization, 2017). Apart from access to financial services (e.g. agricultural loans and crop insurance) allows farmers to invest in more productive and resilient farming practices, enhancing yields and minimizing post-harvest losses. Access to digital payment system ensures the efficient distribution of social protection programs (e.g. food aid and transfer of subsidies), thus in ultimate providing access to nutritious food, especially during times of crisis for vulnerable populations (United Nations, 2020).

Financial inclusion through extending a variety of services aims to improve health outcomes, thus ensuring 'Good Health and Well-being' (i.e. SDG 3). Access to a savings account helps individuals and households as well to manage unexpected medical costs, thus reducing their likelihood to fall into the poverty traps owing to health shocks (Rojas-Suarez, 2016). More specifically, access to affordable health insurance adds a cushion especially for the poor to manage large medical expenditures and avoid drainage of their hard earned savings. Further, the digital payment platforms helps in delivering health services more effectively and in a transparent way to ensure remote consultations and drug distribution, especially in poor areas (World Health Organization, 2020). Through improving the financial resilience of individuals and households and streamlining the access to healthcare, financial inclusion paves the way for increased health and well-being for all.

One of the significant areas where financial inclusion has a substantial contribution is to ensure 'Quality Education' (i.e. SDG 4). Financial inclusion helps to promote fair access to quality education and encourage lifelong learning opportunities through decreasing the financial burden of education. Access to education, especially for the marginalized (i.e. the low-income households) sections of society have to come across a financial constraint (i.e. the expenditure on education). These expenses can be in the form of tuition fees, books, transportation and like and is observed to increase over the passage of time. Access to savings accounts allows parents to plan and invest for their children's education, while education loans can help students pursue higher education or vocational training (UNESCO, 2019). Digital payments further ensure smooth and effective disbursal of scholarships and awards without fuelling for corruption.

Financial inclusion as a miniature form of a broader picture (i.e. social inclusion) helps to ensure 'Gender Equality' (i.e. SDG 5). Women, as a significant part of the social structure are often devoid of being part of the formal financial institutions especially in reference to the developing countries facing social, cultural, and economic barriers (Global Findex Database, 2017). The economic structure of a nation, which frequently becomes the center of attraction, can profoundly change its landscape if women gain access to financial services. Funding opportunities to women-led enterprises empower women entrepreneurs to start and expand businesses, boosting their economic independence and decision-making authority within the household and community (Kabeer, 2005). Access to savings accounts can empower women with greater control over their finances, boosting their bargaining power and minimizing their exposure to economic abuse. Furthermore, mobile money can offer a safe and convenient option for women to receive and handle money, particularly in conditions where physical mobility might be hindered (GSMA, 2020). By strengthening women economically, financial inclusion contributes greatly to gender equality and the empowerment of all women and girls.

Expansion of the micro, small and medium-sized firms (MSMEs) through providing access to financial services encourages job creation and support entrepreneurship (International Labour Organization, 2019). Access to affordable loans allows businesses to invest in new technologies, expand operations, and recruit more workers. Furthermore, access to digital payments helps to lower the cost and further boost the efficiency of corporate transactions, leading to a more vibrant and egalitarian economy. Financial inclusion also fosters formalization of the informal sector by enabling informal firms and workers to use formal financial services, leading to higher tax income and social protection.

Financial inclusion promotes the evolution of resilient infrastructure by allowing individuals and communities to invest in essential services, such as clean energy and water infrastructure, seeking participation through community-led initiatives or public–private partnerships leading to the accomplishment of the 'Industry, Innovation, and Infrastructure' (i.e. SDG 9) (United Nations Capital Development Fund, 2018). Additionally, the insurgence of digital finance especially across the remote locations promotes formation and extension of innovative infrastructure, enabling accessible and efficient payment methods. These steps help to advance the economic diversification and progress through establishing new business models and extensive sectors.

Financial inclusion aims to reduce the urban–rural divide in accessing financial services and empower vulnerable groups through active participation in the economy, thus contributing to 'Reduced Inequalities' (i.e. SDG 10). Financial inclusion ensures delivery of financial services to under-represented populations, including women, rural communities, youth, and persons with disabilities, while directly confronting the gaps in access to economic opportunities and resources (CGAP, 2019). By supplying a stepping stone out of poverty and providing instruments for wealth growth, financial inclusion helps to a more equitable distribution of income and opportunity.

Although financial inclusion shares some obvious linkages with SDGs those come under the social and economic category it has a persistent (however indirect) effect on environmental sustainability (a component of 'Climate Action') (i.e. SDG 13). People and communities can take action against climate change by investing in renewable energy, climate-smart farming and other climate-resilient technologies and practices when they have access to financial services (Green Climate Fund, 2021). Access to micro-insurance protects the vulnerable populations from climate-related disasters, and help to recover from shocks through building resilience. Further, access to digital payment systems smoothen and make transparent the distribution of climate finance along with incentives for adopting sustainable practices.

Financial inclusion helps to mitigate the causes of conflict and instability through promoting economic empowerment and eradicating poverty, thus achieving the aim of 'Peace, Justice, and Strong Institutions' (i.e. SDG 16) (United States Institute of Peace, 2015). Individuals become less likely to be susceptible to illegal activity and more likely to contribute to improved social cohesion when they have a stake in the legal economy and can use authorized financial channels. Additionally, governance and accountability can be improved through establishing a robust and transparent financial system that fosters trust in institutions.

Finally, financial inclusion requires a multi-stakeholder approach involving governments, financial institutions, technology providers, civil society organizations and international development partners which calls for the 'Partnerships for the Goals' (i.e. SDG 17) (United Nations Development Programme, 2022). Country-specific policies and associated regulatory framework helps to maintain consumer safety and investments in promoting financial literacy which are some of the crucial ingredients of successful collaborations. While expansion of accessibility to financial services across the state borders calls for international cooperation along with ensuring free flow of knowledge leading to scaling up of the activities and overcome shared problems.

To summarize, the relationship between financial inclusion and the Sustainable Development Goals is complex and varied. Financial inclusion is not just a goal in and of itself, but it is also a powerful enabler and accelerator for achieving a wide range of development outcomes across the social, economic, and environmental dimensions of sustainability. Financial inclusion empowers individuals, households, and businesses by facilitating access to and promoting the effective use of affordable, useful, and responsible financial services. Recognizing and harnessing this sensitive link is critical to achieving the 2030 Agenda's transformative objective of creating a more inclusive, egalitarian, and sustainable society for all. Continued efforts to increase financial access, improve financial literacy, and develop new financial solutions tailored to the requirements of marginalized groups will be critical to realizing the full potential of financial inclusion as a pillar of sustainable development.

The design of the SDGs is in a way that accomplishment in one global target may lead to advancement in another. In reality, one set of SDGs serves primarily as a vehicle for accomplishing other SDGs. In this connection, one could certainly consider SDG 8 (i.e. Decent employment and economic development) paving the way to effectively achieve SDG 1 (i.e. No Poverty). The structural composition of the sustainable development goals however, can broadly be classified under the three pillars that emphasizes on the social targets (Goals 1–6), economic goals (Goals 7–11), and goals aimed at achieving environmental sustainability (Goals 12–15). The inter-linkages between the developmental goals are expected to have synergic benefits. While poverty eradication (SDG 1), achieving zero hunger (SDG 2) and equitable economic growth (SDG 8) remain top priorities for development, this section attempts to highlight the ways on how these goals can be achieved through ensuring inclusive financial systems.

6.2 Financial Inclusion and Decent Work and Economic Growth (SDG 8)

The significance of finance as a catalyst for growth is a highly contested subject in financial economics. Theoretical foundations indicate that finance functions as a solvent in the economic growth process (Levine, 1997). The results demonstrate a positive, primary correlation between financial development and economic growth. Finance is identified in the research of Gurley and Shaw (1955) and McKinnon (1973) as an essential instrument for economic development. The existence of efficient financial tools, markets, and institutions mitigates frictions by lowering information and transaction costs. Moreover, countries transiting from developing to developed status require an advanced financial system to thrive (Patrick, 1966). Numerous researches investigating the correlation between financial development and economic growth indicate a positive association (Eryilmaz et al., 2015; Goldsmith, 1969; Habibullah & Eng, 2006). Habibullah and Eng (2006) identified results analogous to those of Calderón and Liu (2003), Fase and Abma (2003), and Christopoulos and Tsionas (2004) regarding the correlation between financial growth and economic expansion within a system GMM framework established by Arellano and Bond (1991) and Blundell and Bond (1998). The theoretical structure of the study is based on Pagano's (1993) research, which integrates the works of Romer (1989) and Lucas Jr. (1988). Results suggest that across a panel of 13 Asian nations over 1990–98 financial development enhances growth (Habibullah & Eng, 2006). The traditional role of finance impacts economic growth via capital accumulation (Lewis, 1955; Nurkse, 1962) and technological advancement, particularly productivity enhancement. Following the capital accumulation strategy a steady-state per capita growth can be attained through either exploiting capital externalities or manufacturing capital goods with continuous returns to scale while avoiding the use of non-reproducible variables (Rebelo, 1991; Romer, 1986). While observing 109 countries from developing as well as industrialist countries from 1960–94, economic growth is seen to accelerate with financial deepening through capital aggregation and more slowly through productivity growth (Calderón & Liu, 2003). Through adjustment of the rate of savings or by dispersing the savings among the available capital producing technologies the formal financial system configures the accumulation of the capital. Further, the steady-state growth climbs by change of the pace of technical innovation based on the interventions that concentrates on advanced production processes (Aghion & Howitt, 1990; Grossman & Helpman, 1991). Formal financial system is also regarded to support economic growth through enhancing liquidity of long-term high-return commitments, which typically investors are less willing to get engaged with. This was further highlighted as one of the key motivations driving industrial revolution in England (Hicks, 1969). Even if technological progress was present, the driving force behind this eighteenth century growth was the liquidity factor of the stock market. A well-functioning financial market should ensure appropriate liquidity of capital commitments so that as and when required it can be transformed into purchasing power of the investor. However, the ease and speed of this conversion largely rely on the information asymmetry and transaction cost that acts as a friction to such conversion. The four pillars of a formal financial system (e.g. Financial Instruments, Financial Institutions, Financial Markets and Financial Services) functions as a facilitator to lower these frictions and so boost the rate of savings and long-run growth. The different forms of financial functions (e.g. trading of risks, savings mobilization, proper capital allocation, simplifying the trading of goods and enforcing financial contracts) help ameliorate these frictions on a broader perspective rather focusing on a particular instrument (i.e. money) or an institution (i.e. banks). This rather capitalizes on the functional link between financial structure (i.e. financial instruments, markets and institutions) and the delivery of financial services (Levine, 1997). Furthermore, Goldsmith (1969) illustrates that financial development strategies may greatly improve per capita GDP by assessing 35 nations from 1860 to 1963. Likewise, Eryilmaz et al. (2015) attempted to observe the impact of total credit to GDP and total savings to GDP on annual GDP per capita after taking control of public spending to GDP and export deficit. Employing the two-way random effects model, the study identified comparable evidences across the 23 OECD nations from 1980 to 2012. Similar to the conclusions of past research, Rajan and Zingales (1998) indicate that financial markets supply the needs for development. They claim that financial development and economic growth may both be driven by a single missing component (e.g. the propensity of households to save). Since, savings influence the pace of growth, early financial development unsurprisingly might be connected with growth and subsequent evidence of cross-sectional regressions might fail to contradict the hypothesis so developed. They explain such specifications as conjectural coincidence. Rajan and Zingales (1998) criticizes the size of stock market as an indicator of growth as the markets itself capitalizes growth whereas they also criticize the level of credit extended by financial institutions as an indicator of financial development because the lending by institutions is based on future growth prospects. This study categorizes the industries based on their requirements of external money and develops a comparative view with respect to the scale of financial status the country belongs to. Industrial expansion is largely driven by the introduction of new enterprises or increase in the average size of the present firms. However, entry of a new firm is more dependent on external capital and thus vulnerable to financial changes. Finance-growth association has been further investigated in the study of Demirgüç-Kunt and Maksimovic (1999). This study basically evaluates the proportion of firms where growth rates surpass the growth rate of firms those are backed entirely by internal resources. Results further reveal that this proportion is positively associated to stock market turnover as well as with the measure of legal enforcements. However, Greenwood and Jovanovic (1990) shows financial intermediation and economic growth are endogeneously determined. They suggest that financial organizations can provide larger returns than compared to individuals whereas growth in turn builds costlier financial structures. Financial development brings down the gaps between external and internal finance leading to growth notably for organizations which are relying on external financing. Bencivenga and Smith (1993) asserts the evolution of financial markets as exogeneous that depends mostly on legislations and government regulations. However, financial development cannot just happen quickly (Diamond, 1989). Similar evidences have been detected in a wide panel of 159 nations by Khan and Senhadji (2000) from 1960 to 1999. However, a thorough assessment of the data suggests that throughout the lower income categories economic development is more susceptible to the development of financial sector (Herwartz & Walle, 2014). Efficient financial intermediaries together with superior resource allocation with greater total factor productivity increase quicken economic progress after considering the likelihood of endogeneity across the regressors in 63 nations between 1960 and 1995 (Beck et al., 2000). Furthermore, a positive link with growth in the economy is found when focusing on the exogenous part of financial sector expansions (Levine et al., 2000). However, the reliability of both studies Beck et al. (2000) and Levine et al. (2000) in terms of meeting the data characteristics associated with integration and co-integration came into question owing to the challenges in predicting the present equilibrium over the long-term connection (Christopoulos & Tsionas, 2004). When investigating the equilibrium over the long-term link, Christopoulos and Tsionas (2004) demonstrated a cause-and-effect (uni-directional) connection between financial sector expansion and economic success in a sample of 10 rising economies. The links, however, were constructed on the concept of cross-sectional independence rather than dependency (Bist, 2018). A closer look at the situation throughout the ASEAN region from 1991 to 2011 indicates co-integrating links between economic growth along all the four components of financial sector development (the growth in the insurance and banking industries, as well as the stock and bond markets) (Pradhan et al., 2017). In terms of causality, the study identifies disparities in the magnitude and directions of financial development factors concerning economic growth. The expanding importance of banking services and economic growth has one-way causation, whereas the gains in the stock market, insurance industry, and economic expansion have two-way causality.

Dahiya and Kumar (2020) uncover a strong link between economic development and the usage part of financial inclusion instead of the composite index produced in this regard when employing the Bayesian VAR model across Indian states. After accounting for the probable interruption owing to the liberalization policy in 1991, the aggregate index of financial inclusion is shown to favorably affect economic development both in the long and short-term applying the ARDL and VECM framework respectively during 1980–2014 (Lenka & Sharma, 2017). Furthermore, Sharma (2016) shows a positive link with economic development in a VAR configuration for 2004–13 when the metrics of financial inclusion are studied individually as well as together as an index. The aforementioned analysis discovers the existence of bi-directional causation utilizing the Granger causality test. It does not, however, disclose the co-integrating relationships between the components.

When focusing on low-income nations and relying on both the simple and uni-directional features of causation, the supply leading theory becomes increasingly prominent (Jung, 1986). Hassan et al. (2011) revealed a bi-directional association between financial expansion and economic development in poor and middle-income nations, with the major exception of the two lowest categories, which exhibit uni-directional causality. Nevertheless, an array of examples from the research reveals that, relying on the circumstances, financial performance indicators could possibly hold a damaging or even non-existent impact on economic achievement (Ahmad et al., 2021; Lucas Jr., 1988). Singh (1997) finds a minor correlation between the components when 65 developing nations are studied. Considering the indicators from both the banking and stock market across the BRICS nations for 1993–2014, Guru and Yadav (2019) in a one-step system GMM specification observed that indicators from the banking and stock market both compliments each other's growth and stimulate the growth of the economy. The size of the financial intermediaries, credit to deposit ratio and domestic credit to private sector as an indicator of banking sector development favorably effects the growth of the economy. Similarly, the value of shares traded is also observed to strongly influence the growth in the economy under the presence of banking sector development variables. However, the impact of the turnover ratio is not considerable on the economic growth metrics across the industrialized nations (Mhadhbi, 2014). This was also visible across the GCC nations (Grassa & Gazdar, 2014) as well as across the northern and southern Mediterranean nations (Ayadi et al., 2015). Similar outcomes were reached while considering a broad panel comprising 101 developing and developed nations (Ductor & Grechyna, 2015).

The contemporary research on finance-led growth adds to the data in support of a non-linear connection. Samargandi et al. (2015) identified a U-shaped inversion between finance and economic development while investigating the long-term threshold implications in a panel of 52 nations with modest incomes from 1980 to 2008. Furthermore, Rousseau and Wachtel (2011) and Arcand et al. (2015) show that after reaching the threshold, the influence of financial development on economic growth decreases, even when non-linear components are included. In this regard, Arcand et al. (2015) show that once loan to the private sector crosses a certain level, it begins to have a negative effect on output growth. Similarly, results from the threshold regression model suggests development in finance to greatly influence economic growth (Deidda & Fattouh, 2002) more promisingly across the countries from higher-income group than the poorer ones. However, Law and Singh (2014) suggest that greater financing does not inevitably boost economic advancement, based on data from 87 countries, the bulk of which are in the developed and emerging stages.

It is not unique to use a panel co-integration model to analyze the long-run connection between financial inclusion and growth in the economy. Using both the FMOLS and DOLS estimators, Sethi and Acharya (2018) assert that the combination of the evaluation of financial accessibility (as determined by Sarma, 2012) had a long-term impact on the 31 nations' economic success between 2004 and 2010. The research also applies unconstrained VECM in a fixed-effect model and discovers a two-way causal link between the components. Using the system GMM method, similar evidence was reported across 152 countries for 2004–15 and 44 developing markets for 1990–2018 (Emara & El Said, 2021; Van et al., 2021). Furthermore, Kim and Hassan (2018) apply the dynamic VAR framework in the system GMM using lagged regressors to find that financial inclusion boosts economic development across the 55 OIC member nations. The Impulse Response Function results support the findings, and the panel's Granger causality test validates mutually reinforcing effects.

6.3 Financial Inclusion and No Poverty (SDG 1)

This study presents both cross-sectional and country-specific evidence from extensive research highlighting the interrelations between financial development, economic growth, income distribution, and poverty reduction. Research examining the relationship between financial inclusion and poverty alleviation through macro-level statistics offers a general overview but lacks detailed and thorough insights. Moreover, country-specific research at the household level is specific yet inadequate for constructing a comprehensive overview. Chibba (2009) successfully underscores the significance of formal financial intermediation in reducing poverty incidence and so enhancing the achievement of the Millennium Development Goals (now the SDGs). Traditional poverty-reduction strategies are vital, but they are insufficient to hasten poverty reduction; hence, financial inclusion (FI) may be considered as a chance to explore. Addresses from the United Nations Conference on 'Financial Inclusion in Africa' in June 2006; the World Bank's conference on 'Access to Finance' in March 2007; and many other efforts taken solely or jointly recognize and support the vital role delved within the FI-PR-SDG alliance encouraging a nation's economic and social progress. International collaborations support the premise that economic and social growth cannot be promoted in isolation, leaving a substantial sector of society, particularly the marginalized out from its peripheral. As a result, an endeavor to extend the accessibility of financial services to all parts of society is crucial for boosting development and configuring capability in such a manner that SDGs are better fulfilled (Dugarova & Gülasan, 2017). An analysis of cross-sectional empirical evidences in the literature, whether employing the nations or states as units, offers a generalized viewpoint. Considering the influence of financial inclusion on vulnerability and susceptibility to poverty in Ghanaian households following the Ghana Living Standard Survey 2016/17, Koomson et al. (2020) observe a 27% decrease in the household's likelihood of being poor, as well as a 28% protection against future exposures to poverty. A gender-based comparison demonstrates that male-headed families are relatively on the inefficient side than compared to the female-headed households to cut down poverty as well as susceptible to poverty. Further, the impacts of financial inclusion towards reducing the occurrence of poverty are more noticeable across the rural areas than compared to the metropolitan areas. Literary data reveals that sensitivity to poverty is an area that has been relatively less emphasized (Imai et al., 2010). There can be occasions where now non-poor households slip in to poverty traps in the approaching future while confronting contingencies (Ozughalu, 2016). Coulibaly (2019) believes that before the year's end of 2019, Africa will be 70% of the worlds impoverished. This will further intensify by the end of 2030 when approximately 13 countries would confront extreme poverty. The Maya declaration as agreed by 90 poor countries in the G20 summit in 2010 which had targeted to decrease poverty through financial inclusion (AFI, 2015) had identified that 75% of the countries are financially excluded. The purpose of such declaration was also congruent with the Sustainable Development Goals (SDG 1) (ICSU & ISSC, 2015). Financial inclusion climbed from 51 percent in 2011 to 69 percent in 2017, among nations that signed the declaration. However, the review of account ownership in terms of gender is highly suspicious notably throughout the developing countries for female (59%) than compared to male (79%) (Demirgüç-Kunt et al., 2018). So far as the rural urban gap is concerned, urban people in developing countries have higher access than compared to the rural counterparts. As a result, the relationship between financial inclusion and poverty (usually assessed by susceptibility to poverty) requires greater examination in light of gender and locational differences (Swamy, 2014). The benefits of financial inclusion serve to eliminate poverty by both direct and indirect pathways while remaining within the structure of financial growth. Financial inclusion directly reduces poverty through extending the access to credit, insurance and other services that facilitates the resources required to meet day-to-day consumption, investment, and overall growth of the economy (Rajan & Zingales, 1998). It also assists to boost the entrepreneurial capacities of individuals that augment their income, consumption and thus the living standards (World Bank, 2007). Financial inclusion provides job and through greater social expenditure on education, health and social security that contributes to growth of an economy which is frequently called as an indirect channel to support development (Abosedra et al., 2016). This was also previously a component of policy advice by Perotti (1993). Further, economic liberalization through over utilization of credit cards, ATM's and online banking can adversely boost indebtedness which might invite poverty as a consequence (Lyons & Hunt, 2003). Enhanced accessibility to finance has also some unfavorable consequences on the status of prevailing inequalities (Koomson et al., 2020). As a result, a better technique is offered, emphasizing the ubiquitous availability of financial services to improve economic development (Beck & Demirgüç-Kunt, 2008). This was also consistent with the World Bank's goal of implementing the UFA to eliminate poverty by 2020. Enhanced accessibility of financial services is also praised to ensure sustained coverage of education, health, and insurance goods which mitigates the risk of uncertainty especially for the vulnerable sectors of the community (Brown et al., 2013; Mârza et al., 2015). Certain adaptations that improve access to credit and savings products serve to limit risks and create a cushion for the households to battle poverty (Emerick et al., 2016; Hallegatte et al., 2016). Mobile money as a recent inventive product is one of most significant strategic vehicles that begin on gaining accessibility to varied financial products and thus accomplishing sustained economic growth (Donovan, 2012). Among others, Burgess and Pande (2005) found that extending rural bank branches, savings, and credit throughout Indian provinces may successfully address the problem of poverty reduction. Inoue and Hamori (2012) demonstrate that deposit and credit amounts as a fraction of production may greatly lower poverty head count across an uneven panel of 28 Indian states. Jalilian and Kirkpatrick (2002) analyze the influence of financial expansion on decreasing poverty utilizing cross-country data from developing nations. They observed that financial inclusion helps to build the asset structure of the poor households by giving them the opportunity to invest in innovative technology, health, and education. This in turn will allow them to achieve sustainable living conditions (World Bank, 2001). Further, Jalilian and Kirkpatrick (2005) found that financial development considerably reduces poverty when they simulate the same sample to make accurate judgments on the causal link between the dimensions. They remark that up to a certain degree of economic growth the expansion of the financial sector eradicates poverty through resolving inequality.

Following the introduction of the SDGs (UNDESA, 2016), Boukhatem (2016), utilizing the framework of the GMM estimator in a group of 67 low and middle-income nations, reveals that financial development may drastically decrease poverty. This research, on the other hand, tackles the issues of over-identification and weak tools, resulting in a more trustworthy estimate of the impact. In a similar approach, Kiendrebeogo and Minea (2016) assert that access to financial services reduces poverty undeniably. After controlling for simultaneity and small sample bias, the research suggests that volatility in finances may have a negative impact on the connection between financial success and poverty. Further, this study also admits the presence of non-linearity in predicting poverty on the basis of financial progress. Following the McKinnon's 'Conduit' effect (McKinnon, 1973) and Shaw's 'Intermediation' effect (Shaw, 1973), finance may not have a direct impact on poverty, but it might abolish poverty through economic growth. However, Greenwood and Jovanovic (1990) claims that accessibility to bank's credit notably for the poor can be limited by the high costs of modest loans. This problem could get aggravated when it comes to observe the distribution of income (which appears to be highly skewed) notably throughout the developing countries in favor of the affluent side of the population (World Bank, 1995). As a result, a favorable effect on economic growth may not inevitably translate to a decrease in poverty (Beck et al., 2007). In this connection, Arestis and Caner (2010) were unable to prove a significant influence that capital account liberalization may have on decrease of poverty. Through increasing the allocation of resources in to productive investments and with sufficient monitoring, financial intermediaries can dramatically reduce poverty.

Thus, financial inclusion, as an integral component of a bigger picture (i.e., social inclusion), seems to be extremely crucial in addressing problems of poverty and inequality by unlocking potential possibilities, particularly for the disadvantaged portions of society. Even increasing the heterogeneity by evaluating a larger cross-sectional band of 116 developing nations during 2004–16, shows that access and use of formal financial services should be expanded in order to enhance societies' overall well-being (Omar & Inaba, 2020). Analyzing the influence of financial inclusion on specific development indicators (e.g., income, spending, food security, and so on) with regard to gender inequalities, for example, does not support the exception (Swamy, 2014). The influence of formal financial intermediation on production growth is diverse, leading to an expected result (Bell & Rousseau, 2001). Furthermore, there is minimal data, particularly employing a cross-country arrangement that explores whether financial inclusion helps poverty through improving output development.

6.4 Financial Inclusion and Zero Hunger (SDG 2)

Currently, one of the most significant threats to humanity is undernourishment. Human capital formation and economic development rely on sufficient nutrition (Bryce et al., 2008; Sachs, 2002; Sen, 1995). It is now assured that all forms of malnutrition would be eliminated, at least by 2030, in accordance with commitments 2.1 and 2.2. The commitment 2.1 guarantees that everyone has year-round access to safe, nutritional, and sufficient food. The FAO's studies from 2017 and 2018 indicate that conflict, climatic variability, and other extremes impede efforts to eradicate hunger. The principal barriers to eradicating hunger include the economic recessions outlined in the 2019 report, the repercussions of COVID-19, and the unprecedented desert locust infestations in Eastern Africa in 2020.

Alongside hunger, individuals are progressively compromising their intake of both the quantity and quality of food. Common circumstances leading to undernourishment, or more severely, malnutrition, include meal omission and complete hunger.

There could be a lot of explanations, including increasing food costs, decreasing food supply, lower income, and many more. Warr (2014) identifies the availability and affordability of food as a prominent driver of food insecurity (FI), which contributes to undernourishment by diminishing dietary quality. Income is yet another key variable that impacts food security. By assuring food security, a constant and sufficient income that can afford a nutritious diet helps to prevent the prevalence of undernourishment (PoU) and malnutrition. Requiring money to be adequate is unjustifiable. By providing the opportunity to save, the promotion of financial services can partially ensure income security.

Proper nutrition is a vital component for analyzing the effectiveness of the tactics focused on development. Development projects frequently emphasizes on the relevance of significant economic issues, notably income and pricing, in addressing under nutrition. Surprisingly, appropriate diet is not proven to improve certain markers of health instead improving the productivity (Behrman et al., 1988). Recent strategies have relied on econometric calculations that illustrate the relationships between energy intakes and income, prices, and other factors based on household level data. However, it is the appropriateness of these intakes relative to each person's requirements that is crucial, not the intakes themselves. Except certain extreme instances, human metabolism is capable of adjust with whatever intake of nutrient they consume (Sahn, 1987). Additionally severe intrapersonal volatility throughout the period is perceived so far intakes are concerned (Srinivasan, 1981; Sukhatme, 1982) and thus this measure might become a very poor proxy for health indicators (Behrman et al., 1988).

The role of income and price in the formation of famines has also been underscored by empirical evidences studying the elements that determine individual levels of nutrition (Ravallion, 1987; Sen, 1982). This raises the question of whether the prevalence of undernourishment should be considered transient or an aggregate estimate responding to prices and income. Furthermore, characteristics such as the incidence of corruption, effectiveness, and stability of the administration and the state of violence have proven necessary to decrease undernourishment (Lee et al., 2016; Moore et al., 2006).

Further, food prices have grown gradually since 2006, reaching a maximum in 2008 and causing a global food crisis that has never been overcome (Lee et al., 2016). A spike in food costs exacerbates worries about food security by boosting the possibility of famine and limiting family intake of nutritious foods. The recession globally in 2007, which combined food, fuel, and financial crises, showed a decline in the intake of food score (determining the frequency and diversity of food intakes) due to food price hikes in Haiti, Nepal, and Niger (Christian, 2010[1]). This forced families to experience vulnerability as a result of a logical tradeoff in terms of both food quality and quantity, leading to increasing malnutrition. According to Darnton-Hill and Cogill (2010), these vulnerabilities have an effect on mother and child health by lowering micronutrients and increasing infectious diseases, which may lead to death in severe cases. Similarly, growing food costs have had a major and adverse influence on nutrition, resulting in greater rates of new-born and infant mortality in both developing nations as well as least developed nations (LDCs) between 2001 and 2011 (Lee et al., 2016). These data supplement the assumption that nations having lower income have a bigger share of household spending on food, particularly in LDCs, and hence rely considerably on the import of food from outside. A study of the prior literature especially across the poorer nations demonstrates an incidence of increasing hunger, malnutrition, and death caused by the rising food costs among new-borns and children (FAO, 2008; ADB, 2012). The relatively wealthier part of the society (measured in monetary terms) is further reported to experience massive cutbacks in private expenditure and adopting dramatic measures such as increased usage of food banks (Butler, 2012; Selke, 2011), attempting suicide due to poverty, and dying from extended famine (Gentleman, 2014). When such strategies are used, micronutrient and calorie shortages develop, worsening the incidence of malnutrition (Ortiz et al., 2011). Furthermore, such spikes in the food costs are evident to drastically reverse the successes so far as the Sustainable Development Goals (SDGs) are concerned. The rise in food costs is predicted to raise both child and infant death rates (Goal 3), which

[1] Further read Brinkman et al. (2010) and Darnton-Hill & Cogill (2010).

were lowered by 25 percentage points between 1990 and the year 2015 (Brinkman et al., 2010). A surge in global food commodity prices (as judged by the FAO Food Price Index) happens with a one-period lag (Lee & Park, 2013). Furthermore, rising food prices do not result in enhanced advantages for the marginal farmers (Wise & Harvie, 2009).

Income becomes yet another essential component in determining food security. The presence of a consistent and enough income may assist purchase a healthy meal, therefore reducing the prevalence of undernourishment (PoU) and consequently malnutrition. The sufficiency of income requirement cannot be justified. However, expanding financial services may help to foster economic stability by offering a chance to save. Recent study reveals that the income elasticity of individual energy use in emerging nations is lower than previously anticipated (Behrman et al., 1988; Bouis & Haddad, 1988). Policies designed to raise the income especially for the underprivileged either through accelerating the growth or addressing inequality. Raising the poor's income (by policies targeted at boosting growth and decreasing income disparity) has helped to overcome hunger, which is regarded to be one of the most important drivers of development (Ravallion, 1990). The income sensitivity of calories may range from 0 to infinity. Empirical evidence suggests a size ranging from 0.08 to 0.14 in Philippines (Bouis & Haddad, 1992) to 0.55 among the poorest families in Maharashtra (India) (Subramaniam & Deaton, 1996). Research findings in Alderman (1993) and Strauss and Thomas (1995) similarly shows near-zero estimates with point estimates statistically significant at 0.37 for India (Behrman & Deolalikar, 1987). However, the sensitivity of PoU to income may be overestimated due to the substitution effect between nutritional values and other dietary aspects within food-staple aggregates (Behrman & Deolalikar, 1987). When the non-linear structure of the Engel Curve for calories is taken into account, the poverty demonstrates more sensitivity than the rich (Ravallion, 1990; Reutlinger & Selowsky, 1976; Strauss & Thomas, 1990). Especially among the families with lower income, it was revealed that demand with respect to price of food was more responsive (Green et al., 2013). Furthermore, the sensitivity of food quality to money gives inadequate explanation for such outcomes (Alderman, 1993). Bouis and Haddad (1988) give an alternative interpretation fearing erroneous variables.

Economic downturns, as assessed by per capita GDP, exacerbate infant deaths (Baird et al., 2011). Falling incomes as a result of the economic downturn are communicated to households through lower food and health costs (FAO, 2012), whereas increased income is expected to provide a variety of improved prospects, including better housing, adequate nutrition with a supply of hygienic water and sanitation, and, most importantly, access to health-care services. However, panel studies evaluating a number of economies together reveal that economic improvement may not be translating into improved health conditions unless purposeful governmental actions are made (Cutler et al., 2006; Dreze & Sen, 2002). Even with a long-term view, the findings remain inconclusive (Lee et al., 2016).

Despite being a major economic sector in rising nations, employing a considerable percentage of the working population, agricultural productivity froze between 1960 and 1980 and has been slowly increasing since then. Review of Warr (2014) reveals that increase in agricultural productivity, while expecting the prices to stay constant, improves the incidence of undernourishment. Policies that aim for agricultural protection and minimize substantially the dependency on the imports often accidentally raises food prices via achieving food self-sufficiency, which eventually aggravates undernourishment (Tiwari & Zaman, 2010; Warr, 2014). The speed of expansion in this industry has been regarded slow most likely due to the unsupportive policies and constraint of the resources. (Alston & Pardey, 2015; Mattoo et al., 2020). From having a proportion of 15–20% of overall Official Development Assistance (ODA) dedicated to agriculture in 1970–80 (Cabral & Howell, 2012), it is presently less than 3% in 2007 (Webb, 2010) and has never topped 4% during 2012–20. This unexpected behavior, together with growing demand for food, has rendered the nationwide agricultural research initiatives unviable. Operations such as seed multiplication and providing of loans across the rural belts are on the brink of eradication (Webb, 2010). Furthermore, with a weak Food Supply Chain (FSC), poor and undeveloped nations are already contending with the issue of food insecurity, which has been compounded by the implementation of COVID-19 (Béné, 2020). Food shortages and supply chain disruptions have a direct effect on the health and well-being of a third of the world's population (Khan et al., 2022). High fluctuation in transaction costs leads to more disruptions in FSC, reducing food supply (Hobbs, 2020; Reardon et al., 2020). However, the influence on undernourishment seems to be transitory. Several difficulties, including crop losses, environmental disasters, insufficient food storage, and outmoded FSC technology, must be solved to assure food supply for billions of undernourished people (Baležentis et al., 2021; Tian et al., 2016; Webb, 2010).

6.5 Conclusions

Financial inclusion, broadly defined as having access to and using reasonably priced, usable, and responsible financial products and services, is critical for social empowerment, economic growth and poverty reduction (World Bank, 2021). It is provided by official financial institutions such as banks, microfinance organizations, and mobile money providers and covers a wide range of services such as savings, credit, insurance, payments, and transfers. In contrast, a lack of financial inclusion leads to financial exclusion, which occurs when individuals and households, particularly those in underserved areas, are denied access to or use of these essential services, perpetuating cycles of poverty and vulnerability.

Poverty is often defined as a lack of assets, money, and opportunity. These deprivations are frequently addressed by providing a diverse range of financial services and in this light, access to financial services becomes the most straightforward and obvious strategy to combat poverty. Individuals who open savings accounts become more resilient to shocks and channel their money into productive activities, often allowing them to acquire assets (Demirgüç-Kunt et al., 2018). Furthermore, microcredit to small enterprises opens up potential income and work opportunities for the unemployed young and populace (Morduch & Haley, 2002). Secured payment systems play an important role in this regard by lowering the possible risks of handling cash, particularly when receiving remittances and social assistance, as well as reducing the prevalence of corruption (CGAP, 2015). Financial inclusion, achieved via good resource management, assists the poor in emerging from extreme poverty and emphatically eliminating multidimensional poverty.

Aside from poverty, one of the most critical challenges that financial inclusion addresses is achieving 'Zero Hunger' (SDG 2). Financial inclusion aims to eliminate hunger and malnutrition by ensuring food security and increasing agricultural productivity. Small farmers are often debarred of accessing quality seeds, fertilizers, innovative ways of irrigation, and advanced agricultural inputs on financial grounds. Other than accessing the conventional financial products (e.g. agricultural loans and crop insurance) farmers can devise mechanism of resilient farming practices to improve productivity and thus yields. Access to nutritious food especially during the crisis for the vulnerable can be instrumental through allowing access to a digital payment system catering to efficient distribution of social protection programs (e.g. Food aid and subsidy transfers).

In a nutshell, the relationship between financial inclusion and the SDGs is multifaceted and diverse. Financial inclusion does not work in isolation and acts as a potent enabler and accelerator for achieving a wide range of development outcomes in the social, economic, and environmental dimensions of sustainability. Individuals, households, and enterprises benefit from financial inclusion as it calibrates the conventional financial services into more affordable and usable services. Recognizing and leveraging this delicate relationship is crucial to accomplishing the 2030 Agenda's transformative goal of establishing a more inclusive, egalitarian and sustainable society for all. Continued efforts to extend financial access, promote financial literacy and develop new financial solutions targeted to the needs of marginalized groups will be required to realize financial inclusion's full potential as a pillar of sustainable development.

The SDGs are designed in such a way that achieving one global goal may result in progress towards another. In practice, one set of SDGs is mostly used as a vehicle to achieve other SDGs. In this regard, one may surely see SDG 8 (Decent employment and economic development) as paving the path to efficiently achieving SDG 1 (No Poverty). However, the structural makeup of the sustainable development goals can be broadly grouped into three pillars that emphasize social aims (Goals 1–6), economic goals (Goals 7–11), and environmental sustainability goals (Goals 12–15). The interactions between the developmental goals are predicted to produce synergistic effects. Poverty eradication (SDG 1), zero hunger (SDG 2), and equitable economic growth (SDG 8) continue to be major development targets.

References

Abosedra, S., Shahbaz, M., & Nawaz, K. (2016). Modelling causality between financial deepening and poverty reduction in Egypt. *Social Indicators Research, 126*(3), 955–969. https://doi.org/10.1007/s11205-015-0929-2

AFI. (2015). Maya declaration report: Commitments into action. Alliance for Financial Inclusion. Retrieved on 5/9/2021 from http://www.afi-global.org/sites/default/files/publications/maya_report_2015-final.pdf

Aghion, P., & Howitt, P. (1990). *A model of growth through creative destruction.*

Ahmad, M., Majeed, A., Khan, M. A., Sohaib, M., & Shehzad, K. (2021). Digital financial inclusion and economic growth: Provincial data analysis of China. *China Economic Journal, 14*(3), 291–310.

Alderman, H. (1993). New research on poverty and malnutrition: What are the implications for policy? *Including the poor*, World Bank, 115–132.

Alliance for Financial Inclusion. (2017). *The Maya Declaration: Advancing financial inclusion for all.* https://www.afi-global.org/publications/the-maya-declaration-advancing-financial-inclusion-for-all-2/

Alston, J. M., & Pardey, P. G. (2015). Agricultural R&D, food prices, poverty, and malnutrition redux. In *The fight against hunger and malnutrition: The role of food, agriculture, and targeted policies* (pp. 208–239) Oxford University Press.

Arcand, J. L., Berkes, E., & Panizza, U. (2015). Too much finance? *Journal of Economic Growth, 20*(2), 105–148. https://doi.org/10.1007/s10887-015-9115-2

Arellano, M. & Bond, S. R. (1991). Some test of specification for panel data: Monte Carlo evidence and application to employment equations. *Review of Economic Studies, 58*, 277–297. https://www.jstor.org/stable/2297968

Arestis, P., & Caner, A. (2010). Capital account liberalisation and poverty: How close is the link? *Cambridge Journal of Economics, 34*(2), 295–323. https://doi.org/10.1093/cje/bep062

Asian Development Bank. (2012). *Food security and poverty in Asia and the pacific: Key challenges and policy issues.* Manila. https://www.adb.org/publications/food-security-and-poverty-asia-and-pacific-key-challenges-and-policy-issues

Ayadi, R., Arbak, E., Naceur, S. B., & De Groen, W. P. (2015). Financial Development, Bank Efficiency, and Economic growth across the Mediterranean. In R. Ayadi, M. Dabrowski, L. De Wulf (Eds.), *Economic and social development of the Southern and Eastern Mediterranean countries* (pp. 219–233). Springer. https://doi.org/10.1007/978-3-319-11122-3_14

Baird, S., Friedman, J., & Schady, N. (2011). Aggregate income shocks and infant mortality in the developing world. *Review of Economics and statistics, 93*(3), 847–856. https://www.jstor.org/stable/23016081

Baležentis, T., Li, T., & Chen, X. (2021). Has agricultural labor restructuring improved agricultural labor productivity in China? A decomposition approach. *Socio-Economic Planning Sciences, 76*, 100967. https://doi.org/10.1016/j.seps.2020.100967

Beck, T., Levine, R., & Loayza, N. (2000). Finance and the sources of growth. *Journal of Financial Economics, 58*(1–2), 261–300. https://doi.org/10.1016/s0304-405x(00)00072-6

Beck, T., & Demirgüç-Kunt, A. (2008). Access to finance: An unfinished agenda. *World Bank Economic Review, 22*(3), 383–396. https://doi.org/10.1093/wber/lhn021

Beck, T., Demirguc-Kunt, A., & Levine, R. (2007). Finance, inequality, and the poor. *Journal of Economic Growth, 12*(1), 27–49. https://doi.org/10.1007/s10887-007-9010-6

Behrman, J. R., & Deolalikar, A. B. (1987). Will developing country nutrition improve with income? A case study for rural South India. *Journal of political Economy, 95*(3), 492–507. https://doi.org/ https://doi.org/10.1086/261469

Behrman, J. R., Deolalikar, A. B., & Wolfe, B. L. (1988). Nutrients: impacts and determinants. *The World Bank Economic Review, 2*(3), 299–320. https://www.jstor.org/stable/3989840

Bell, C., & Rousseau, P. L. (2001). Post-independence India: A case of finance-led industrialization? *Journal of Development Economics, 65*(1), 153–175. https://doi.org/10.1016/s0304-3878(01)00132-8

Bencivenga, V. R., & Smith, B. D. (1993). Some consequences of credit rationing in an endogenous growth model. *Journal of Economic Dynamics and Control, 17*(1–2), 97–122. https://doi.org/10.1016/S0165-1889(06)80006-0

Béné, C. (2020). Resilience of local food systems and links to food security - A review of some important concepts in the context of COVID-19 and other shocks. *Food Security, 12*(4), 805–822. https://doi.org/10.1007/s12571-020-01076-1

Bist, J. P. (2018). Financial development and economic growth: Evidence from a panel of 16 African and non-African low-income countries. *Cogent Economics & Finance, 6*(1), 1–17. https://doi.org/10.1080/23322039.2018.1449780

Blundell, R. W., & Bond, S. R. (1998). Initial conditions and moment restrictions in dynamic panel data models. *Journal of Econometrics, 87*, 29–52.

Bouis, H. E., & Haddad, L. J. (1988). *Comparing calorie-income elasticities using calories derived from reported food purchases and a twenty-four hour recall of food intakes: An application using Philippine data*. University of Warwick.

Bouis, H. E., & Haddad, L. J. (1992). Are estimates of calorie-income elasticities too high? A recalibration of the plausible range. *Journal of Development Economics, 39*(2), 333–364. https://doi.org/10.1016/0304-3878(92)90043-9

Boukhatem, J. (2016). Assessing the direct effect of financial development on poverty reduction in a panel of low-and middle-income countries. *Research in International Business and Finance, 37*, 214–230. https://doi.org/10.1016/j.ribaf.2015.11.008

Brown, J. K., Zelenska, T. V., & Mobarak, M. A. (2013). *Barriers to adoption of products and technologies that aid risk management in developing countries*. Retrieved from August 11, 2022, from http://hdl.handle.net/10986/16365

Brinkman, H. J., De Pee, S., Sanogo, I., Subran, L., & Bloem, M. W. (2010). High food prices and the global financial crisis have reduced access to nutritious food and worsened nutritional status and health. *The Journal of Nutrition, 140*(1), 153S-161S. https://doi.org/10.3945/jn.109.110767

Bryce, J., Coitinho, D., Darnton-Hill, I., Pelletier, D., Pinstrup-Andersen, P., & Maternal and Child Undernutrition Study Group. (2008). Maternal and child undernutrition: Effective action at national level. *The Lancet, 371*(9611), 510–526.

Burgess, R., & Pande, R. (2005). Do rural banks matter? Evidence from the Indian social banking experiment. *American Economic Review, 95*(3), 780–795. https://www.jstor.org/stable/4132740

Butler, P. (2012, June 25). Demand for food parcels explodes as welfare cuts and falling pay hit home—early-warning indicator should set alarm bells ringing about poverty levels, government told. *The Guardian*. https://www.theguardian.com/society/2012/jun/25/breadline-britain-growth-food-parcels

Cabral, L., & Howell, J. (2012). Measuring aid to agriculture and food security. *ODI Briefing Paper*, (72).

Calderón, C., & Liu, L. (2003). The direction of causality between financial development and economic growth. *Journal of Development Economics, 72*(1), 321–334.

CGAP. (2015). *The opportunities and challenges of digitizing payments for the poor*. https://www.cgap.org/sites/default/files/CGAP-FocusNote-Opportunities-and-Challenges-of-Digitizing-Payments-for-the-Poor-Oct-2015.pdf

CGAP. (2019). *Financial inclusion and the SDGs*. https://www.cgap.org/topics/financial-inclusion-and-sdgs

Chibba, M. (2009). Financial inclusion, poverty reduction and the millennium development goals. *The European Journal of Development Research, 21*(2), 213–230. https://doi.org/10.1057/ejdr.2008.17

Christian, P. (2010). Impact of the economic crisis and increase in food prices on child mortality: Exploring nutritional pathways. *The Journal of Nutrition, 140*(1), 177S-181S. https://doi.org/10.3945/jn.109.111708

Christopoulos, D. K., & Tsionas, E. G. (2004). Financial development and economic growth: Evidence from panel unit root and Cointegration tests. *Journal of Development Economics, 73*(1), 55–74. https://doi.org/10.1016/j.jdeveco.2003.03.002

Coulibaly, B. S. (2019). *Foresight Africa: Top priorities for the continent in 2019*. Retrieved from September 12, 2023 from Africa Growth initiative, Brookings Institute website. https://www.brookings.edu/wp-content/uploads/2019/01/BLS18234_BRO_book_007_WEB.pdf

Cutler, D., Deaton, A., & Lleras-Muney, A. (2006). The determinants of mortality. *Journal of Economic Perspectives, 20*(3), 97–120.

Dahiya, S., & Kumar, M. (2020). Linkage between financial inclusion and economic growth: An empirical study of the emerging Indian economy. *Vision, 24*(2), 184–193. https://doi.org/10.1177/0972262920923891

Darnton-Hill, I., & Cogill, B. (2010). Maternal and young child nutrition adversely affected by external shocks such as increasing global food prices. *The Journal of Nutrition, 140*(1), 162S-169S. https://doi.org/10.3945/jn.109.111682

Deidda, L., & Fattouh, B. (2002). Non-linearity between finance and growth. *Economics Letters, 74*(3), 339–345. https://doi.org/10.1016/S0165-1765(01)00571-7

Demirgüç-Kunt, A., & Maksimovic, V. (1999). *Financial constraints, uses of funds, and firm growth: An international comparison*. The World Bank.

Demirgüç-Kunt, A., Klapper, L., Singer, D., Ansar, S., & Hess, J. (2018). *The global Findex database 2017: Measuring financial inclusion and the Fintech revolution*. World Bank. Retrieved from July 18, 2022 from http://documents.worldbank.org/curated/en/332881525873182837/pdf/126033-PUB-PUBLIC-pubdate-4-19-2018.pdf

Diamond, D. W. (1989). Reputation acquisition in debt markets. *Journal of Political Economy, 97*(4), 828–862. https://www.jstor.org/stable/1832193

Donovan, K. (2012). *Mobile money for financial inclusion*. Retrieved from December 1, 2021 from http://elibraryworldbank.org/doi/abs/10.1596/9780821389911_ch04

Dreze, J., & Sen, A. (2002). *India: Development and participation*. Oxford University Press on Demand.

Ductor, L., & Grechyna, D. (2015). Financial development, real sector, and economic growth. *International Review of Economics & Finance, 37*, 393–405. https://doi.org/10.1016/j.iref.2015.01.001

Dugarova, E., & Gülasan, N. (2017). *Global trends: Challenges and opportunities in the implementation of the sustainable development goals*. Academic Press.

Emara, N., & El Said, A. (2021). Financial inclusion and economic growth: The role of governance in selected MENA countries. *International Review of Economics & Finance, 75*, 34–54. https://doi.org/10.1016/j.iref.2021.03.014

Emerick, K., de Janvry, A., Sadoulet, E., & Dar, M. H. (2016). Technological innovations, downside risk, and the modernization of agriculture. *American Economic Review, 106*(6), 1537–1561. https://www.jstor.org/stable/43861131

Eryılmaz, F., Bakır, H., & Mercan, M. (2015). Financial development and economic growth: Panel data analysis. In Ö. Olgu, H. Dinçer & Ü. Hacıoğlu (Eds.), *Handbook of research on strategic developments and regulatory practice*

in global finance (pp. 233–245). IGI Global. https://doi.org/10.4018/978-1-4666-7288-8.ch015
Fase, M. M. G., & Abma, R. C. N. (2003). Financial environment and economic growth in selected Asian countries. *Journal of Asian Economics, 14*, 11–21.
Food and Agriculture Organization. (2008). *The state of food insecurity in the world 2008: High food prices and food security—Threats and opportunities*. Rome. https://www.fao.org/publications/card/en/c/3d431db5-34c8-5e01-9ad9-b2b0a2b9b918/
Food and Agriculture Organization. (2012). *The state of food insecurity in the world 2012: Economic growth is necessary but not sufficient to accelerate reduction of hunger and malnutrition*. Rome.
Food and Agriculture Organization. (2017). *The future of food and agriculture—Trends and challenges*. FAO.
Gentleman, A. (2014, February 28). Vulnerable man starved to death after benefits were cut—44- year-old died months after sickness and housing benefits were stopped following Atos fitness-for-work assessment. *The Guardian*. https://www.theguardian.com/society/2014/feb/28/man-starved-to-death-after-benefits-cut.
Global Findex Database. (2017). *Global Findex Database 2017: Measuring financial inclusion and the Fintech revolution*. World Bank. https://globalfindex.worldbank.org/
Goldsmith, R. W. (1969). *Financial structure and development*. Yale University Press.
Green Climate Fund. (2021). *Financial products*. https://www.greenclimate.fund/projects/financial-products
Greenwood, J., & Jovanovic, B. (1990). Financial development, growth, and the distribution of income. *Journal of political Economy, 98*(5, Part 1), 1076–1107. https://www.jstor.org/stable/2937625
Grassa, R., & Gazdar, K. (2014). Financial development and economic growth in GCC countries: A comparative study between Islamic and conventional finance. *International Journal of Social Economics, 41*(6), 493–514. https://doi.org/10.1108/IJSE-12-2012-0232
Green, R., Cornelsen, L., Dangour, A. D., Turner, R., Shankar, B., Mazzocchi, M., & Smith, R. D. (2013). The effect of rising food prices on food consumption: Systematic review with meta-regression. *Bmj, 346*. https://doi.org/10.1136/bmj.f3703
Grossman, G. M., & Helpman, E. (1991). Quality ladders in the theory of growth. *The Review of Economic Studies, 58*(1), 43–61.
GSMA. (2020). *The mobile gender gap report 2020*. GSMA. https://www.gsma.com/mobilefordevelopment/wp-content/uploads/2020/03/GSMA-Mobile-Gender-Gap-Report-2020.pdf

Gurley, J. G., & Shaw, E. S. (1955). Financial aspects of economic development. *The American Economic Review, 45*(4), 515–538. http://www.jstor.org/stable/1811632

Guru, B. K., & Yadav, I. S. (2019). Financial development and economic growth: Panel evidence from BRICS. *Journal of Economics, Finance and Administrative Science, 24*(47), 113–126. https://doi.org/10.1108/JEFAS-12-2017-0125

Habibullah, M. S., & Eng, Y. K. (2006). Does financial development cause economic growth? A panel data dynamic analysis for the Asian developing countries. *Journal of the Asia Pacific Economy, 11*(4), 377–393. https://doi.org/10.1080/13547860600923585

Hallegatte, S., Vogt-Schilb, A., Bangalore, M., & Rozenberg, J. (2016). *Unbreakable: Building the resilience of the poor in the face of natural disasters*. World Bank Publications.

Hassan, M. K., Sanchez, B., & Yu, J. S. (2011). Financial development and economic growth: New evidence from panel data. *The Quarterly Review of Economics and Finance, 51*(1), 88–104. https://doi.org/10.1016/j.qref.2010.09.001

Herwartz, H., & Walle, Y. M. (2014). Determinants of the link between financial and economic development: Evidence from a functional coefficient model. *Economic Modelling, 37*, 417–427. https://doi.org/10.1016/j.econmod.2013.11.029

Hicks, J. R. (1969). *A theory of economic history*.

Hobbs, J. E. (2020). Food supply chains during the COVID-19 pandemic. *Canadian Journal of Agricultural Economics/revue Canadienne D'agroeconomie, 68*(2), 171–176. https://doi.org/10.1111/cjag.12237

ICSU, & ISSC. (2015). Review of the sustainable development goals: The science perspective. *Paris: International Council for Science (ICSU)*. Retrieved from September 5, 2021 from http://www.genderportal.eu/resources/review-sustainable-development-goals-science-perspective

Imai, K. S., Arun, T., & Annim, S. K. (2010). Microfinance and household poverty reduction: New evidence from India. *World Development, 38*(12), 1760–1774. https://doi.org/10.1016/j.worlddev.2010.04.006

Inoue, T., & Hamori, S. (2012). How has financial deepening affected poverty reduction in India? Empirical analysis using state-level panel data. *Applied Financial Economics, 22*(5), 395–408. https://doi.org/10.1080/09603107.2011.613764

International Labour Organization. (2019). *The future of work: A centenary initiative*. ILO.

Jalilian, H., & Kirkpatrick, C. (2002). Financial development and poverty reduction in developing countries. *International Journal of Finance and Economics, 7*(2), 97–108. https://doi.org/10.1002/ijfe.179

Jalilian, H., & Kirkpatrick, C. (2005). Does financial development contribute to poverty reduction? *The Journal of Development Studies, 41*(4), 636–656. https://doi.org/10.1080/00220380500092754

Jung, W. S. (1986). Financial development and economic growth: International evidence. *Economic Development and Cultural Change, 34*(2), 333–346. https://doi.org/10.1086/451531

Kabeer, N. (2005). Gender equality and women's empowerment: A critical analysis of the third millennium development goal. *Gender & Development, 13*(1), 13–24.

Khan, M. S., & Senhadji A. S. (2000). Financial development and economic growth: An overview. *IMF Working Papers, 2000*(209), 1–24. https://doi.org/10.5089/9781451874747.001

Khan, S. A. R., Razzaq, A., Yu, Z., Shah, A., Sharif, A., & Janjua, L. (2022). Disruption in food supply chain and undernourishment challenges: An empirical study in the context of Asian countries. *Socio-Economic Planning Sciences, 82*, Article 101033. https://doi.org/10.1016/j.seps.2021.101033

Kiendrebeogo, Y., & Minea, A. (2016). Financial development and poverty: Evidence from the CFA Franc Zone. *Applied Economics, 48*(56), 5421–5436. https://doi.org/10.1080/00036846.2016.1176114

Kim, D. W., Yu, J. S., & Hassan, M. K. (2018). Financial inclusion and economic growth in OIC countries. *Research in International Business and Finance, 43*, 1–14. https://doi.org/10.1016/j.ribaf.2017.07.178

Koomson, I., Villano, R. A., & Hadley, D. (2020). Effect of financial inclusion on poverty and vulnerability to poverty: Evidence using a multidimensional measure of financial inclusion. *Social Indicators Research, 149*(2), 613–639. https://doi.org/10.1007/s11205-019-02263-0

Law, S. H., & Singh, N. (2014). Does too much finance harm economic growth? *Journal of Banking & Finance, 41*, 36–44. https://doi.org/10.1016/j.jbankfin.2013.12.020

Lee, H. H., & Park, C. Y. (2013). International transmission of food prices and volatilities: A panel analysis. *Asian Development Bank Economics Working Paper Series*, (373).

Lee, H. H., Lee, S. A., Lim, J. Y., & Park, C. Y. (2016). Effects of food price inflation on infant and child mortality in developing countries. *The European Journal of Health Economics, 17*(5), 535–551.

Lenka, S. K., & Sharma, R. (2017). Does financial inclusion spur economic growth in India? *The Journal of Developing Areas, 51*(3), 215–228. https://www.jstor.org/stable/10.2307/26416941

Levine, R. (1997). Financial development and economic growth: Views and agenda. *Journal of Economic Literature, 35*(2), 688–726. http://www.jstor.org/stable/2729790

Levine, R., Loayza, N., & Beck, T. (2000). Financial intermediation and growth: Causality and causes. *Journal of Monetary Economics, 46*(1), 31–77. https://doi.org/10.1016/S0304-3932(00)00017-9

Lewis, W. A. (1955). *Theory of economic growth*. Unwin University Books.

Lucas, R. E., Jr. (1988). On the mechanics of economic development. *Journal of Monetary Economics, 22*(1), 3–42. https://doi.org/10.1016/0304-3932(88)90168-7

Lyons, A. C., & Hunt, J. (2003). *The credit practices and financial education needs of community college students*. Retrieved on July 10, 2021 from https://ssrn.com/abstract=2265596

Mattoo, A., Rocha, N., & Ruta, M. (2020). *The evolution of deep trade agreements* (Policy Research Working Paper No. 9283). World Bank.

McKinnon, R. I. (1973). *Money and capital in economic development*. Brookings Institution. https://www.brookings.edu/book/money-and-capital-in-economic-development/

Mârza, B., Angelescu, C., & Tindeche, C. (2015). Agricultural insurances and food security. The new climate change challenges. *Procedia Economics and Finance, 27*, 594–599. https://doi.org/10.1016/S2212-5671(15)01038-2

Mhadhbi, K. (2014). Financial development and economic growth: A dynamic panel data analysis. *International Journal of Econometrics and Financial Management, 2*(2), 48–58. https://doi.org/10.12691/ijefm-2-2-1

Morduch, J., & Haley, M. (2002). *Analysis of the effects of microfinance on the poor: A critical review*. USAID.

Moore, S., Teixeira, A. C., & Shiell, A. (2006). The health of nations in a global context: Trade, global stratification, and infant mortality rates. *Social Science & Medicine, 63*(1), 165–178. https://doi.org/10.1016/j.socscimed.2005.12.009

Nurkse, R. (1962). *Problems of capital formation in underdeveloped countries*. Blackwell.

Omar, M. A., & Inaba, K. (2020). Does financial inclusion reduce poverty and income inequality in developing countries? A panel data analysis. *Journal of Economic Structures, 9*(1), 1–25. https://doi.org/10.1186/s40008-020-00214-4

Ortiz, I., Chai, J., & Cummins, M. (2011). *Escalating food prices: The threat to poor households and policies to safeguard a recovery for all* (UNICEF Social and Economic Policy Working Paper). UNICEF.

Ozughalu, U. M. (2016). Relationship between household food poverty and vulnerability to food poverty: Evidence from Nigeria. *Social Indicators Research, 125*, 567–587. https://doi.org/10.1007/s11205-014-0845-x

Pagano, M. (1993). Financial markets and growth: An overview. *European Economic Review, 37*(2–3), 613–622.

Patrick, H. T. (1966). Financial development and economic growth in underdeveloped countries. *Economic Development and Cultural Change, 14*(2), 174–189. https://doi.org/10.1086/450153

Perotti, R. (1993). Political equilibrium, income distribution, and growth. *The Review of Economic Studies, 60*(4), 755–776. https://www.jstor.org/stable/2298098

Pradhan, R. P., Arvin, M. B., Bahmani, S., Hall, J. H., & Norman, N. R. (2017). Finance and growth: Evidence from the ARF countries. *The Quarterly Review of Economics and Finance, 66*, 136–148. https://doi.org/10.1016/j.qref.2017.01.011

Rajan, R. G., & Zingales, L. (1998). Financial dependence and growth. *American Economic Review, 88*(3), 559–586. https://www.jstor.org/stable/116849

Ravallion, M. (1987). *Markets and famines*. Oxford University Press.

Ravallion, M. (1990). Income effects on undernutrition. *Economic development and cultural change, 38*(3), 489–515. https://doi.org/10.1086/451812

Reardon, T., Mishra, A., Nuthalapati, C. S., Bellemare, M. F., & Zilberman, D. (2020). COVID-19's disruption of India's transformed food supply chains. *Economic and Political Weekly, 55*(18), 18–22.

Rebelo, S. (1991). Long-run policy analysis and long-run growth. *Journal of Political Economy, 99*(3), 500–521. https://doi.org/10.1086/261764

Reutlinger, S., & Selowsky, M. (1976). *Malnutrition and poverty. Magnitude and policy options*. Published for the World Bank by the Johns Hopkins University Press.

Rojas-Suarez, L. (2016). *Financial inclusion and financial stability: An unconventional view*. Center for Global Development. https://www.cgdev.org/publication/financial-inclusion-and-financial-stability-unconventional-view

Romer, P. M. (1986). Increasing returns and long-run growth. *Journal of Political Economy, 94*(5), 1002–1037. https://doi.org/10.1086/261420

Romer, P. (1989) Capital accumulation and the theory of long-run growth, In: R. Barro (Ed) *Modern Business Cycle Theory* (Cambridge: Harvard University Press).

Rousseau, P. L., & Wachtel, P. (2011). What is happening to the impact of financial deepening on economic growth? *Economic Inquiry, 49*(1), 276–288. https://doi.org/10.1111/j.1465-7295.2009.00197.x

Sachs, J. (2002). *Macroeconomics and health*. WHO.

Sahn, D. (1987). *Causes and implications of seasonal variability in household food security*. Food Policy Research Institute.

Samargandi, N., Fidrmuc, J., & Ghosh, S. (2015). Is the relationship between financial development and economic growth monotonic? Evidence from a sample of middle-income countries. *World Development, 68*, 66–81. https://doi.org/10.1016/j.worlddev.2014.11.010

Sarma, M. (2012). *Index of financial inclusion—A measure of financial sector inclusiveness*.

Selke, S. (Ed.). (2011). *Food banks in Germany: Aspects of a social movement between food redistribution and poverty intervention*. Springer Science & Business Media.

Sethi, D., & Acharya, D. (2018). Financial inclusion and economic growth linkage: Some cross country evidence. *Journal of Financial Economic Policy, 10*(3), 369–385. https://doi.org/10.1108/JFEP-11-2016-0073

Sen, A. (1982). *Poverty and famines: An essay on entitlement and deprivation*. Oxford University Press.

Sen, A. (1995). *Inequality reexamined*. Harvard University Press.

Sharma, D. (2016). Nexus between financial inclusion and economic growth: Evidence from the emerging Indian economy. *Journal of Financial Economic Policy, 8*(1), 13–36. https://doi.org/10.1108/JFEP-01-2015-0004

Shaw, E. S. (1973). *Financial deepening in economic development*. Oxford University Press.

Singh, A. (1997). Financial liberalization, stock markets and economic development. *The Economic Journal, 107*(442), 771–782. https://www.jstor.org/stable/2957801

Srinivasan, T. (1981). Malnutrition: Some measurement and policy issues. *Journal of Development Economics, 8*(1), 3–19. https://doi.org/10.1016/0304-3878(81)90043-2

Strauss, J., & Thomas, D. (1990). *The shape of the calorie-expenditure curve (No. 595)* (Center Discussion Paper).

Strauss, J., & Thomas, D. (1995). Human resources: Empirical modeling of household and family decisions. *Handbook of Development Economics, 3*, 1883–2023. https://doi.org/10.1016/s1573-4471(05)80006-3

Subramanian, S., & Deaton, A. (1996). The demand for food and calories. *Journal of Political Economy, 104*(1), 133–162. https://doi.org/10.1086/262020

Sukhatme, P. V. (1982). Newer concepts in nutrition and their implications for policy. Maharashtra association for the cultivation of science research institute.

Swamy, V. (2014). Financial inclusion, gender dimension, and economic impact on poor households. *World Development, 56*, 1–15. https://doi.org/10.1016/j.worlddev.2013.10.019

Tian, J., Bryksa, B. C., & Yada, R. Y. (2016). Feeding the world into the future-food and nutrition security: The role of food science and technology. *Frontiers in Life Science, 9*(3), 155–166. https://doi.org/10.1080/21553769.2016.1174958

Tiwari, S., & Zaman, H. (2010). *The impact of economic shocks on global undernourishment* (*World Bank Policy Research Working Paper*, (5215)).

UNDESA. (2016). *The sustainable development goals report 2016*. UN DESA. https://unstats.un.org/sdgs/report/2016/

UNESCO. (2019). *Global education monitoring report 2019: Migration, displacement and education—Building bridges, not walls*. UNESCO.

United Nations. (2020). *The sustainable development goals report 2020*. United Nations. https://unstats.un.org/sdgs/report/2020/The-Sustainable-Development-Goals-Report-2020.pdf

United Nations Capital Development Fund. (2018). *Financing local development for the SDGs*. UNCDF. https://www.uncdf.org/article/3588/financing-local-development-for-the-sdgs

United Nations Development Programme. (2022). *SDG 17: Partnerships for the goals*. UNDP. https://www.undp.org/sustainable-development-goals/partnerships-goals-sdg-17

United States Institute of Peace. (2015). *Why economic growth is not enough: Peacebuilding and the importance of economic development*. USIP. https://www.usip.org/publications/2015/05/why-economic-growth-not-enough-peacebuilding-and-importance-economic

Van, L. T. H., Vo, A. T., Nguyen, N. T., & Vo, D. H. (2021). Financial inclusion and economic growth: An international evidence. *Emerging Markets Finance and Trade, 57*(1), 239–263.

Warr, P. (2014). Food insecurity and its determinants. *Australian Journal of Agricultural and Resource Economics, 58*(4), 519–537. https://doi.org/10.1111/1467-8489.12073

Webb, P. (2010). Medium-to long-run implications of high food prices for global nutrition. *The Journal of Nutrition, 140*(1), 143S-147S. https://doi.org/10.3945/jn.109.110536

Wise, T. A., & Harvie, A. (2009). Boom for whom? Family farmers saw lower on-farm income despite high prices. *GDAE Policy Brief*, (09–02).

World Bank. (1995). *World development report: Workers in an intergrating world*. World Bank.

World Bank. (2001). *World development report 2000–2001: Attacking poverty*. Retrieved on October 15, 2021 from World Bank Group. Oxford University Press. Website. https://openknowledge.worldbank.org/handle/10986/11856

World Bank. (2007). *Finance for all? Policies and pitfalls in expanding access*. The World Bank.

World Bank. (2021). *Financial inclusion*. https://www.worldbank.org/en/topic/financialinclusion

World Health Organization. (2020). *Digital health: A guide for action*. WHO.

CHAPTER 7

Financial Inclusion and Economic Growth (SDG 8)

7.1 Introduction

Financial inclusion, often referred as access to affordable and appropriate financial services for all individuals and households, is critical to achieving Sustainable Development Goal (SDG) 8, which seeks to "promote sustained, inclusive, and sustainable economic growth, full and productive employment and decent work for all" (UNCDF, 2025). By incorporating excluded individuals into the formal financial system, financial inclusion increases economic participation, decreases inequities, and enhances resilience against economic shocks. This section investigates the complex connection between financial inclusion and SDG 8, investigating how access to financial services boosts economic productivity, encourages entrepreneurship, and boosts financial systems in emerging economies.

Financial inclusion directly aids to sustainable GDP development (Target 8.1) by allowing people and enterprises to invest, save and manage risks efficiently. For instance, access to financing allows micro, small, and medium-sized firms (MSMEs) to increase operations, adopt new tools and generate jobs, hence enhancing productivity (Target 8.2) (FSD, 2025). In Kenya, mobile money services like M-Pesa have changed financial access, boosting small companies and providing employment prospects in rural areas. In a similar way, the World Bank indicates that financial inclusion can bridge the $5.7 trillion funding gap faced by MSMEs in emerging

S. Ghosh et al., *Financial Inclusion*, Sustainable Development Goals Series, https://doi.org/10.1007/978-981-95-3385-5_7

nations, which is crucial for diversification of economies and innovation (FSD, 2025).

Target 8.3 underlines the need for policies that boost entrepreneurship and eliminate informal employment. Financial inclusion facilitates this by providing affordable credit, insurance and payment systems to informal workers and small businesses. For example, India's Jan Dhan Yojana initiative, which opened over 400 million bank accounts for the unbanked, has significantly reduced reliance on informal lenders and increased participation in the formal economy (FSD, 2025). Additionally, digital financial services (DFS) have enabled informal sector workers to build credit histories, access government subsidies, and transition into regulated employment.

A core target under SDG 8 (Target 8.10) emphasizes on empowering the domestic financial institutions to broaden accessibility to banking, insurance and financial services. The Global Findex 2021 reports that 76% of adults now have a financial account, yet disparities persist, particularly in low-income countries where only 32% of adults can access emergency funds. Financial inclusion initiatives - such as agent banking, mobile money, and simplified KYC (Know Your Customer) requirements - have been instrumental in closing these gaps. In Sub-Saharan Africa, mobile money adoption has reduced the gender gap in account ownership by 4 percentage points, demonstrating its role in inclusive growth (FSD, 2025).

While not explicitly a target under SDG 8, financial inclusion intersects with SDG 5 (Gender Equality) and SDG 10 (Reduced Inequalities) by empowering women and marginalized groups. Studies show that women with savings accounts are more likely to invest in businesses, manage household finances, and withstand economic shocks. In Kenya, tailored financial products for women have increased female entrepreneurship rates by 18%, contributing to broader economic resilience. However, barriers such as lack of collateral, discriminatory norms and low financial literacy still hinder progress, necessitating targeted interventions.

Financial inclusion also supports climate adaptation (SDG 13) by enabling investments in sustainable agriculture, renewable energy, and climate-smart infrastructure. For instance, 41.1% of rural Kenyans invest in green initiatives compared to 26.2% of urban dwellers, highlighting the potential of inclusive finance to drive environmental sustainability. Insurance products for smallholder farmers further mitigate climate-related risks, ensuring stable incomes and food security (FSD, 2025). Further, extending financial support to the job seekers and entrepreneurs through the Prakerja program in Indonesia is a classical exhibition of promoting decent work and economic growth (T20 India, 2023).

Despite its benefits, financial inclusion suffers from some serious drawbacks. These restrictions impede the economic growth and often hinder the speed of development.

Among the major hindrances, rural populations often lack access to internet and further due to non-availability of IDs, they remain involuntarily excluded from accessing the means of digital finance. In Kenya, 16.7% of borrowers in 2024 were found to default on their loan repayments, which took place due to irresponsible lending practices and overstressing on debt from lenders part. Weak consumer protection layoffs are yet another important impediment of growth leading to exploitation by predatory lenders (FSD, 2025).

Financial inclusion is a catalyst for achieving SDG 8, unlocking economic potential, reducing inequalities, and fostering resilience. By addressing systemic barriers and prioritizing inclusive policies, governments and institutions can ensure that financial systems serve as engines of sustainable development. As the UNCDF emphasizes, "leaving no one behind" in financial access is not just a moral imperative but an economic necessity (UNCDF, 2025).

This section attempts to characterize the association between financial inclusion and economic development (SDG 8). Since, the literature shows mixed evidence on the causative connections, this study has considered both the probable propositions (1) Whether, Economic Development improves the extent of Financial Inclusion (The case of Reverse Causality) in Sect. 7.2.2, and subsequently (2) Whether, Financial Inclusion can accelerate the pace of Economic Development in Sect. 7.2.1.

7.2 Theoretical Connections Between Financial Inclusion and Economic Development (SDG 8)

7.2.1 The Conventional Connection

Further, considering a sample of 26 states and 4 union territories of India for 2002–2019 to analyze the impact of financial inclusion[1] (measured as natural logarithm of savings, credit and branches) and Gross Fixed Capital Formation (indicating the investment) on the output growth (measured as natural logarithm of RNSDP, RealIP, and per capita RNSDP) after considering a dummy for the financial crisis of 2007.[2] All the three unit-root tests across the panel structure in Table 7.1suggests the data to contain a unit-root at the level and subsequently become stationary at their first differences and therefore confirming the existence of a co-integrating vector (i.e. a linear combination of the unit-root variables). Majority of the six between and within-group estimates of co-integrating vector are significant using Kao's residual co-integration test (Pedroni, 2004) in Table 7.2. Long-run estimates of elasticity are all significant and also showing a convergence across both the parametric alternative (i.e. DOLS) and non-parametric alternative (i.e. FMOLS) in Table 7.3.

The calculated coefficients of the ECM factor (lagged one) and the dynamics in the short term are shown in Table 7.4. ECM refers to the previous period's divergence from a long-run equilibrium, which determines its short-run dynamics. The term for error correction calculates the rate at which a dependent variable (in this example, ΔlnRNSDP, ΔlnrealIP, and ΔlnpercapitaRNSDP) returns to equilibrium with respect to other variables. It is clear from the table that the short-run dynamics for one period lagged variables of indicators of financial outreach are not all significant except for the $\Delta lnSavingsac_{t-1}$ implies that in the short-run savings account growth significantly affect the overall growth of lnrealIP and lnpercapita RNSDP across the states in India. The Global Financial crisis[3] (D_{GFC}) as a dummy remains negative and significant indicating the slowdown of the economy during 2008 in India.

[1] Preliminary investigation of the associations (at level and 1st difference) using Pearson correlation is provided in Table 7.1.2 in the appendix.

[2] Data description and summary statistics are provided in Table 7.1.1 in the appendix.

[3] To incorporate the shock of the GFC on three key economic development and growth indicators.

Table 7.1 Panel Unit-root tests

	At Level				*At 1st difference*			
	LLC	*ADF*	*PP*	*Order of integration*	*LLC*	*ADF*	*PP*	*Order of integration*
lnRNSDP	47.09	0.01	0.02	I (1)	−3.88***	135.02***	130.11***	I (0)
lnrealIP	13.12	1.67	0.87	I (1)	−8.23***	167.81***	319.23***	I (0)
lnpercapitaRNSDP	17.06	0.49	0.03	I (1)	−2.39***	79.95***	160.90***	I (0)
lnOffices	7.15	2.98	0.05	I (1)	−4.03***	56.01	69.16	I (1) or I (0)
lnSavingsac	11.89	0.77	0.02	I (1)	−3.31***	58.34	78.47**	I (0)
lnCreditac	15.78	0.94	0.43	I (1)	−5.87***	119.57***	218.71***	I (0)
lnGFCF	0.86	39.83	33.92	I (1)	-23.76***	496.08***	519.76***	I (0)

Note (i) ***, ** and * shows 1%, 5% and 10% level of significance
(ii) Schwarz Information Criterion is used as the Lag selection criteria; Test specification: no intercept
(iii) I (1) and I (0) indicate the order of integration [i.e. Non-stationary for I (1) and Stationary for I (0) respectively]

A preliminary investigation of the growth in production and indicators of financial inclusion shows a co-movement. The production during the period of study across the states and UTs grew on an average by 6–8% and the growth across the indicators of financial inclusion on an average was observed roughly near 5–11%. This was further supported by a significant correlation between the variables at level. The long-run association as observed between growth in output and indicators of financial inclusion under both the parametric and non-parametric alternatives are valid as well as complimentary to each other.[4] However, short-run dynamics remains inconclusive due to absence of data for a longer period of time (Pierse & Snell, 1995). Implications to the present study converges with earlier research of Apergis et al. (2007) which was an extension of the Christopoulos and Tsionas (2004) considering a cross-sectional portfolio of 10 developing nations across 1970–2000. However, literary evidence in support of supply leading hypothesis (i.e. the finance-led growth) is mixed. Watchel and Rousseau (1994) and Rousseau (1998) find evidence

[4] Further read Habibullah and Eng (2006) and Eryilmaz et al. (2015) for relevant discussion.

in favor of this hypothesis whereas Arestis et al. (2001); Luintel and Khan (1999), and Arestis and Demetriades (1997) find evidence of simultaneous causality.

The present study provides a microscopic view of the hypothesis by framing the heterogeneity across the states and UTs of India. An insight from the data shows that in the long-run financial inclusion promotes economic growth. However, short-run dynamics are inconclusive and are quite reasonable due to the fact that improvement in outreach of financial services takes certain time to materialize in to output growth. It is quite reasonable to believe that opening of savings account or provision of credit does not influence output instantaneously especially across the provinces of India where it is unlikely to get a loan just after applying for the same (Ghosh & Sahu, 2024). Further, using a single lag to represent the equation reduces the degrees of freedom and thus in short-run results became inconclusive.[5] Thus, in the long-run it is quite certain that improvement in availability (i.e. a probable growth of bank branches) and subsequent increment in the usage of conventional savings and credit provided by the banking industry will raise the output levels. The Government of India through implementing the National Strategy for Financial Inclusion (NFSI) for 2019–2024 assumes a proactive role to improve the status of financial inclusion. Under such strategy, initiatives such as the Pradhan Mantri Jan Dhan Yojana (PMJDY) which focuses to open no-cost bank accounts, Pradhan Mantri Suraksha Bima Yojana (PMSBY) covering a one-year death cum disability insurance at a premium of Rs. 12/ year, and Pradhan Mantri Jeevan Jyoti Beema Yojana (PMJJBY) to cover a life at a premium of Rs. 330/ year are worth mentioning.

7.2.2 *The Case of Reverse Causality*

This study examines 27 Indian states and 5 territories of the union from 2002–2019 to determine the impact of productivity (measured in per capita state-wise Net Domestic Product) on savings behavior (measured in

[5] Reviewing the causal connections across the VAR or VECM structure is likely to have more than one lag. This gives the justification of proceeding with the framework of the Granger Causality test (Lopez & Weber, 2017).

Table 7.2 Results for panel Co-integration analysis

	Model with lnRNSDP and lnGFCF			*Model with lnrealIP and lnGFCF*			*Model with lnpercapitaRNSDP and lnGFCF*		
	With lnOffices	*With lnSavingsac*	*With lnCreditac*	*With lnOffices*	*With lnSavingsac*	*With lnCreditac*	*With lnOffices*	*With lnSavingsac*	*With lnCreditac*
Panel A									
Group PP	−8.03***	−4.96***	−6.13***	−10.21***	−7.76***	−10.87***	−10.25***	−5.17***	−10.73***
	(0.002)	(0.001)	(0.001)	(0.003)	(0.001)	(0.002)	(0.001)	(0.002)	(0.001)
Group ADF	−5.87***	−6.39***	−4.85***	−7.12***	−3.82**	−5.79***	−7.07***	−5.61***	−6.32***
	(0.001)	(0.002)	(0.002)	(0.002)	(0.04)	(0.002)	(0.001)	(0.002)	(0.001)
Panel B									
Panel	13.11***	7.81***	16.36***	2.38**	−1.57	−1.68	2.04**	2.59***	4.19***
v-statistic	(0.001)	(0.001)	(0.001)	(0.049)	(0.83)	(0.87)	(0.048)	(0.001)	(0.001)
Panel	2.04	0.39	−0.78	0.85	0.23	0.71	0.91	0.86	0.72
pp-statistic	(0.79)	(0.69)	(0.21)	(0.73)	(0.34)	(0.68)	(0.72)	(0.63)	(0.81)
Panel	−4.68***	−5.68***	−8.68***	−7.88***	−10.37***	−12.01***	−7.88***	−5.45***	−12.02***
ADF-statistics	(0.001)	(0.001)	(0.003)	(0.001)	(0.002)	(0.002)	(0.001)	(0.002)	(0.001)
Panel	−4.71***	−5.87***	−9.16***	−7.03***	−8.70***	−10.15***	−7.31***	−5.72***	−10.42***
ρ-statistic	(0.002)	(0.001)	(0.001)	(0.001)	(0.002)	(0.001)	(0.002)	(0.002)	(0.002)

Note (i) *p*-values are shown within the parenthesis
(ii) *, ** and *** presents the level of significance at 10%, 5% and 1% level respectively
Source Authors Calculation

Table 7.3 Estimation of the long-run association

(1)	*(2) Panel DOLS*	*(3) Panel FMOLS*	*(4) Panel DOLS*	*(5) Panel FMOLS*	*(6) Panel DOLS*	*(7) Panel FMOLS*
Panel A	*Dependent variable: lnRNSDP*					
lnOffices	0.768*** (25.71)	0.897*** (41.27)				
lnSavingsac			0.512*** (18.16)	0.479*** (26.59)		
lnCreditac					0.711*** (16.14)	0.744*** (21.73)
lnGFCF	0.093*** (5.60)	0.106*** (8.61)	0.039 (1.52)	0.053*** (3.14)	0.101*** (4.28)	0.085*** (4.31)
Panel B	*Dependent variable: lnRealIP*					
lnOffices	0.782*** (8.90)	0.885*** (8.40)				
lnSavingsac			0.374*** (8.29)	0.432*** (13.48)		
lnCreditac					0.625*** (9.64)	0.691*** (12.94)
lnGFCF	0.182*** (4.19)	0.204*** (7.70)	0.050 (1.23)	0.157*** (5.18)	0.150*** (4.37)	0.180*** (5.90)
Panel C	*Dependent variable: lnpercapitaRNSDP*					
lnOffices	0.669*** (17.93)	0.754*** (27.71)				
lnSavingsac			0.378*** (17.04)	0.380*** (23.47)		
lnCreditac					0.565*** (13.94)	0.579*** (18.56)
lnGFCF	0.053*** (2.89)	0.081*** (6.45)	0.019 (0.90)	0.036** (2.36)	0.022*** (3.34)	0.064*** (3.60)

Note (i) *t*-statistics is shown within the parenthesis
(ii) *, ** and *** presents the level of significance at 10%, 5% and 1% level respectively
Source Authors Calculation

Table 7.4 Panel error correction model

	Dependent variable: $\Delta lnRNSDP$			*Dependent variable:* $\Delta lnrealIP$			*Dependent variable:* $\Delta lnpercapitaRNSDP$		
(1)	*(2)*	*(3)*	*(4)*	*(5)*	*(6)*	*(7)*	*(8)*	*(9)*	*(10)*
$\Delta lnRNSDP_{t-1}$	−0.209*** (−4.48)	−0.202*** (−4.40)	−0.003*** (−4.44)						
$\Delta lnrealIP_{t-1}$				−0.110** (−2.26)	−0.101** (−2.08)	−0.112** (−2.31)			
$\Delta lnpercapitaRNSDP_{t-1}$							−0.100** (−2.10)	−0.097** (−2.03)	−0.102 (0.048)
$\Delta lnOffices_{t-1}$	0.002 (0.02)			0.053 (0.20)			−0.103 (−1.24)		
$\Delta lnSavingsac_{t-1}$		0.015 (0.44)			0.194* (1.75)			0.060* (1.62)	
$\Delta lnCreditac_{t-1}$			0.003 (0.14)			0.029 (0.45)			−0.004 (−0.16)
$\Delta lnGFCF_{t-1}$	−0.001 (−0.26)	−0.097 (−0.32)	−0.002 (−0.34)	0.003 (0.19)	0.001 (0.05)	0.003 (0.19)	−0.007 (−1.35)	−0.005 (−1.08)	
ECM_{t-1}	0.006 (1.37)	−0.002* (−1.95)	−0.003** (−2.19)	−0.006 (−0.66)	−0.006 (−0.67)	−0.004 (−0.43)	0.002*** (2.59)	0.007* (1.86)	0.003*** (2.72)
D_{GFC}	−0.021*** (−3.41)	−0.022*** (−4.03)	−0.021*** (−4.44)	−0.051** (−2.56)	−0.066*** (−3.62)	−0.047*** (−3.03)	−0.011* (−1.66)	−0.022*** (−3.41)	−0.017*** (−3.31)
Constant	−0.098*** (−3.42)	−0.097*** (−4.23)	0.097*** (18.27)	0.113*** (7.98)	0.104*** (6.89)	0.112*** (7.78)	0.073*** (14.73)	0.068*** (12.89)	0.073*** (13.97)
R^2	0.083	0.088	0.090	0.0364	0.044	0.036	0.0420	0.038	0.044

(continued)

Table 7.4 (continued)

	Dependent variable: $\Delta lnRNSDP$			*Dependent variable: $\Delta lnrealIP$*			*Dependent variable: $\Delta lnpercapitaRNSDP$*		
(1)	*(2)*	*(3)*	*(4)*	*(5)*	*(6)*	*(7)*	*(8)*	*(9)*	*(10)*
F—Statistics	7.083	7.706	7.706	2.953	3.609	2.926	3.426	3.084	3.569
Observarions	398	398	398	397	397	397	397	397	397

Note (i) *t*-statistics are provided in the parenthesis
(ii) *, ** and *** shows the significance at 10%, 5% and 1% level respectively
Source Authors Calculation

per capita number of savings accounts), after controlling for population, inflation, and government spending on health and education.[6]

The descriptive statistics shows that the variables across the panel are distributed non-normally (significant JB statistics) (see Table 7.5) and a contemporaneous correlation (partial correlation) exists between NSA and GDP at 1% significance (see Table 7.5.2 in the appendix). Further, there seems no linear association between the explanatory variables leading to biased estimates of the elasticities (see Table 7.5.3 in the appendix). The F- statistic values for all the three static panel regressions (i.e. Pooled, Fixed and Random effects) show a good fit. However, the Restricted F, BP Lagrange Multiplier, and Hausman test (see Table 7.7) suggests the Fixed Effect specification to be a perfect fit among the available alternatives. The assumptions underlying the model are duly satisfied following Woolridge (2016) in Table 7.5.6 in the appendix. With no perfect linear connection between the variables that explain the result (Table 7.5.2 provided in the appendix) and the error component meeting the mean reversal criterion (as shown in Fig. 7.1), we infer that the fixed effects estimator produces unbiased results. Additionally, with a constant variance (Fig. 7.1) and absence of autocorrelation among the lagged values (Table 7.5.4)[7] of the residual term, we find the fixed effect specification to be the appropriate model to estimate the parameter unbiasedly. We further find the estimate of the parameters to be efficient using the Between-effects panel specification in Table 7.5.5 provided in the appendix. The time-averaged idiosyncratic error terms ($\overline{u}_i$) obtained from applying the fixed effect model shows an insignificant dependency on the explanatory variables and consequently validates the robustness of the estimated parameters (see Fig. 7.1). Additionally, it is observed from Table 7.5.6 in the appendix that the expected values of the idiosyncratic error term [i.e. E (u_{it})] given the values of the independent variables (x_i's) and the unobserved effects (α_i's) in all time periods are significantly

[6] A preliminary description of the variables with its respective sources is provided in Table 7.5.1 in the appendix.

[7] The autocorrelation among residuals (*please consider the Ljung-Box Q-statistics provided in Table–7.5.4*) shows the presence of correlation at least for one of the pairs up to a certain lag at 1 percent level. But, since the test is of a 'Portmanteau' type and the alternative hypothesis is found to be very loosely designed, we may avoid the presence of correlation among the lagged terms of error component.

equal to zero. This basically validates the elasticities (i.e. the $\hat{\beta}'$s) to be unbiased and free from the issue of possible endogenous conditions.

Findings from the sample show that per capita state-wise Net Domestic Product has a positive influence on increasing the income and thus the number of savings accounts in the economy. The impact (though less important) of other variables in the model on increasing the number of

Table 7.5 Summary statistics

Variable	*Mean*	*Standard deviation*	*Skewness*	*Kurtosis*	*Jarque–Bera Statistics*	*Probability*
NSA	0.852	0.369	0.941	5.273	75.463	0.000***
POP	41,205,538	46,540,344	1.770	6.715	228.204	0.000***
CPI	122.902	12.069	−0.523	2.144	15.832	0.000***
EDU	21.558	12.358	7.329	82.303	56,366.880	0.000***
HEL	5.008	1.392	1.956	8.204	367.372	0.000***
GDP	87,600	51,648.92	1.601	5.929	163.227	0.000***

Note *** Indicates significance at 1 percent level
Source Authors Calculation

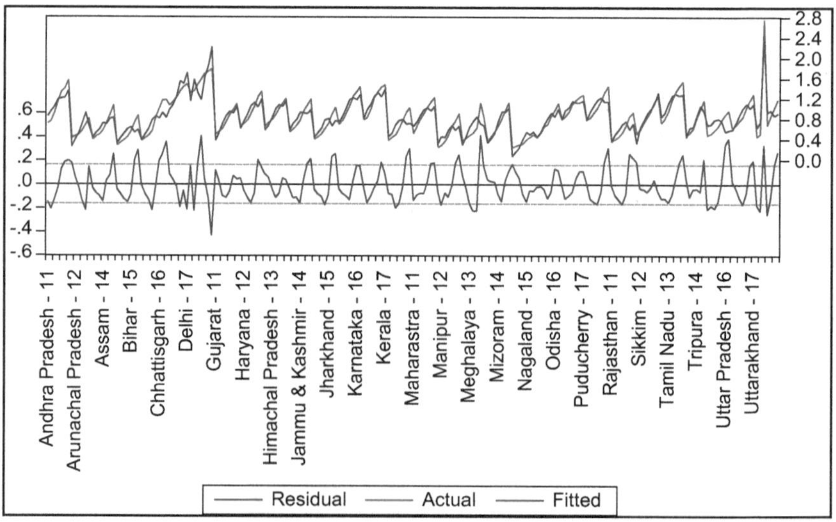

Fig. 7.1 Fitness of the fixed effect model (*Source* Calculated by Authors)

savings accounts are consistent with theoretical constructs. Analysis of the controlling factors further shows that an increase of 14.49 points of HEL and 100 points of CPI with respect to its base year includes an additional individual in the framework of the formal finance. Expenditure on health accounts almost 15% of the annual family income and turns "financially catastrophic" as they exceed the relative ability to pay (Wyszewianski, 1986). Especially considering the case of India where a significant share of population lives below the international poverty lines, it appears obvious that they are unable to protect themselves not only from illness involving high costs/ catastrophic illnesses [e.g. Acquired Immune Deficiency Syndrome (AIDS) and Cancer] but also illnesses from acute urinary tract infection and a strep throat in absence of an adequate health insurance plan. Therefore, increased investment from government's part either by way of building up new hospitals or modifying the old ones or even by way of extending social health insurance schemes (e.g. Ayushman Bharat, Rashtriya Swasthya Bima Yojana etc.) for the poor offloads to some extent the burden of catastrophic health care expenses on them resulting in to a positive association with their propensity to save (Berki, 1986) and hence promote financial inclusion. Similarly, results show that enhanced allotment of funds to education from the total state budget promote savings by reducing the burden of individuals. However, the relation appears to be non-significant.

Anecdotal evidence suggests a negative association between inflation and the behavior of saving for an investor. Quite genuinely it is argued that as the price level rises the purchasing power of money deteriorates and hence has an adverse effect on savings. However, findings of the present study suggest a positive relation between inflation and financial inclusion at least for the sample considered across the provinces of India. Similar evidence has been documented earlier by Chaturvedi et al. (2011) while investigating the interdependence of economic growth, savings, and inflation in a framework of simultaneous equation using two-stage least square comprising nations across the South and South-eastern part of the Asian continent. Applying the panel data econometrics across the nations of Asia for 1981–2011, Patra et al. (2015) affirms inflation to positively affect the rate of savings.

The relation between inflation and financial inclusion may have a dual characteristic. Primarily, as inflation rises the rate of savings is expected to shrink due to the fall in purchasing power of money but in long run it may be so that the savings rate increases gradually, hence relation may be of non-linear type. In this regard, the present study refrains from exploring such form of relationship between constructs, following the inconclusive views of Dash and Kumar (2018) in Indian context (Table 7.6).

Table 7.6 Panel regression results (Dependent variable: NSA)

	Pooled regression model		*Fixed effect model*		*Random effect model*	
	Coefficient	*t-Stat*	*Coefficient*	*t-Stat*	*Coefficient*	*t-Stat*
Intercept	−0.925	−4.847***	−0.505	−2.15**	−0.975	−6.979***
POP	8.96E–10	2.194**	−2.24E–08	−4.23***	−2.95E–10	−0.439
CPI	0.011	7.632***	0.010	6.46***	0.011	10.734***
EDU	−0.001	−0.468	0.0003	0.33	0.0002	0.177
HEL	−0.006	−0.390	0.069	1.97*	0.009	0.494
GDP	4.56E-06	11.398***	8.33E-06	2.63**	5.00E-06	8.410***
F-Statistics	48.17***		40.85***		37.99***	
R-squared	0.544		0.834		0.485	
Sum of squared residuals	12.842		4.662		8.305	
Std. Error of regression	0.252		0.164		0.203	

Note ***, ** and * indicate significance at 1, 5 and 10 percent level respectively
Source Authors Calculation

Table 7.7 Tests of specifications

	Null Hypothesis	*Appropriate test*	*Statistic*
Pooled regression model vs. Fixed effect model	$u_i = 0$	Restricted F Test	F (29173) = 10.47***
Pooled Regression model vs. Random effect model	$\sigma^2{}_i = 0$	Breusch-Pagan Lagrange Multiplier Test	χ^2 = 50.54***
Fixed effect model vs. Random effect model	Difference in coefficients are not systematic	Hausman Test	χ^2 (5) = 111.13***

Note *** Indicate significance at 1% level
Source Calculated by Authors

7.3 Empirical Observations

Beginning the voyage of studying literature is intended to reveal the rationale why the degree of inclusion has become an essential sign of growth. Following the Arrow-Debreu model with substantial amount of contingent contracts financial intermediation is not needed when there is no transaction cost and asymmetric information. Even after substantial reduction of the contingent contracts (the unrealistic assumptions) through using the Arrow certificates contemporary research focuses establishing general equilibrium in incomplete markets (Friesen, 1979). The presence of intermediaries becomes essential when imperfections or some sort of frictions enters the model. Due to frictions, the market operations become less efficient thus, exchange of finance becomes costly. Financial intermediaries facilitate these exchanges making it affordable and thus reducing the frictions through mobilizing funds. Fundamentally, these frictions arise due to technological and sometimes improper incentivizing the asymmetrically distributed information. Technologically generated frictions restrict the individual to access the economies of scale. These frictions are relaxed to some extent by way of mobilizing the savings, facilitating the hedging, diversifying, and pooling of risks, efficient allocation of resources and through screening the management's performance by taking corporate control. Financial system growth may be defined as the development of the magnitude, efficacy, and long-term viability of financial markets, as well as increased accessibility to financial markets, which can have a variety of economic advantages. A strong financial market, for example, channels an economy's savings to lucrative investments (Stiglitz & Weiss, 1983), lowers information costs, resulting in improved allocation of capital (Greenwood & Jovanovic, 1990) and lowers the cost of governing corporations (Bencivenga & Smith, 1993). There have been innumerous indicators those give a good proxy of different aspects of the financial system. However, initially the monetary aggregates (e.g. M1 and M2) were largely been used to characterize the development of financial structures. These indicators were incompetent for the reasons that they had portrayed the ability of the financial system to smoothly provide the services of transaction rather than the channelizing of savings to the borrowers instead. Alternatively using the liquid liabilities of the banking system (i.e. M3) overcomes the shortcomings but suffers from the problem of endogeneity due to the inclusion of M2 within it. In terms of development status, developing countries are primarily driven by banking sector development, whereas industrialized countries are slowed by non-banking sector development (for example, bond and stock market developments). To investigate the type of relationship across inclusion and additional measures of development, the investigation draws on many studies, including King and Levine (1993), which seek to determine the relationship within financial advancements, growth, and particular reasons for growth, as well as to test Schumpeter's claims made 80 years ago. The study's findings show that financial development elements are strongly associated with growth characteristics. However, a panel across the nations from Latin America shows that credit from banks reduces GDP where, De Gregorio and Guidotti (1995) suspects the lax regulatory environment being a promising reason behind such phenomena. The importance of finance as an instrument for growth is one of the most debated topics in financial economics. The theoretical foundations show that funding acts as a solvent in the process of economic growth (Levine, 1997). The findings indicate a favorable, first-order link between the development of finance and economic growth. Finance is also found in the studies of Gurley and Shaw (1955) and McKinnon (1973) as an indispensable tool for economic development. Presence of well-functioning financial instruments, markets, and institutions reduces the frictions by way of curtailing the information and transaction costs. Furthermore, nations in the process of transitioning from developing to developed status need a sophisticated financial system to flourish (Patrick, 1966). A number of studies exploring the relationship between financial development and economic expansions in the forefront show a positive relation (Eryilmaz et al., 2015; Goldsmith, 1969; Habibullah & Eng, 2006). Habibullah and Eng (2006) discovered findings that are similar to the results of Calderón and Liu (2003), Fase and Abma (2003), and Christopoulos and Tsionas (2004) when considering the association across financial growth and economic expansion in a system GMM framework constructed by Arellano and Bover (1995) and Blundell and Bond (1998). The theoretical framework of the study relies on the study of Pagano (1993), which combines Romer (1989) and Lucas Jr (1988). Results show that across a panel of 13 Asian countries over 1990–98 financial development promotes growth (Habibullah & Eng, 2006). The conventional function of finance influences economic growth through capital accumulation (Lewis, 1955; Nurkse, 1962) and technological innovation or more specifically productivity growth. Following the capital accumulation approach a steady-state per capita growth can be achieved through either using capital externalities or producing capital goods with constant returns to scale while avoiding the use of non-reproducible factors (Rebelo, 1991; Romer, 1986). While observing 109 countries from developing as well as industrialist countries from 1960–1994, economic growth is found to accelerate with financial deepening through capital aggregation and more steadily through productivity growth (Calderón & Liu, 2003). Through alteration of the rate of savings or by redistributing the savings among the available capital producing technologies the formal financial system configures the accumulation of the capital. Further, the steady-state growth escalates through alteration of the rate of technological innovation based on the interventions that focuses on advanced production processes (Aghion & Howitt, 1990; Grossman & Helpman, 1991). Formal financial system is also believed to promote economic growth through augmenting liquidity of long-term high-return commitments, which often investors are less likely to get involved with. This was further identified as one of the major drivers behind industrial revolution in England (Hicks, 1969). Even though technological innovation was present, the thriving force behind this eighteenth century growth was the liquidity factor of the capital market. A well-functioning financial market should ensure adequate liquidity of capital commitments so that as and when required it can be converted into purchasing power of the investor. However, the ease and speed of this conversion largely depends on the information asymmetry and transaction cost that acts as a friction to such conversion. The four pillars of a formal financial system (e.g. Financial Instruments, Financial Institutions, Financial Markets, and Financial Services) acts as a facilitator to reduce these frictions and thus enhance the rate of savings and long-run growth. The different forms of financial functions (e.g. trading of risks, savings mobilization, proper capital allocation, simplifying the trading of goods and enforcing financial contracts) help ameliorate these frictions on a broader perspective rather focusing on a particular instrument (i.e. money) or an institution (i.e. banks). This rather capitalizes on the functional relationship between financial structure (i.e. financial instruments, markets and institutions) and the provisions of financial services (Levine, 1997). Furthermore, Goldsmith (1969) demonstrates that financial development policies may significantly boost per capita GDP by evaluating 35 nations during 1860–1963. Likewise, Eryilmaz et al. (2015) attempted to observe the impact of total credit to GDP and total savings to GDP on annual GDP per capita after taking control of public spending to GDP and export deficit. Employing the two-way random effects model, the study discovered comparable evidences across the 23 OECD nations from 1980 to 2012. Similar to the conclusions of previous research, Rajan and Zingales (1998) suggest that financial markets supply the needs for development. They argue that financial development and economic growth may both be influenced by a common omitted factor (e.g. the propensity of households to save). Since, savings influence the rate of growth, initial financial development unsurprisingly might be correlated with growth and consequential evidence of cross-sectional regressions might unable to refute the hypothesis so formed. They explain such specifications as conjectural coincidence. Rajan and Zingales (1998) criticizes the size of stock market as an indicator of growth as the markets itself capitalizes growth whereas they also criticize the level of credit extended by financial institutions as an indicator of financial development because the lending by institutions is based on future growth prospects. This study categorizes the industries based on their requirements of external finance and draws a comparative view with respect to the scale of financial status the country belongs to. Industrial expansion is majorly driven by the entry of new firms or increase in the average size of the existent firms. However, entry of a new firm is more dependent on external financing and thus sensitive to financial developments. Finance-growth correlation has been further tested in the study of Demirgüç-Kunt and Maksimovic (1999). This study basically considers the proportion of firms where growth rates surpass the growth rate of firms those are backed only by internal resources. Results further show that this proportion is positively related to stock market turnover as well as with the measure of legal enforcements. However, Greenwood and Jovanovic (1990) shows financial intermediation and economic growth are endogeneously determined. They argue that financial institutions can generate more returns than compared to individuals whereas growth in turn builds costly financial structures. Financial development cuts down the differences between external and internal financing leading to growth especially for firms those are reliant on external financing. Bencivenga and Smith (1993) contends the development of financial markets as exogeneous that depends primarily on legislations and government regulations. However, financial development cannot just happen instantaneously (Diamond, 1989). Similar evidences have been observed in a wide panel of 159 nations by Khan and Senhadji (2000) from 1960 to

1999. However, a thorough study of the data reveals that across the lower income groups economic development is more sensitive to the development of financial sector (Herwartz & Walle, 2014). Efficient financial intermediaries along with improved resource allocation with greater total factor productivity growth expedite economic expansion after considering the possibility of endogeneity across the regressors in 63 nations between 1960 and 1995 (Beck et al., 2000). Furthermore, a positive relationship with growth in the economy is shown when focusing on the exogenous part of financial sector expansions (Levine et al., 2000). However, estimation of the equilibrium considering a relatively longer period of study criticizes the reliability of both the studies [Beck et al. (2000) and Levine et al. (2000)] while meeting the data characteristics associated with integration and co-integration (Christopoulos & Tsionas, 2004). Further, Christopoulos and Tsionas (2004) observes a uni-directional connection between financial sector expansion and economic success while considering a sample of 10 emerging economies and emphasizing on the equilibrium over the long-term connection. The linkages, however, were observed on the premise of assuming independent cross-sectional units instead of an absolute dependency (Bist, 2018). Further, Pradhan et al. (2017) finds co-integrating links between economic growth and across the four components of financial sector development (i.e. growth across the Banking, Insurance, Stock and the Bond markets) considering the ASEAN region during 1991–2011. However, significant differences are found in the magnitude and directions of financial development factors concerning economic growth when emphasized on causality. The growing importance of banking services and economic growth has one-way causation, but the increases in the stock market, insurance industry, and economic expansion have two-way causality.

Dahiya and Kumar (2020) with the Bayesian VAR model across Indian states finds a substantial correlation between economic development and the use of financial services instead of the composite index produced in this respect. After accounting for the probable interruption owing to the liberalization policy in 1991, the aggregate index of financial inclusion is shown to favorably affect economic development both in the long and short-term applying the ARDL and VECM framework respectively during 1980–2014 (Lenka & Sharma, 2017). Furthermore, Sharma (2016) sees a positive relationship with economic development in a VAR configuration for 2004–2013 when the measures of financial inclusion are examined individually as well as together as an index. The aforementioned investigation discovers the existence of bi-directional causation using the Granger causality test. It does not, however, disclose the co-integrating links between the components.

When concentrating on low-income countries and capitalizing on both the simple and uni-directional aspects of causation, the supply leading theory becomes more dominant (Jung, 1986). Across the poor and middle-income countries, Hassan et al. (2011) observes a bi-directional relationship between financial growth and economic development. However, significant exception remains across the two lowest categories (in terms of income), exhibiting a uni-directional causation. Nevertheless, an array of examples from the literature demonstrates that, based on the context, financial performance indicators could possibly possess a detrimental or even non-existent impact on economic achievement (Ahmad et al., 2021; Lucas Jr., 1988). Singh (1997) finds a modest connection between the components when 65 developing countries are considered. Considering the indicators from both the banking and stock market across the BRICS nations for 1993–2014, Guru and Yadav (2019) in a one-step system GMM specification observed that indicators from the banking and stock market both compliments each other's growth and stimulate the growth of the economy. The size of the financial intermediaries, credit to deposit ratio and domestic credit to private sector as an indicator of banking sector development positively influences the growth of the economy. Similarly, the value of shares traded is also found to significantly influence the growth in the economy under the presence of banking sector development variables. However, the impact of the turnover ratio is not significant on the economic growth indicators across the developed nations (Mhadhbi, 2014). This was also evident across the GCC nations (Grassa & Gazdar, 2014) as well as across the northern and southern Mediterranean nations (Ayadi et al., 2015). Similar conclusions were reached while considering a large panel comprising 101 developing and developed nations (Ductor & Grechyna, 2015).

The current research on finance-led growth adds to the evidence in support of a non-linear connection. Samargandi et al. (2015) found a U-shaped inversion between finance and economic development when studying the long-term threshold implications in a panel of 52 nations with moderate incomes from 1980 to 2008. Furthermore, Rousseau and Wachtel (2011) and Arcand et al. (2015) show that after crossing the threshold, the influence of financial development on economic growth disappears, even when non-linear factors are included. In this context,

Arcand et al. (2015) show that once credit to the private sector exceeds a certain limit, it begins to have a negative influence on production growth. Similarly, results from the threshold regression model suggests development in finance to substantially influence economic growth (Deidda & Fattouh, 2002) more promisingly across the countries from higher-income group than the lower ones. However, Law and Singh (2014) argue that more financing does not inevitably enhance economic progress, based on data from 87 countries, the majority of which are in the developed and emerging stages.

It is not innovative to use a panel co-integration model to assess the long-run association between financial inclusion and growth in the economy. Using both the FMOLS and DOLS estimators, Sethi and Acharya (2018) claim that the combination of the evaluation of financial accessibility (as judged by Sarma, 2012) had a long-term impact on the 31 countries' economic progress between 2004 and 2010. The research also employs unconstrained VECM in a fixed effect model and discovers a two-way causal relationship between the constructs. Using the system GMM technique, similar evidence was reported across 152 countries for 2004–2015 and 44 developing markets for 1990–2018 (Emara & El Said, 2021; Van et al., 2021). Furthermore, Kim and Hassan (2018) use the dynamic VAR framework in the system GMM using lagged regressors to find that financial inclusion promotes economic development among the 55 OIC member nations. The Impulse Response Function results support the findings, and the panel's Granger causality test validates mutually reinforcing effects.

7.4 Conclusions

While capitalizing on the conventional connection, the study's findings show that, from 2002 to 2019, the three criteria used to measure financial inclusion had a substantial long-term impact on output growth across all Indian provinces. Nevertheless, research into the immediate impacts yields conflicting findings. According to a preliminary analysis of the stationarity properties for the indicators of financial inclusion and economic progress, the variables exhibit non-stationary behavior at level and then move apart before becoming stationary at their first differences. Furthermore, the non-stationary variables are confirmed to be co-integrated and may eventually share a relationship by Pedroni's residual-based panel co-integration test (Pedroni, 1999).

Co-integrated vector estimates using between- and within-group estimators (i.e., FMOLS and DOLS approaches) demonstrate the long-term effects of financial inclusion indicators on specific developmental characteristics. On the other hand, it seems that the lnRNSDP and lnpercapitaRNSDP variables in the VECM setup are adjusting extremely slowly towards equilibrium. From a policy perspective, the study appears to be especially important since it suggests that in order to support long-term economic development, basic financial services should be made more widely available and used throughout Indian provinces. These findings are consistent with other studies by Pradhan (2010), Apergis et al. (2007), and Sethi and Acharya (2018), which emphasized the role that financial outreach plays in creating economic momentum. More banking locations and more people using credit and savings accounts will probably boost the engagement of funds, which might lead to a bigger exposure to productivity and a rise in the economy's production growth. The results are favorable to the finance-led growth theory and support Patrick's supply-led growth hypothesis, especially in developing nations like India.

In this connection, access to financial services through digital means provides availability of the basic financial services (e.g. savings, credit, insurance, and payment) especially to the underserved and unbanked population (Mazer & McKee, 2021). Especially in the areas where banking infrastructure is inadequate, mobile money services (e.g. Kenya's m-Pesa) allow to overcome the conventional barriers to the users (Jack & Suri, 2011). Through the mobile money accounts, rural dwellers are able to manage risks and investments more promptly, thus investible corpus is channelized towards the education and healthcare uplifting the growth of the economy (Garg, 2022). Further, the digital platforms aid to access the microloans to entrepreneurs more conveniently, thus allowing seamless growth of the business along with creation of employment (Suri & Jack, 2016). On a broader scale, digital platforms allow to reduce frictions, thus improving the efficiency of the financial transactions through reducing the transaction costs and encouraging faster and cheaper cross-border payments (Narayan & Smith, 2020). However, this growing reliance on digital means involves the potential risks of cyber security, mostly concerning the privacy of data that is often stored on centralized servers. Data breaches which can be one of the obvious phenomena can potentially lead to significant financial losses for the consumers and financial institutions (Chen et al., 2019). Lack of awareness on cyber security can potentially lead to phishing and identity theft which increases the likelihood of fraud (Ayo et al., 2022). Often financial institutions are exposed to cyber attacks like ransom ware and Distributed Denial of Service (DDoS) attacks leading to hefty losses and disruption of the services (Hafner, 2020). Especially across the developing nations, a robust security system of the data is one among the serious challenges that arises due to the financial and technical constraints (Ashraf et al., 2021). Among other barriers to seamless digital transactions especially across the rural areas is the poor connectivity of the internet. Often underdeveloped communication infrastructure creates hindrances restricting the access to mobile banking and online payments (Aker & Mbiti, 2010). Further, inadequate digital training programs especially across the rural areas where educational attainments are relatively lesser makes financial services ineffective (Vargas & Orozco, 2019). Apart from tackling these challenges comprehensive cyber security measures in the form of multi-factor authentication, encrypting the data and fraud detection system are in place (Xia & Zhang, 2022). Public private partnerships have built extensive broadband connections, improved the mobile network coverage and through electricity grids have extended the access to underserved areas (World Bank, 2021). However, the fundamental question of whether households with lower incomes are more unbanked throughout Indian provinces, particularly given that 20–30% of the country's population lives below the international poverty line, serves as the impetus for looking into the reverse causality case. The study demonstrates a contemporaneous link between per capita savings account and per capita real GDP, with the former showing a positive correlation and the latter showing a significant partial correlation (at the 1% level). Additionally, an examination of the association's performance from 2002 to 2019 using a panel configuration consisting of 27 states and 5 union territories reveals that government spending on health care, inflation, and economic growth all indicate a markedly favorable expansion of formal finance. The results of our research validate the empirical findings of Zang and Kim (2007), which indicate that increased productivity would have a favorable impact on financial inclusion. It would also confirm the long-standing theories of Robinson and Lucas. It is evident that growth led finance exists when the standard preconditions of the residuals derived are met. Nonetheless, given the compelling evidence of growth led finance, the study does not refute the presence of the finance-led growth hypothesis. Growth may also be influenced by finance, or there may be a reciprocal relationship between the two.

References

Aghion, P., & Howitt, P. (1990). A model of growth through creative destruction.

Ahmad, M., Majeed, A., Khan, M. A., Sohaib, M., & Shehzad, K. (2021). Digital financial inclusion and economic growth: Provincial data analysis of China. *China Economic Journal, 14*(3), 291–310.

Aker, J. C., & Mbiti, I. M. (2010). Mobile phones and economic development in Africa. *Journal of Economic Perspectives, 24*(3), 207–232. https://doi.org/10.1257/jep.24.3.207

Apergis, N., Filippidis, I., & Economidou, C. (2007). Financial deepening and economic growth linkages: a panel data analysis. *Review of World Economics*, *143*, 179–198. https://www.jstor.org/stable/40441131

Arcand, J. L., Berkes, E., & Panizza, U. (2015). Too much finance? *Journal of Economic Growth, 20*(2), 105–148. https://doi.org/10.1007/s10887-015-9115-2

Arellano, M., & Bover, O. (1995). Another look at the instrumental variable estimation of error-components models. *Journal of Econometrics*, *68*(1), 29–51. https://doi.org/10.1016/0304-4076(94)01642-D

Arestis, P., & Demetriades, P. (1997). Financial development and economic growth: assessing the evidence. *The economic journal*, *107*(442), 783–799. https://www.jstor.org/stable/2957802

Arestis, P., Demetriades, P. O., & Luintel, K. B. (2001). Financial development and economic growth: the role of stock markets. *Journal of money, credit and banking*, 16–41. https://www.jstor.org/stable/2673870

Ashraf, S., Mehmood, M., & Khan, I. (2021). Cybersecurity in digital financial services: Challenges and solutions. *Journal of Financial Regulation and Compliance, 29*(4), 561–574. https://doi.org/10.1108/JFRC-09-2020-0241

Ayadi, R., Arbak, E., Naceur, S. B., & De Groen, W. P. (2015). Financial development, bank efficiency, and economic growth across the Mediterranean. In R. Ayadi, M. Dabrowski, & L. De Wulf (Eds.), *Economic and social development of the Southern and Eastern Mediterranean Countries* (pp. 219–233). Springer. https://doi.org/10.1007/978-3-319-11122-3_14

Ayo, C. K., Akter, S., & Aljoza, M. (2022). Consumers' trust and adoption of digital financial services: Insights from developing economies. *International Journal of Bank Marketing, 40*(5), 1–22. https://doi.org/10.1108/IJBM-07-2021-0299

Beck, T., Levine, R., & Loayza, N. (2000). Finance and the sources of growth. *Journal of Financial Economics, 58*(1–2), 261–300. https://doi.org/10.1016/s0304-405x(00)00072-6

Bencivenga, V. R., & Smith, B. D. (1993). Some consequences of credit rationing in an endogenous growth model. *Journal of Economic Dynamics and Control, 17*(1–2), 97–122. https://doi.org/10.1016/S0165-1889(06)80006-0

Berki, S. E. (1986). A look at catastrophic medical expenses and the poor. *Health Affairs, 5*(4), 138–145. https://doi.org/10.1377/hlthaff.5.4.138

Bist, J. P. (2018). Financial development and economic growth: Evidence from a panel of 16 African and non-African low-income countries. *Cogent Economics & Finance, 6*(1), 1–17. https://doi.org/10.1080/23322039.2018.1449780

Blundell, R. W., & Bond, S. R. (1998). Initial conditions and moment restrictions in dynamic panel data models. *Journal of Econometrics, 87*, 29–52.

Calderón, C., & Liu, L. (2003). The direction of causality between financial development and economic growth. *Journal of Development Economics, 72*(1), 321–334.

Chaturvedi, V., Dholakia, R. H., & Kumar, B. (2011). Inter-Relationship between Economic Growth, Savings and Inflation in Asia. *SSRN Electronic Journal*. https://doi.org/10.2139/ssrn.1212096

Chen, Z., Zhang, M., & Wu, Y. (2019). A survey of data security and privacy protection in digital finance. *IEEE Access, 7*, 57356–57369. https://doi.org/10.1109/ACCESS.2019.2913491

Christopoulos, D. K., & Tsionas, E. G. (2004). Financial development and economic growth: Evidence from panel unit root and cointegration tests. *Journal of Development Economics, 73*(1), 55–74. https://doi.org/10.1016/j.jdeveco.2003.03.002

Dahiya, S., & Kumar, M. (2020). Linkage between financial inclusion and economic growth: An empirical study of the emerging Indian economy. *Vision, 24*(2), 184–193. https://doi.org/10.1177/0972262920923891

Dash, S. K., & Kumar, L. (2018). Does inflation affect savings non-linearly? Evidence from India. *Margin: The Journal of Applied Economic Research, 12*(4), 431–457. https://doi.org/10.1177/0973801018786155

Demirgüç-Kunt, A., & Maksimovic, V. (1999). *Financial constraints, uses of funds, and firm growth: An international comparison*. The World Bank.

De Gregorio, J., & Guidotti, P. E. (1995). Financial development and economic growth. *World Development, 23*(3), 433–448. https://doi.org/10.1016/0305-750X(94)00132-I

Deidda, L., & Fattouh, B. (2002). Non-linearity between finance and growth. *Economics Letters, 74*(3), 339–345. https://doi.org/10.1016/S0165-1765(01)00571-7

Diamond, D. W. (1989). Reputation acquisition in debt markets. *Journal of Political Economy, 97*(4), 828–862. https://www.jstor.org/stable/1832193

Ductor, L., & Grechyna, D. (2015). Financial development, real sector, and economic growth. *International Review of Economics & Finance, 37*, 393–405. https://doi.org/10.1016/j.iref.2015.01.001

Emara, N., & El Said, A. (2021). Financial inclusion and economic growth: The role of governance in selected MENA countries. *International Review of Economics & Finance, 75*, 34–54. https://doi.org/10.1016/j.iref.2021.03.014

Eryılmaz, F., Bakır, H., & Mercan, M. (2015). Financial development and economic growth: Panel data analysis. In Ö. Olgu, H. Dinçer, & Ü. Hacıoğlu (Eds.), *Handbook of research on strategic developments and regulatory practice in global finance* (pp. 233–245). IGI Global. https://doi.org/10.4018/978-1-4666-7288-8.ch015

Fase, M. M. G., & Abma, R. C. N. (2003). Financial environment and economic growth in selected Asian countries. *Journal of Asian Economics, 14*, 11–21.

Friesen, P. H. (1979). The Arrow-Debreu model extended to financial markets. *Econometrica: Journal of the Econometric Society*, 689–707. https://doi.org/10.2307/1910415

FSD Kenya. (2025). *Financial inclusion and health: A balancing act for Kenya's future*. https://www.fsdkenya.org/blogs-publications/financial-inclusion-and-health-a-balancing-act-for-kenyas-future/

Garg, S. (2022). Digital financial inclusion and poverty alleviation: A study of mobile banking services. *Journal of Financial Services Marketing, 27*(2), 98–113. https://doi.org/10.1057/s41264-022-00095-1

Ghosh, S., Sahu, T. N., & Mallik, G. (2024). Does financial inclusion stimulate India's economy: A panel data analysis. *Journal of Public Affairs, 24*(1), Article e2891. https://doi.org/10.1002/pa.2891

Goldsmith, R. W. (1969). *Financial structure and development*. Yale University Press.

Greenwood, J., & Jovanovic, B. (1990). Financial development, growth, and the distribution of income. *Journal of political Economy, 98*(5, Part 1), 1076–1107. https://www.jstor.org/stable/2937625

Grassa, R., & Gazdar, K. (2014). Financial development and economic growth in GCC countries: A comparative study between Islamic and conventional finance. *International Journal of Social Economics, 41*(6), 493–514. https://doi.org/10.1108/IJSE-12-2012-0232

Grossman, G. M., & Helpman, E. (1991). Quality ladders in the theory of growth. *The Review of Economic Studies, 58*(1), 43–61.

Gurley, J. G., & Shaw, E. S. (1955). Financial aspects of economic development. *The American Economic Review, 45*(4), 515–538. http://www.jstor.org/stable/1811632

Guru, B. K., & Yadav, I. S. (2019). Financial development and economic growth: Panel evidence from BRICS. *Journal of Economics, Finance and Administrative Science, 24*(47), 113–126. https://doi.org/10.1108/JEFAS-12-2017-0125

Habibullah, M. S., & Eng, Y. K. (2006). Does financial development cause economic growth? A panel data dynamic analysis for the Asian developing countries. *Journal of the Asia Pacific Economy, 11*(4), 377–393. https://doi.org/10.1080/13547860600923585

Hafner, L. (2020). Cyber risks in financial services: A global overview. *Journal of Financial Crime, 27*(3), 710–721. https://doi.org/10.1108/JFC-06-2020-0129

Hassan, M. K., Sanchez, B., & Yu, J. S. (2011). Financial development and economic growth: New evidence from panel data. *The Quarterly Review of Economics and Finance, 51*(1), 88–104. https://doi.org/10.1016/j.qref.2010.09.001

Herwartz, H., & Walle, Y. M. (2014). Determinants of the link between financial and economic development: Evidence from a functional coefficient model. *Economic Modelling, 37*, 417–427. https://doi.org/10.1016/j.econmod.2013.11.029

Hicks, J. R. (1969). A theory of economic history.

Jack, W., & Suri, T. (2011). Mobile money: The economics of M-PESA. *National Bureau of Economic Research Working Paper Series*, No. 16721. https://doi.org/10.3386/w16721

Jung, W. S. (1986). Financial Development and economic growth: International evidence. *Economic Development and Cultural Change, 34*(2), 333–346. https://doi.org/10.1086/451531

Khan, M. S., & Senhadji A. S. (2000). Financial development and economic growth: An overview. *IMF Working Papers, 2000*(209), 1–24. https://doi.org/10.5089/9781451874747.001

Kim, D. W., Yu, J. S., & Hassan, M. K. (2018). Financial inclusion and economic growth in OIC countries. *Research in International Business and Finance, 43*, 1–14. https://doi.org/10.1016/j.ribaf.2017.07.178

King, R. G., & Levine, R. (1993). Finance and growth: Schumpeter might be right. *The Quarterly Journal of Economics*, 717–737. https://www.jstor.org/stable/2118406

Law, S. H., & Singh, N. (2014). Does too much finance harm economic growth? *Journal of Banking & Finance, 41*, 36–44. https://doi.org/10.1016/j.jbankfin.2013.12.020

Lenka, S. K., & Sharma, R. (2017). Does financial inclusion spur economic growth in India?. *The Journal of Developing Areas, 51*(3), 215–228. https://www.jstor.org/stable/https://doi.org/10.2307/26416941

Levine, R. (1997). Financial development and economic growth: Views and agenda. *Journal of Economic Literature, 35*(2), 688–726. http://www.jstor.org/stable/2729790

Levine, R., Loayza, N., & Beck, T. (2000). Financial intermediation and growth: Causality and causes. *Journal of Monetary Economics, 46*(1), 31–77. https://doi.org/10.1016/S0304-3932(00)00017-9

Lewis, W. A. (1955). *Theory of economic growth*. Unwin University Books.

Lopez, L., & Weber, S. (2017). Testing for Granger causality in panel data. *The Stata Journal, 17*(4), 972–984. https://doi.org/10.1177/1536867X1801700

Lucas, R. E., Jr. (1988). On the mechanics of economic development. *Journal of Monetary Economics, 22*(1), 3–42. https://doi.org/10.1016/0304-3932(88)90168-7

Luintel, K. B., & Khan, M. (1999). A quantitative reassessment of the finance–growth nexus: Evidence from a multivariate VAR. *Journal of Development Economics, 60*(2), 381–405. https://doi.org/10.1016/S0304-3878(99)00045-0

Mazer, R., & McKee, K. (2021). Digital financial inclusion: A conceptual framework and its implications for economic growth. *Development Policy Review, 39*(2), 205–223. https://doi.org/10.1111/dpr.12432

McKinnon, R. I. (1973). *Money and capital in economic development*. Brookings Institution. https://www.brookings.edu/book/money-and-capital-in-economic-development/

Mhadhbi, K. (2014). Financial development and economic growth: A dynamic panel data analysis. *International Journal of Econometrics and Financial Management, 2*(2), 48–58. https://doi.org/10.12691/ijefm-2-2-1

Narayan, L., & Smith, S. (2020). Digital financial services: Opportunities and challenges. *Financial Services Review, 29*(1), 75–88. https://doi.org/10.12816/003234

Nurkse, R. (1962). *Problems of capital formation in underdeveloped countries*. Blackwell.

Pagano, M. (1993). Financial markets and growth: An overview. *European Economic Review, 37*(2–3), 613–622.

Patra, S. K., Murthy, S., & Biswal, A. (2015). Savings-growth-inflation nexus in Asia: Panel data approach. *Journal of Economics and Finance. e-ISSN, 2321*(5933), 75–85.

Patrick, H. T. (1966). Financial development and economic growth in underdeveloped countries. *Economic Development and Cultural Change, 14*(2), 174–189. https://doi.org/10.1086/450153

Pedroni, P. (1999). Critical values for cointegration tests in heterogeneous panels with multiple regressors. *Oxford Bulletin of Economics and Statistics, 61*(S1), 653–670.

Pedroni, P. (2004). Panel cointegration: Asymptotic and finite sample properties of pooled time series tests with an application to the PPP hypothesis. *Econometric Theory, 20*(3), 597–625. https://doi.org/10.1017/S0266466604203073

Pierse, R. G., & Snell, A. J. (1995). Temporal aggregation and the power of tests for a unit root. *Journal of Econometrics, 65*(2), 333–345. https://doi.org/10.1016/0304-4076(93)01589-E

Pradhan, R. P. (2010). Financial deepening, foreign direct investment and economic growth: Are they cointegrated. *International Journal of Financial Research, 1*(1), 37–43. https://doi.org/10.5430/ijfr.v1n1p37

Pradhan, R. P., Arvin, M. B., Bahmani, S., Hall, J. H., & Norman, N. R. (2017). Finance and growth: Evidence from the ARF countries. *The Quarterly Review of Economics and Finance, 66*, 136–148. https://doi.org/10.1016/j.qref.2017.01.011

Rajan, R. G., & Zingales, L. (1998). Financial dependence and growth. *American Economic Review, 88*(3), 559–586. https://www.jstor.org/stable/116849

Rebelo, S. (1991). Long-run policy analysis and long-run growth. *Journal of Political Economy, 99*(3), 500–521. https://doi.org/10.1086/261764

Romer, P. M. (1986). Increasing returns and long-run growth. *Journal of Political Economy, 94*(5), 1002–1037. https://doi.org/10.1086/261420

Romer, P. (1989) Capital accumulation and the theory of long-run growth. In R. Barro (Ed.), *Modern business cycle theory*. Harvard University Press.

Rousseau, P. L. (1998). The permanent effects of innovation on financial depth: Theory and US historical evidence from unobservable components models. *Journal of Monetary Economics, 42*(2), 387–425. https://doi.org/10.1016/S0304-3932(98)00027-0

Rousseau, P. L., & Wachtel, P. (2011). What is happening to the impact of financial deepening on economic growth? *Economic Inquiry, 49*(1), 276–288. https://doi.org/10.1111/j.1465-7295.2009.00197.x

Samargandi, N., Fidrmuc, J., & Ghosh, S. (2015). Is the relationship between financial development and economic growth monotonic? Evidence from a sample of middle-income countries. *World Development, 68*, 66–81. https://doi.org/10.1016/j.worlddev.2014.11.010

Sarma, M. (2012). *Index of financial inclusion—A measure of financial sector inclusiveness*.

Sethi, D., & Acharya, D. (2018). Financial inclusion and economic growth linkage: Some cross country evidence. *Journal of Financial Economic Policy, 10*(3), 369–385. https://doi.org/10.1108/JFEP-11-2016-0073

Sharma, D. (2016). Nexus between financial inclusion and economic growth: Evidence from the emerging Indian economy. *Journal of Financial Economic Policy, 8*(1), 13–36. https://doi.org/10.1108/JFEP-01-2015-0004

Singh, A. (1997). Financial liberalization, stock markets and economic development. *The Economic Journal, 107*(442), 771–782. https://www.jstor.org/stable/2957801
Stiglitz, J. E., & Weiss, A. (1983). Incentive effects of terminations: Applications to the credit and labor markets. *The American Economic Review, 73*(5), 912–927. https://www.jstor.org/stable/1814662
Suri, T., & Jack, W. (2016). The long-run poverty and gender impacts of M-Pesa. *Science, 354*(6311), 1288–1292. https://doi.org/10.1126/science.aaf0013
T20 India. (2023). *Ensuring progress on SDGs through digital transformation and financial inclusion.* Think20 India. https://t20ind.org/research/ensuring-progress-on-sdgs-through-digital-transformation-and-financial-inclusion/
UNCDF. (2025). *Financial Inclusion and the SDGs.* https://www.uncdf.org/financial-inclusion-and-the-sdgs
Van, L. T. H., Vo, A. T., Nguyen, N. T., & Vo, D. H. (2021). Financial inclusion and economic growth: An international evidence. *Emerging Markets Finance and Trade, 57*(1), 239–263.
Vargas, S., & Orozco, M. (2019). Digital financial literacy in developing economies. *Global Financial Inclusion Review, 10*(4), 33–44. https://doi.org/10.1787/gfi-2019-12
Wachtel, P., & Rousseau, P. (1994). *Financial intermediation and economic growth: A historical comparison of the US, UK and Canada* (No. 94–04).
Wooldridge, J. M. (2016). *Introductory econometrics: A modern approach.* Nelson Education.
World Bank. (2021). *The state of financial infrastructure in emerging economies.* World Bank Group. https://www.worldbank.org/financial-infrastructure
Wyszewianski, L. (1986). Financially catastrophic and high-cost cases: Definitions, distinctions, and their implications for policy formulation. *Inquiry*, 382–394. https://www.jstor.org/stable/29771827
Xia, Y., & Zhang, W. (2022). The role of cybersecurity in the success of digital financial inclusion. *Journal of Financial Technology and Innovation, 5*(2), 221–238. https://doi.org/10.1093/jfticn2022
Zang, H., & Kim, Y. C. (2007). Does financial development precede growth? Robinson and Lucas might be right. *Applied Economics Letters, 14*(1), 15–19. https://doi.org/10.1080/13504850500425469

CHAPTER 8

Financial Inclusion and Poverty Alleviation (SDG 1)

8.1 Introduction

Addressing poverty and inequality is one of the key policy goals pursued via increased access to different financial services, particularly for the most vulnerable segments of society (Sahoo et al., 2017). The majority of research found a negative relationship between poverty and inequality as well as accessibility to legally authorized banking services (Burgess & Pande, 2005; Galor & Zeira, 1993; Honohan, 2004; Jalilian & Kirkpatrick, 2002). Studies in this connection are motivated by empirical evidences of strong correlations between the distribution of income and the aggregate economic activity (measured by income per capita) (Kravis, 1960). The distribution of income is more concentrated for countries which are financially sounder (Lydall, 1968). Consistent evidence was reported in the findings of Torsten and Tabellini (1991) which exhibits that equity increases both the level of income as well as the growth in the economy. However, the extent of influence of both poverty and inequality on the access to financial services varies largely across the contexts being studied (Beck et al., 2007; Boukhatem, 2016; Buckland et al., 2005; Clarke et al., 2006; Inoue & Hamori, 2012; Jeanneney & Kpodar, 2011). The growth of financial sector promotes economic development albeit qualifications exist on the preferential influence that the sectoral growth may have on the wealthier part of the population (Clarke et al., 2006). Experience has shown that financial markets are driven by selection bias

S. Ghosh et al., *Financial Inclusion*, Sustainable Development Goals Series, https://doi.org/10.1007/978-981-95-3385-5_8

and, in many cases, moral hazards owing to the necessity for collateral from borrowers, resulting in the poor being deprived of cash. Thus, initially the development of financial sector might increase the inequality among the society but in the long run, this momentum is reversed (Banerjee & Newman, 1993). Under static equilibrium condition, the existence of inequality due to capital market imperfections ensures the enforcement of employment contracts. Thus, weaker sections of the society prefer wage work over self-employment and the wealthier part takes the charge of investment through entrepreneurship. Contraction in inequality promotes income growth of the poor by 40%, so far as the long-run impact of financial development is concerned while the remaining 60% is contributed towards aggregate economic growth (Beck et al., 2007). The rate of growth in per capita income along with the growth in the Lorenz curve explains the changes in the distribution of income. This also remains consistent when alteration is made with the absolute measure of poverty (Datt & Ravallion, 1992; Kakwani, 1993). Financial inclusion, as a strategic approach to eliminating inequality by giving access to financial services, promotes poverty reduction. However, addressing inequality (measured by any of the usual Pigou-Dalton criterion) certainly does not mean a reduction of poverty (Datt & Ravallion, 1992). Referring to the conduit effect as proposed by McKinnon et al. (2011) observe that financial development directly and strongly compresses the extent of poverty through distributional effects than compared to the indirect effect through economic growth.

Considering a representative group of countries that were developing from 1966 to 2000, the results suggest that those on the lower end of the economic spectrum gain greatly from enhanced payment and savings options provided by the banking sector. However, availability of credit remains as still one of the major concerns. The motive of finance as insisted by Keynes (1937) and Mckinnon (1973) through his conduit effect not only acts as a motivation for the financial intermediation that might address poverty rather complements the alternate wing of savings opportunities leading to investment. Following Kanbur (2001), growth in the per capita income certainly reduces poverty as illustrated by a majority of literary evidences, yet the dispute is on exploring alternative policies. Financial instability is yet another an important condition that has detrimental effects on the relationship between financial development and eradication of poverty (Loayza & Ranciere, 2006). Especially for the poor, those are more vulnerable to such movements which often lead to macro-economic instability in the form of fluctuations in growth and inflation.

8.2 Financial Inclusion and Poverty—Some Theoretical Insights

This study finds cross-sectional as well as country-specific evidences in a large body of research that emphasizes the interconnections among financial development, economic expansion, income distribution, and poverty alleviation. Studies that investigate the link between financial inclusion and eliminating poverty using macro-level data present a broad picture but are incapable of providing nuanced and comprehensive insights. Further, country-specific research at the household level is particular but insufficient for establishing a broad picture. Chibba (2009) effectively highlights the importance of formal financial intermediation in lowering the occurrence of poverty and so better accomplishing the Millennium Development Goals (now the SDGs). Traditional poverty-reduction methods are necessary, but they are insufficient to expedite poverty reduction; so, financial inclusion (FI) may be seen as an opportunity to explore. Addresses from the United Nations Conference on 'Financial Inclusion in Africa' in June 2006; the World Bank's conference on 'Access to Finance' in March 2007; and many other efforts taken solely or jointly recognize and support the vital role delved within the FI-PR-SDG alliance encouraging a nation's economic and social progress. International partnerships support the concept that economic and social growth cannot be promoted in isolation, leaving a significant portion of society, particularly the marginalized out from its periphery. As a result, an effort to broaden the accessibility of financial services to all segments of society is vital for promoting development and configuring capability in such a way that SDGs are better realized (Dugarova & Gülasan, 2017). An examination of cross-sectional empirical evidences in the literature, whether using the nations or states as units, yields a generalized viewpoint. Considering the influence of financial inclusion on vulnerability and susceptibility to poverty in Ghanaian households following the Ghana Living Standard Survey 2016/17, Koomson et al. (2020) observe a 27% decrease in the household's likelihood of being poor, as well as a 28% protection against future exposures to poverty. A gender-based comparison shows that male-headed households are relatively on the inefficient side than compared to the female-headed households to cut down poverty as well as vulnerability to poverty. Further, the effects of financial inclusion towards curtailing the incidence of poverty are more prominent across the rural areas than compared to the urban areas. Literary evidence shows that vulnerability to poverty is an area that has been relatively less emphasized (Imai et al., 2010). There might be instances where presently non-poor households fall in to poverty traps in the coming future while encountering contingencies (Ozughalu, 2016). Coulibaly (2019) predicts that before the year's end of 2019, Africa will be 70% of the worlds impoverished. This would further aggravate by the end of 2030 when nearly 13 countries would face extreme poverty. The Maya declaration as signed by 90 developing countries in the G20 summit of 2010 which had aimed to reduce poverty through financial inclusion (AFI, 2015) had identified that 75% of the countries are financially excluded. The aim of such declaration was also consistent with the Sustainable Development Goals (SDG 1) (ICSU & ISSC, 2015). Financial inclusion increased from 51 percent in 2011 to 69 percent in 2017, among nations that signed the declaration. However, the review of account ownership in terms of gender is quite suspicious especially across the developing countries for female (59%) than compared to male (79%) (Demirgüç-Kunt et al., 2018). So far as the rural urban divide is concerned, urban dwellers in developing countries have greater access than compared to the rural counterparts. As a result, the relationship between financial inclusion and poverty (typically assessed by susceptibility to poverty) warrants more investigation in light of gender and locational disparities (Swamy, 2014). The benefits of financial inclusion serve to reduce poverty by both direct and indirect routes while remaining within the structure of financial growth. Financial inclusion directly reduces poverty through broadening the access to credit, insurance and other services that facilitates the resources required to meet day-to-day consumption, investment, and overall growth of the economy (Rajan & Zingales, 1998). It also aids to improve the entrepreneurial capabilities of individuals that augment their income, consumption and thus the living standards (World Bank, 2007). Financial inclusion creates job and through increased social spending on education, health, and social protection that leads to growth of an economy which is often referred as an indirect channel to foster development (Abosedra et al., 2016). This was also previously a part of policy recommendation by Perotti (1993). Further, economic liberalization through over usage of credit cards, ATM's and internet banking can adversely increase indebtedness which might invite poverty as a consequence (Lyons & Hunt,

2003). Enhanced accessibility to finance has also some adverse effects on the status of prevailing inequalities (Koomson et al., 2020). As a result, a better method is proposed, emphasizing the ubiquitous availability of financial services to boost economic development (Beck & Demirgüç-Kunt, 2008). This was also aligned with the World Bank's goal of implementing the UFA to reduce poverty by 2020. Enhanced accessibility of financial services is also acclaimed to ensure sustainable coverage of education, health, and insurance products which mitigates the risk of uncertainty especially for the marginalized sections of the society (Brown et al., 2013; Mârza et al., 2015). Certain customizations that improve access to credit and savings products help to mitigate risks and provide a cushion for the households to combat poverty (Emerick et al., 2016; Hallegatte et al., 2016). Mobile money as a recent innovative product is one of most prominent strategic vehicles that embark on achieving accessibility to varied financial products and thus achieving sustainable economic growth (Donovan, 2012). Among others, Burgess and Pande (2005) discover that expanding rural bank branches, savings, and credit throughout Indian provinces may successfully address the problem of poverty reduction. Inoue and Hamori (2012) demonstrate that deposit and credit amounts as a proportion of production may considerably decrease poverty head count across an unbalanced panel of 28 Indian states. Jalilian and Kirkpatrick (2002) examine the influence of financial growth on decreasing poverty utilizing cross-country data from developing nations. They found that financial inclusion helps to strengthen the asset structure of the poor households by giving them the opportunity to invest in newer technologies, health, and education. This in turn will help them to achieve sustainable living standards (World Bank, 2001). Further, Jalilian and Kirkpatrick (2005) discover that financial development greatly reduces poverty when they simulate the same sample to make reliable conclusions on the causal link between the dimensions. They observe that up to a certain level of economic growth the expansion of the financial sector eradicates poverty through addressing inequality. However, studies so far, focusing to draw a generalized perspective across a large cross-sectional sample suffers from the problem of structural homogeneity. While encountering the assumption of structural homogeneity, Ram (1999) stratified the sample countries in to three separate groups on the basis of their average growth performance. Results show significant differences across the groups so far as the correlations are concerned between financial development and indicators of growth. Further, the presence of heterogeneity is confirmed while classifying the regions considered in the model prescribed by Levine et al., (2000) (Anderson & Tarp, 2003). Results show that progressive exclusion of the regions from the sample space makes the financial development indicators insignificant. Country specific studies are also in support of the views that financial development promotes economic growth. However, the functional form differs across the cross-sectional units (Demetriades & Hussein, 1996). Following the establishment of the SDGs (UNDESA, 2016), Boukhatem (2016), using the framework of the GMM estimator in a group of 67 low and middle-income countries, shows that financial development may dramatically decrease poverty. This research, on the other hand, tackles the difficulties of over-identification and poor instruments, resulting in a more trustworthy estimate of the impact. In a similar manner Kiendrebeogo and Minea (2016) assert that access to financial services reduces poverty indisputably. After correcting for simultaneity and small sample bias, the research concludes that instability in finances may have a negative impact on the connection between financial progress and poverty. Further, this study also acknowledges the presence of non-linearity while predicting poverty on the basis of financial development. Following the McKinnon's 'Conduit' effect (McKinnon, 1973) and Shaw's 'Intermediation' effect (Shaw, 1973), finance may not have a direct impact on poverty, but it might eradicate poverty through economic growth. However, Greenwood and Jovanovic (1990) asserts that accessibility to bank's credit especially for the poor might be hindered by the high costs of small loans. This problem might get aggravated when it comes to observe the distribution of income (which appears to be highly skewed) especially across the developing countries in favor of the richer side of the population (World Bank, 1995). As a result, a beneficial effect on economic growth may not necessarily lead to a decrease in poverty (Beck et al., 2007). In this connection, Arestis and Caner (2010) were unable to establish a significant effect that capital account liberalization may have on reduction of poverty. Through improving the allocation of savings in to productive investments and with adequate monitoring, financial intermediaries can significantly reduce poverty. However, availability of data can lead to selection bias of the sample which may comprise countries with strong financial backbone instead of the targeted group which should have taken the countries that are in their developing stage (Kiendrebeogo & Minea, 2016). However, financial development often represented by the broad money stock (M2 and M3) (further read

Akhter & Daly, 2009; Perez-Moreno, 2011) following the Mckinnon's 'Conduit effect' can be questioned on the ground that broad money with reference to developing countries are held outside the banking system. Park and Mercado (2015), who constructed a panel of 37 emerging Asian countries, and more recently Neaime and Gaysset (2018), who considered eight nations from the Middle East and Africa, support financial inclusion to reduce poverty and inequality (albeit early). Some research; however, acknowledge that aggregate data is inadequate to capture the influence at the household level (Beck & Demirgüç-Kunt, 2008).

A recent assessment of the literature reveals the ineffectiveness of empirical evidences demonstrating financial inclusion as a method to relieve poverty (Mader, 2018). Furthermore, there are research investigations that deny that economic progress reduces the extent of poverty (Todaro & Smith, 2012). Studies such as Dawood et al. (2019) in the Indonesian context, Koomson et al. (2020) in Ghana, Uddin et al. (2014) in Bangladesh, Li (2018) in China, and Aideyan (2009); Seck et al. (2017); Dimova and Adebowale (2018) in the Nigerian context deserve mention. According to the research, financial development or inclusion decreases the incidence of poverty to some amount, however the functional form seems to be varied in the situations studied.

Thus, financial inclusion, as an integral component of a larger picture (i.e., social inclusion), seems to be immensely crucial in addressing problems of poverty and inequality by unlocking potential possibilities, particularly for the disadvantaged segments of society. Even amplifying the variability by analyzing a longer cross-sectional band of 116 developing countries from 2004 to 2016 implies that access and use of formal financial services should be promoted in order to maximize societies' overall wellbeing (Omar & Inaba, 2020). Analyzing the influence of financial inclusion on specific development indicators (e.g., income, spending, food security, and so on) with regard to gender differences, for example, does not support the exception (Swamy, 2014). The influence of formal financial intermediation on production growth is varied, leading to a predictable conclusion (Bell & Rousseau, 2001). Furthermore, there is limited data, particularly using a cross-country arrangement that investigates whether financial inclusion improves poverty through enhancing production growth. However, Majid et al. (2017) provide a typical illustration of the research topic that we want to answer in the current work in the Indonesian setting.

8.3 Relationship Between Financial Inclusion and Poverty Eradication (SDG 1)

Financial inclusion as a significant contributor of the inclusive development mechanism can be viewed as a relevant and complementary measure for poverty eradication (Chibba, 2009). In this respect, using Sample 3, this study aims to determine the relationship between financial inclusion and output growth, as well as to examine whether inclusion-led output development reduces poverty.[1] Table 7.1.2 in the appendix provides a primary study of the link between the variables using Pearsons' correlation. The results show ΔlnGDP to negatively (although not significant) correlate with the measure of poverty.

Further, the growth in the measure of financial inclusion is also found to share a significant positive correlation with ΔlnGDP. Such characterization gives us reasonable evidence that financial inclusion may improve the output growth and may also reduce poverty through improving the output growth. Using the two-stage least squares method the study finds financial inclusion to improve the output growth and reduce poverty significantly across the 67 countries from 2004–18.

Critical modeling of the errors either using the parametric autoregression to estimate its ARMA structure in the test regression (i.e. ADF) as well as adjusting its form for any possible serial correlation and heteroscedasticity (i.e. PP) for the variables of interest, results in Table 8.1 suggests the data to be trend stationary [i.e. I (0)] in level, hence available to applications of the usual asymptotic theory.

Assuming poverty as a linear function of output growth and other set of macro-economic variables those are controlled for, we present a generalized form of the basic equation (i.e. Eq. 8.1).

$$\begin{aligned} \mathrm{POV}_{it} = {} & \beta_0 + \beta_1 \Delta\mathrm{lnGDP}_{it'} + \beta_2 \Delta\mathrm{UNEMP}_{it} \\ & + \beta_3 \mathrm{INF}_{it} + \beta_4 \Delta\mathrm{lnPOP}_{it} + \beta_5 \Delta\mathrm{DSE}_{it} + \mathrm{u}_{it'} \end{aligned} \tag{8.1}$$

Using the OLS estimator, we observe the output growth to increase poverty in Table 8.2 across the 67 countries for 2004–18. Economic inspection of such influence suggests that the uneven distribution of income may configure the output growth causing poverty to fall

[1] Table 7.1.1 in the Appendix contains summary statistics and an explanation of the variables utilised.

Table 8.1 Panel unit-root tests

At Level			
	ADF	*PP*	*Order of integration*
POV	521.568***	658.582***	I(0)
ΔlnGDP	270.716***	264.170***	I(0)
ΔUNEMP	639.723***	610.067***	I(0)
INF	224.284***	241.162***	I(0)
ΔlnPOP	317.601***	320.274***	I(0)
ΔGSE	707.287***	749.410***	I(0)
ΔIFI	736.290***	799.895***	I(0)

Notes (i)***, ** and * shows 1%, 5% and 10% level of significance
(ii) Schwarz Information Criterion is used as the Lag selection criteria; Include in test equation: no intercept
(iii) I (1) and I (0) indicate the order of integration [i.e. Non-stationary for I (1) and Stationary for I (0) respectively]
Source Authors Calculation

(Churchill & Marisetty, 2020). Hence, inequality appears as a theoretically important variable that has been omitted from our model. Such omission can make the model biased and inconsistent. We therefore address this problem through instrumenting the values of output growth (i.e. $\Delta\text{lnGDP'}_{it}$) in Eq. 8.1 with the estimated values of output growth (i.e. $\Delta\text{lnGDP}_{it} - hat$) obtained from Eq. 8.2.

$$\Delta\text{lnGDP}_{it} = \beta_0 + \beta_1\Delta\text{IFI}_{it} + \beta_2\Delta\text{UNEMP}_{it} + \beta_3\text{INF}_{it} + \beta_4\Delta\text{lnPOP}_{it} + \beta_5\Delta\text{DSE}_{it} + \text{u}_{it} \tag{8.2}$$

Results in panel B of Table 8.2 shows the output growth to significantly reduce the poverty level for the complete sample. Estimates of elasticity using the IV estimators allows us to validate the theoretical construct, assuming the effects of growth in unemployment, inflation, population, states health expenditure to GDP unchanged. Further, the panel shows higher *F*-statistics values for the IV estimator as compared to the OLS estimator, suggesting IV estimator a better fit. The relation and elasticities for the control variables show a convergence with the earlier findings besides share a valid economic reasoning.

A further stratification of the sample based on average output (i.e. median of per capita GDP) and average non-performing loans to gross loans shows the output growth increasing the incidence of poverty,

Table 8.2 Effect of economic growth on poverty

Method	*OLS*					*GMM (IV)*				
Dep. Var	POV_{it}									
Sample		*Income*		*Fragility*			*Income*		*Fragility*	
Class	*Consolidated*	*High*	*Low*	*High*	*Low*	*Full*	*High*	*Low*	*High*	*Low*
$\Delta \ln GDP_{it}$	0.439	0.412*	0.564	0.601	0.285	−4.307**	0.042	−5.084*	−1.223	−7.357**
	(0.304)	(0.232)	(0.598)	(0.512)	(0.319)	(2.006)	(0.975)	(2.899)	(3.597)	(2.838)
$\Delta UNEMP_{it}$	−0.009	0.001	−0.019**	0.003	−0.011	−0.047**	−0.005	−0.031	0.003	−0.103**
	(0.006)	(0.006)	(0.009)	(0.013)	(0.010)	(0.022)	(0.016)	(0.028)	(0.015)	(0.042)
INF_{it}	0.395***	0.231**	0.425***	0.179*	0.340***	0.393***	0.238***	0.403***	0.126	0.347***
	(0.096)	(0.087)	(0.114)	(0.098)	(0.088)	(0.060)	(0.059)	(0.086)	(0.125)	(0.094)
$\Delta \ln POP_{it}$	−2.828	−1.557	−4.566	0.023	−3.058	−4.548***	−1.715**	−6.345***	0.251	−7.801***
	(1.704)	(1.051)	(3.804)	(1.094)	(2.834)	(1.259)	(0.806)	(2.218)	(1.161)	(2.675)
ΔGSE_{it}	0.019	0.020	0.022	0.054*	−0.032	−0.043	0.013	−0.045	0.050	−0.148***
	(0.034)	(0.038)	(0.043)	(0.028)	(0.033)	(0.035)	(0.031)	(0.047)	(0.031)	(0.056)
C	18.644***	5.286***	34.482***	19.977***	17.124***	26.205***	5.804***	44.104***	21.532***	32.529***
	(2.635)	(0.966)	(8.398)	(2.409)	(4.398)	(3.667)	(1.542)	(6.795)	(3.697)	(6.578)
Panel Obs	696	340	356	223	473	692	340	352	222	470
Hausman χ-square	49.34***	7.35†	13.64**	32.73***	35.98***	NA	NA	8.11†	1.30†	NA
F-Statistics	4.43***	2.45*	3.50**	1.90	4.51***	11.20***	4.17***	7.47***	1.64	5.20***

Notes (i) ***, ** and * represents significance at 1%, 5% and 10% level
(ii) Standard errors are given in parenthesis ()
(iii) † represents selection of the Random effects model based on Hausman specification test
Source Calculated by Authors

though not significantly (except in case of countries belonging to output growth above the average) using the OLS estimator. However, these estimates become significantly negative for the countries with below average output and lower fragility applying the IV estimator. The measure of model fitness (i.e. the F-statistics) also got improved using the IV estimator for all the classifications (except the high-fragility group). Since, the F-statistic values for the high-fragility group is not significant, we avoid discussing the results due to inappropriate specification.

Among the theoretically relevant variables that we specify in the model, we observe an upward shift in the unemployment rates reduces poverty significantly across the low-fragility group and for the complete sample. Population growth significantly reduces poverty incidence across all the classifications. Except for the high-income group, it is observed that, as the ratio of health expenditure to GDP increases there is a steeper fall in the incidence of poverty across all the groups. Further, it is observed that a surge in inflation escalates the presence of poverty across all the classifications considered.

While estimating the output growth using the OLS estimator following Eq. 8.2, we find the FE model to appear statistically sound (i.e. significant values of F-statistic) across all the classifications except in the high-fragility group. Results in Table 8.3 shows that financial inclusion scores on an average are able to significantly increase the output growth across the classifications. Additionally, we observe a fall in the output growth due to a positive change in the unemployment rates, health expenditure to GDP of the states and population growth. Further, inflation increases per capita output growth across the selected sample though not significantly.

The objectives of this section are to determine whether financial inclusion increases output growth and whether it decreases poverty by increasing production growth. Table 8.3 demonstrates that progress in financial inclusion has been adequate to boost output growth across all classes when employing the OLS estimator. Moreover, an examination of the vectors linked to the model's predictor variables confirms the anticipated economic behavior. Growth in the population, unemployment, and health spending relative to GDP all considerably are downsizing output growth. In contrast, inflation, or the rise in consumer prices, somewhat accelerates the growth of output.

Previous research by Beck et al. (2000), Christopoulos and Tsionas (2004), Zhang et al. (2012), Beck et al. (2007), and many others has revealed similar evidence. A significant body of research, both theoretical

Table 8.3 Impact of financial inclusion on economic growth

Method	*OLS*				
Dep. Var	$\Delta lnGDP_{it}$				
Sample		*Income*		*Fragility*	
Class	*Consolidated*	*High*	*Low*	*High*	*Low*
ΔIFI_{it}	0.003***	0.017***	0.003***	0.002	0.004***
	(0.001)	(0.004)	(0.001)	(0.001)	(0.001)
$\Delta UNEMP_{it}$	−0.008***	−0.012***	−0.002	0.001	−0.012***
	(0.002)	(0.003)	(0.003)	(0.004)	(0.003)
INF_{it}	0.002	0.011	−0.002	−0.025	0.004
	(0.008)	(0.013)	(0.009)	(0.017)	(0.009)
$\Delta lnPOP_{it}$	−0.369**	−0.447*	−0.326	0.117	−0.623***
	(0.174)	(0.250)	(0.252)	(0.179)	(0.203)
ΔGSE_{it}	−0.013***	−0.021***	−0.012***	−0.004	−0.015***
	(0.003)	(0.007)	(0.004)	(0.009)	(0.004)
C	1.599***	1.387***	1.718***	0.866***	1.986***
	(0.279)	(0.256)	(0.553)	(0.303)	(0.319)
Panel Obs	698	346	352	225	469
Hausman χ-square	10.33*	10.36*	11.23**	7.27†	9.58*
F-Statistics	7.52***	8.57***	4.37***	1.22	8.17***

Notes (i) ***, ** and * represents significance at 1%, 5% and 10% level
(ii) Standard errors are given in parenthesis ()
(iii) † represents selection of the Random effects model based on Hausman specification test
Source Authors Calculation

and empirical, demonstrates that long-term, sustainable economic growth requires an effective and functional financial system. Nearly a century ago, Schumpeter (1911) made the crucial claim that financial institutions, by acting as a middleman between savers and borrowers, are essential to economic growth. Similar to this, contemporary financial theorists endorse the idea that the banking sector enhances economic growth by addressing potential knowledge asymmetries through arbitrage (Levine, 1997). Higher levels of financial development are strongly and robustly connected with quicker rates of economic growth in the present and the future, the accumulation of physical capital, and increases in economic efficiency, following the cross-country econometric analysis (King & Levine, 1993). Enabling the simultaneous equation framework to establish causal relationships among 63 nations between 1960 and 1995 following Levine et al. (2000), the increase of financial intermediaries and economic expansion are strongly positively correlated.

But the presence of structural heterogeneity is a real issue with the cross-sectional approach. While analyzing the financial development growth relationships, categorizing the entire sample reveals significant diversity among the categories (Andersen & Tarp, 2003; Ram, 1999). Similar evidence of a positive association between financial development and growth is further extended by country-specific situations, which, however, show that the functional form is unique (Demetriades & Hussein, 1996; Demetriades et al., 1998; Luintel & Khan, 1999). Because of the structural, institutional, and policy variations among the economies taken into consideration, a significant body of research supporting the finance-led growth hypothesis takes this cross-sectional heterogeneity into account (Jalilian & Kirkpatrick, 2005). The relationship between financial development and economic growth may be hampered by major flaws in the capital market caused by asymmetric knowledge, according to research that has been released (Stiglitz, 1993, 2000). To make the OLS estimator asymptotically efficient, we therefore loosen the requirements of independent and equal distributions for the regressors and errors. This relaxation makes it possible to adopt a broader viewpoint and assert greater theoretical rigor for the outcomes.

Poverty seems to be a crucial policy target that calls into question the effectiveness of development and economic growth. Therefore, answering the question of how to end poverty through financial inclusion may involve both direct and indirect self-examination (Akhter & Daly, 2009). Using the OLS estimator, the results in panel A of Table 8.2 examine how output growth affects the direct path of reducing poverty. The results display contradiction with the traditional hypothesis, which holds that poverty increases with economic growth, even after standard errors have been adjusted for potential cross-sectional variation. This disparity makes sense in light of the facts found in a large body of literature. The distribution of income is largely responsible for the economic reasoning behind output growth that is driving up the rate of poverty, as was found earlier in studies by Kuznets (1955), Paukert (1973), Summers et al. (1984), and Lindert and Williamson (1985), and later in more recent studies by Goudie and Ladd (1999), World Bank (2001), McKay (2002), Ravallion and Chen (2003), and Jalilian and Kirkpatrick (2005). In summary, the relationship between decreasing poverty and economic growth necessitates taking into account the distributional impacts that make growth generally more or less pro-poor over time (Jalilian & Kirkpatrick, 2005). Put differently, the rate of growth affects how income is distributed, and the starting disparities in income, assets, and opportunities that allow the poor to expand also affect how poverty is reduced (World Bank, 2001). Examining the indirect link between the effects of financial inclusion on production growth and poverty seems like a reasonable approach to address the issue of access to opportunities. Panel B of Table 8.2's results demonstrate how financial inclusion lowers poverty by enhancing production growth. While using a common set of explanatory variables in various specifications to assess poverty and production growth may result in mistakes of various kinds, statistically robust results are unlikely to be produced. Nonetheless, a system of equations (also referred to as the TSLS arrangement) can be established with the aid of a significant number of observations showing a weak association between the dependent variables of output growth and poverty (Jalilian & Kirkpatrick, 2005).

Financial institutions typically opt to place their money in locations that are supported by collateral and where they have a history of familiarity and goodwill with the organizations. This helps to improve the productivity of endowed resources and the accumulation of wealth, both of which support economic growth (Beck et al., 2000). Because of this unequal income distribution and the existence of fixed monitoring expenses, the growth status of those with access to the formal financial system differs dramatically from that of those without accessibility (Banerjee, 2003). Using a range of parameters, economists have modeled how flaws in the loan market affect income inequality as the economy develops. Even though we make reference to Greenwood and Jovanovic's (1990) stylized u-shaped relationship, we particularly highlight the later phase where the decreasing cost of utilizing financial services guarantees greater accessibility for all societal segments. This increased accessibility, or better expressed, financial inclusion, boosts credit availability, which boosts production growth while concurrently lowering the rate of poverty.

8.4 Conclusions

This chapter attempts to estimate whether financial inclusion, expressed as a composition of availability, accessibility, and usage dimensions, promotes output growth as well as reduces the poverty incidence by improving the output growth. To accomplish this, the section considers a sample of 67 countries for 2004–18 where data seems to be reasonably available. Additionally, the selected sample is further categorized into high-income and low-income countries based on their positional averages (i.e. median of per capita GDP) and high-fragility and low-fragility countries based on the median values of non-performing loans to gross loans ratio. Such segregations will help comparing the variation of estimates across the groups based on either the economic capacity or fragility.

Finally, based on the results from the empirical analysis, we highlight the three major contributions. Firstly, a positive change in the financial inclusion scores leads to an improvement in the output growth across all the classifications. This suggests that an improvement in the availability, accessibility, or usage dimensions will certainly increase the output growth at least for the sample considered irrespective of the economic capacity or fragility status of the country. Secondly, an improvement in the financial inclusion-led output growth reduces the poverty incidence across all the classifications. Such effects remain stronger across the low-income and the low-fragility groups as compared to the high-income and the high-fragility groups. This especially highlights the importance of financial inclusion for improving the output growth and reducing poverty incidence across the low-income countries. Finally, an improvement in the output growth increases the incidence of poverty though not significantly across all the classifications. This sort of finding, however, seems inconsistent with the underlying theory especially when the measure of inequality is not in place. By contrast, financial inclusion improves the relationship. The increment in the poverty incidence changes to a negative one when financial inclusion is allowed to intermediate, hence, creating a convergence with the underlying theory.

The scope of financial inclusion-led-poverty reduction mainly through improving the outreach of services offered—that is, improving the availability and accessibility dimensions—have recently emerged across the countries from the low-income group. This section finds output growth to increase the incidence of poverty across all the classifications. However, considering the intermediary role played by the measure of financial inclusion into account, the positive relationship between output growth and poverty incidence changes to a negative one, thus, converges with the theoretical proposition.

These results are important from the policy perspective, especially for the countries from the lower-income group. It suggests to extend the availability, accessibility, and usage of financial resources to the excluded part and therefore addressing the problems on poverty eradication. By improving the financial accessibility for the excluded sections, given the adequate regulatory and supervisory measures are in place, policy makers can increase the output growth leading to a fall in the poverty incidence.

References

Abosedra, S., Shahbaz, M., & Nawaz, K. (2016). Modelling causality between financial deepening and poverty reduction in Egypt. *Social Indicators Research, 126*(3), 955–969. https://doi.org/10.1007/s11205-015-0929-2

AFI. (2015). *2015 Maya declaration report: Commitments into action. Alliance for Financial Inclusion*. Retrieved on 5 September 2021 from http://www.afi-global.org/sites/default/files/publications/maya_report_2015-final.pdf

Aideyan, O. (2009). Microfinance and poverty reduction in Rural Nigeria. *Savings and Development, 33*(3), 293–317. https://www.jstor.org/stable/41406498

Akhter, S., & Daly, K. J. (2009). Finance and poverty: Evidence from fixed effect vector decomposition. *Emerging Markets Review, 10*(3), 191–206. https://doi.org/10.1016/j.ememar.2009.02.005

Andersen, T. B., & Tarp, F. (2003). Financial liberalization, financial development and economic growth in LDCs. *Journal of International Development: THe Journal of the Development Studies Association, 15*(2), 189–209. https://doi.org/10.1002/jid.971

Arestis, P., & Caner, A. (2010). Capital account liberalisation and poverty: How close is the link? *Cambridge Journal of Economics, 34*(2), 295–323. https://doi.org/10.1093/cje/bep062

Banerjee, A. V. (2003). Contracting constraints, credit markets, and economic development. *Econometric Society Monographs, 37*, 1–46.

Banerjee, A. V., & Newman, A. F. (1993). Occupational choice and the process of development. *Journal of Political Economy, 101*(2), 274–298. https://www.jstor.org/stable/2138820

Beck, T., & Demirgüç-Kunt, A. (2008). Access to finance: An unfinished agenda. *World Bank Economic Review, 22*(3), 383–396. https://doi.org/10.1093/wber/lhn021

Beck, T., Demirguc-Kunt, A., & Levine, R. (2007). Finance, inequality, and the poor. *Journal of Economic Growth, 12*(1), 27–49. https://doi.org/10.1007/s10887-007-9010-6

Beck, T., Levine, R., & Loayza, N. (2000). Finance and the sources of growth. *Journal of Financial Economics, 58*(1–2), 261–300. https://doi.org/10.1016/s0304-405x(00)00072-6

Bell, C., & Rousseau, P. L. (2001). Post-independence India: A case of finance-led industrialization? *Journal of Development Economics, 65*(1), 153–175. https://doi.org/10.1016/s0304-3878(01)00132-8

Boukhatem, J. (2016). Assessing the direct effect of financial development on poverty reduction in a panel of low-and middle-income countries. *Research in International Business and Finance, 37*, 214–230. https://doi.org/10.1016/j.ribaf.2015.11.008

Brown, J. K., Zelenska, T. V., & Mobarak, M. A. (2013). *Barriers to adoption of products and technologies that aid risk management in developing countries.* Retrieved on 11 August 2022 from http://hdl.handle.net/10986/16365

Buckland, J., Guenther, B., Boichev, G., Geddie, H., & Mutch, M. (2005). There are no banks here: Financial & insurance exclusion in Winnipeg's North End. *Winnipeg Inter-City Research Alliance (WIRA).*

Burgess, R., & Pande, R. (2005). Do rural banks matter? Evidence from the Indian Social Banking Experiment. *American Economic Review, 95*(3), 780–795. https://www.jstor.org/stable/4132740

Chibba, M. (2009). Financial inclusion, poverty reduction and the millennium development goals. *The European Journal of Development Research, 21*(2), 213–230. https://doi.org/10.1057/ejdr.2008.17

Christopoulos, D. K., & Tsionas, E. G. (2004). Financial development and economic growth: Evidence from panel unit root and cointegration tests. *Journal of Development Economics, 73*(1), 55–74. https://doi.org/10.1016/j.jdeveco.2003.03.002

Churchill, S. A., & Marisetty, V. B. (2020). Financial inclusion and poverty: A tale of forty-five thousand households. *Applied Economics, 52*(16), 1777–1788. https://doi.org/10.1080/00036846.2019.1678732

Clarke, G., Xu, L., & Zou, H. (2006). Finance and income inequality: What do the data tell us? *Southern Economic Journal, 72*(3), 578–596.

Coulibaly, B. S. (2019). *Foresight Africa: Top priorities for the continent in 2019.* Retrieved on 12 September 2023 from Africa Growth initiative, Brookings Institute website. https://www.brookings.edu/wp-content/uploads/2019/01/BLS18234_BRO_book_007_WEB.pdf

Datt, G., & Ravallion, M. (1992). Growth and redistribution components of changes in poverty measures: A decomposition with applications to Brazil and India in the 1980s. *Journal of Development Economics, 38*(2), 275–295.

Dawood, T. C., Pratama, H., Masbar, R., & Effendi, R. (2019). Does financial inclusion alleviate household poverty? Empirical evidence from Indonesia. *Economics & Sociology, 12*(2), 235–252. https://doi.org/10.14254/2071-789X.2019/12-2/14

Demetriades, P. O., & Hussein, K. A. (1996). Does financial development cause economic growth? Time-series evidence from 16 countries. *Journal of Development Economics, 51*(2), 387–411. https://doi.org/10.1016/s0304-3878(96)00421-X

Demetriades, P. O., Devereux, M. P., & Luintel, K. B. (1998). Productivity and financial sector policies: Evidence from South East Asia. *Journal of Economic Behavior & Organization, 35*(1), 61–82. https://doi.org/10.1016/s0167-2681(97)00099-1

Demirgüç-Kunt, A., Klapper, L., Singer, D., Ansar, S., & Hess, J. (2018). *The global findex database 2017: Measuring financial inclusion and the fintech revolution.* Retrieved on 18 July 2022 from http://documents.worldbank.org/curated/en/33288 15258 73182 837/pdf/12603 3-PUB-PUBLIC-pubdate-4-19-2018.pdf

Dimova, R., & Adebowale, O. (2018). Does access to formal finance matter for welfare and inequality? Micro level evidence from Nigeria. *The Journal of Development Studies, 54*(9), 1534–1550. https://doi.org/10.1080/00220388.2017.1371293

Donovan, K. (2012). *Mobile money for financial inclusion.* Retrieved on 1 December 2021 from http://elibraryworldbank.org/doi/abs/10.1596/9780821389911_ch04

Dugarova, E., & Gülasan, N. (2017). *Global trends: Challenges and opportunities in the implementation of the sustainable development goals.* Academic Press

Emerick, K., de Janvry, A., Sadoulet, E., & Dar, M. H. (2016). Technological innovations, downside risk, and the modernization of agriculture. *American Economic Review, 106*(6), 1537–1561. https://www.jstor.org/stable/43861131

Galor, O., & Zeira, J. (1993). Income distribution and macroeconomics. *Review of Economic Studies, 60*(1), 35–52. https://www.jstor.org/stable/2297811

Goudie, A., & Ladd, P. (1999). Economic growth, poverty and inequality. *Journal of International Development, 11*(2), 177.

Greenwood, J., & Jovanovic, B. (1990). Financial development, growth, and the distribution of income. *Journal of Political Economy, 98*(5, Part 1), 1076–1107. https://www.jstor.org/stable/2937625

Hallegatte, S., Vogt-Schilb, A., Bangalore, M., & Rozenberg, J. (2016). *Unbreakable: Building the resilience of the poor in the face of natural disasters.* World Bank Publications.

Honohan, P. (2004). Financial development, growth and poverty: How close are the links? *World Bank Policy Research Working Paper 3203.* http://documents.worldbank.org/curated/en/518641468763230741/Financial-development-growth-and-poverty-how-close-are-the-links

ICSU, & ISSC. (2015). Review of the sustainable development goals: The science perspective. *Paris: International Council for Science (ICSU).* Retrieved on 5 September 2021 from http://www.genderportal.eu/resources/review-sustainable-development-goals-science-perspective

Imai, K. S., Arun, T., & Annim, S. K. (2010). Microfinance and household poverty reduction: New evidence from India. *World Development, 38*(12), 1760–1774. https://doi.org/10.1016/j.worlddev.2010.04.006

Inoue, T., & Hamori, S. (2012). How has financial deepening affected poverty reduction in India? Empirical analysis using state-level panel data. *Applied Financial Economics, 22*(5), 395–408. https://doi.org/10.1080/09603107.2011.613764

Jalilian, H., & Kirkpatrick, C. (2002). Financial development and poverty reduction in developing countries. *International Journal of Finance and Economics, 7*(2), 97–108. https://doi.org/10.1002/ijfe.179

Jalilian, H., & Kirkpatrick, C. (2005). Does financial development contribute to poverty reduction? *The Journal of Development Studies, 41*(4), 636–656. https://doi.org/10.1080/00220380500092754

Jeanneney, S., & Kpodar, K. (2011). Financial development and poverty reduction: Can there be a benefit without a cost? *Journal of Development Studies, 47*(1), 143–163. https://doi.org/10.1080/00220388.2010.506918

Kakwani, N. (1993). Poverty and economic growth with application to Cote d'Ivoire. *Review of Income and Wealth, 39*, 121–139. https://doi.org/10.1111/j.1475-4991.1993.tb00443.x

Kanbur, R. (2001). Economic policy, distribution and poverty: The nature of disagreements. *World Development, 29*(6), 1083–1094.

Keynes, J. M. (1937). The "ex-ante" theory of the rate of interest. *The Economic Journal, 47*(188), 663–669. https://www.jstor.org/stable/2225323

Kiendrebeogo, Y., & Minea, A. (2016). Financial development and poverty: Evidence from the CFA Franc Zone. *Applied Economics, 48*(56), 5421–5436. https://doi.org/10.1080/00036846.2016.1176114

King, R. G., & Levine, R. (1993). Finance and growth: Schumpeter might be right. *The Quarterly Journal of Economics*, 717–737. https://www.jstor.org/stable/2118406

Koomson, I., Villano, R. A., & Hadley, D. (2020). Effect of financial inclusion on poverty and vulnerability to poverty: Evidence using a multidimensional measure of financial inclusion. *Social Indicators Research, 149*(2), 613–639. https://doi.org/10.1007/s11205-019-02263-0

Kravis, I. B. (1960). International differences in the distribution of income. *The Review of Economics and Statistics*, 408–416.https://doi.org/10.2307/1925690

Kuznets, S. (1955). Economic growth and income inequality. *The American Economic Review, 45*(1), 1–28.

Levine, R. (1997). Financial development and economic growth: Views and agenda. *Journal of Economic Literature, 35*(2), 688–726. http://www.jstor.org/stable/2729790

Levine, R., Loayza, N., & Beck, T. (2000). Financial intermediation and growth: Causality and causes. *Journal of Monetary Economics, 46*(1), 31–77. https://doi.org/10.1016/S0304-3932(00)00017-9

Li, L. (2018). Financial inclusion and poverty: The role of relative income. *China Economic Review, 52*, 165–191. https://doi.org/10.1016/j.chieco.2018.07.006

Lindert, P. H., & Williamson, J. G. (1985). Growth, equality, and history. *Explorations in Economic History, 22*(4), 341–377. https://doi.org/10.1016/0014-4983(85)90001-4

Loayza, N. V., & Ranciere, R. (2006). Financial development, financial fragility, and growth. *Journal of Money, Credit and Banking*, 1051–1076. https://www.jstor.org/stable/3838993

Luintel, K. B., & Khan, M. (1999). A quantitative reassessment of the finance–growth nexus: Evidence from a multivariate VAR. *Journal of Development Economics, 60*(2), 381–405. https://doi.org/10.1016/S0304-3878(99)00045-0

Lydall, H. F. (1968). Technical progress in Australian manufacturing. *The Economic Journal, 78*(312), 807–826. https://doi.org/10.2307/2229179

Lyons, A. C., & Hunt, J. (2003). *The credit practices and financial education needs of community college students*. Retrieved on 10 July 2021 from https://ssrn.com/abstract=2265596

Mader, P. (2018). Contesting financial inclusion. *Development and Change, 49*(2), 461–483. https://doi.org/10.1111/dech.12368

Majid, M. S. A., Dewi, S., Aliasuddin, & Kassim, S. H. (2017). Does financial development reduce poverty? Empirical evidence from Indonesia. *Journal of the Knowledge Economy*. https://doi.org/10.1007/s13132-017-0509-6

Mârza, B., Angelescu, C., & Tindeche, C. (2015). Agricultural insurances and food security. The new climate change challenges. *Procedia Economics and Finance, 27*, 594–599. https://doi.org/10.1016/S2212-5671(15)01038-2

McKay, A. (2002). Economic growth and distribution of poverty in developing countries. In *Handbook on development policy and management*. Edward Elgar.

McKinnon, R. I. (1973), *Money and capital in economic development*. Brookings Institution. https://www.brookings.edu/book/money-and-capital-in-economic-development/

Nations, U. (2016). *The sustainable development goals 2016*. ESocialSciences.

Neaime, S., & Gaysset, I. (2018). Financial inclusion and stability in MENA: Evidence from poverty and inequality. *Finance Research Letters, 24*, 230–237. https://doi.org/10.1016/j.frl.2017.09.007

Omar, M. A., & Inaba, K. (2020). Does financial inclusion reduce poverty and income inequality in developing countries? A panel data analysis. *Journal of Economic Structures, 9*(1), 1–25. https://doi.org/10.1186/s40008-020-00214-4

Ozughalu, U. M. (2016). Relationship between household food poverty and vulnerability to food poverty: Evidence from Nigeria. *Social Indicators Research, 125*, 567–587. https://doi.org/10.1007/s11205-014-0845-x

Park, C. Y., & Mercado, R. (2015), Financial inclusion, poverty, and income inequality in developing Asia, Asian Development Bank Economics Working Paper Series (426).

Paukert, F. (1973). Income distribution at different levels of development: A survey of evidence. *International Labor Review, 108*, 97–125.

Perez-Moreno, S. (2011). Financial development and poverty in developing countries: A causal analysis. *Empirical Economics, 41*, 57–80. https://doi.org/10.1007/s00181-010-0392-5

Perotti, R. (1993). Political equilibrium, income distribution, and growth. *The Review of Economic Studies, 60*(4), 755–776. https://www.jstor.org/stable/2298098

Rajan, R. G., & Zingales, L. (1998). Financial dependence and growth. *American Economic Review, 88*(3), 559–586. https://www.jstor.org/stable/116849

Ram, R. (1999). Financial development and economic growth: Additional evidence. *Journal of Development Studies, 35*(4), 164–174. https://doi.org/10.1080/00220389908422585

Ravallion, M., & Chen, S. (2003). Measuring pro-poor growth. *Economics Letters, 78*(1), 93–99. https://doi.org/10.1016/s0165-1765(02)00205-7

Sahoo, A. K., Pradhan, B. B., & Sahu, N. C. (2017). Determinants of financial inclusion in tribal districts of Odisha: An empirical investigation. *Sage, 47*(1), 45–64.

Schumpeter, J. A. (1911). *The theory of economic development*. Harvard University Press.

Seck, O., Naiya, I. I., & Muhammad, A. D. (2017). *Effect of financial inclusion on household consumption in Nigeria*. IRTI Working Paper Series WP/2017/03

Shaw, E. S. (1973). *Financial deepening in economic development*. Oxford University Press.

Stiglitz, J. E. (1993). The role of the state in financial markets. *The World Bank Economic Review, 7*(suppl_1), 19–52.

Stiglitz, J. E. (2000). The contributions of the economics of information to twentieth century economics. *The Quarterly Journal of Economics, 115*(4), 1441–1478.

Summers, R., Kravis, I. B., & Heston, A. (1984). Changes in the world income distribution. *Journal of Policy Modeling, 6*(2), 237–269. https://doi.org/10.1016/0161-8938(84)90014-0

Swamy, V. (2014). Financial inclusion, gender dimension, and economic impact on poor households. *World Development, 56*, 1–15. https://doi.org/10.1016/j.worlddev.2013.10.019

Todaro, M., & Smith, S. (2012). Economic Development. *Pearson Addision Wesley*. https://doi.org/10.1177/089124240001400105

Torsten, P., & Tabellini, G. (1991). *'Is inequality harmful for growth? Theory and evidence* (No. 91–155). Working Paper.

Uddin, G. S., Shahbaz, M., Arouri, M., & Teulon, F. (2014). Financial development and poverty reduction nexus: A cointegration and causality analysis in Bangladesh. *Economic Modelling, 36*, 405–412. https://doi.org/10.1016/j.econmod.2013.09.049

UN DESA. (2016). The sustainable development goals report 2016. UN DESA. https://unstats.un.org/sdgs/report/2016/

World Bank. (1995). *World development report: Workers in an integrating world*. World Bank.

World Bank. (2001). *World development report 2000–2001: Attacking poverty*. Retrieved on 15 October 2021 from World Bank Group. New York: Oxford University Press website. https://openknowledge.worldbank.org/handle/10986/11856

World Bank. (2007). *Finance for all? Policies and pitfalls in expanding access*. The World Bank.

Zhang, J., Wang, L., & Wang, S. (2012). Financial development and economic growth: Recent evidence from China. *Journal of Comparative Economics, 40*(3), 393–412. https://doi.org/10.1016/j.jce.2012.01.001

CHAPTER 9

Financial Inclusion and Zero Hunger (SDG 2)

9.1 Introduction

These days, one of the biggest hazards to humanity is undernourishment. Both human capital formation and economic development are dependent on adequate nutrition (Bryce et al., 2008; Sachs, 2002; Sen, 1995). It is now certain that all kinds of malnutrition will be eradicated, at least by 2030, in accordance with commitments 2.1 and 2.2. The commitment 2.1 guarantees that everyone has year-round access to safe, nourishing, and sufficient food. According to the FAO's 2017 and 2018 reports, conflict, climate variability, and other extremes deter attempts to end hunger. Among the major obstacles to ending hunger are the economic downturns detailed in the 2019 edition, the effects of COVID-19, and the extraordinary outbreaks of desert locust in Eastern Africa in 2020. However, there are a lot of threats to reaching SDG 2, or "Zero Hunger."

Even after accounting for the effects of COVID-19, estimates indicate that by 2030, there would be more than 840 million undernourished people worldwide (FAO, 2020). Approximately, 83–132 million more could make this worse if the impact is taken into account.

In addition to hunger, people are increasingly sacrificing their consumption of both quantity and quality food. Among the common situations that result in undernourishment, or more seriously, malnutrition, are missing meals and total hunger.

S. Ghosh et al., *Financial Inclusion*, Sustainable Development Goals Series, https://doi.org/10.1007/978-981-95-3385-5_9

There could be a number of causes, including increased food costs, decreased food supply, reduced income, and many more. But according to Warr (2014), one of the most evident causes of food insecurity (FI) is the availability and cost of food, which paves the path to undernourishment by lowering the quality of diet. Income is yet another significant element that affects food security. Further, the prevalence of undernourishment (PoU) and malnutrition can be safely prevented through taking guarantee on food security through ensuring a stable and sufficient income that helps afford a nutritious diet. It is not justified to require income to be sufficient. But by offering them the chance to save, financial services promotion can help to some degree guarantee income security.

In this context, we focus on the availability of financial services, particularly the provisioning of the basic banking services like the demography-based availability of branches and ATMs, which have the goal of achieving fundamental income stabilization (Ghosh & Sahu, 2020). Understanding the importance of the drivers and better achieving the goals will be made possible by reflecting on the elements that affect the PoU and estimating their impact.

9.2 Financial Inclusion and the Prevalence of Undernourishment (PoU)

Proper nutrition is a key factor for assessing the effectiveness of the strategies emphasizing on development. Development programs usually emphasizes on the relevance of important economic factors, notably income and pricing, in addressing undernutrition. Surprisingly, adequate nutrition is not found to improve certain indicators of health instead increasing the productivity (Behrman et al., 1988). Recent techniques have relied on econometric estimations that exhibit the connections between energy intakes and income, prices, and other factors based on household level data. However, it is the appropriateness of these intakes relative to each person's requirements that is important, not the intakes themselves. Except certain extreme situations, human metabolism is capable to adjust with whatever intake of nutrient they consume (Sahn, 1987). Additionally extreme intrapersonal volatility across the time is perceived so far intakes are concerned (Srinivasan, 1981; Sukhatme, 1982) and thus this measure might become a very poor proxy for health indicators (Behrman et al., 1988).

Fortunately, the FAO's PoU metric is accessible and works well in this scenario. The importance of wealth and price in the genesis of famines has also been highlighted by empirical evidences exploring the factors that influence individual levels of nutrition (Ravallion, 1987; Sen, 1982). This raises the question of whether the PoU should be considered temporary or an aggregate estimate responsive to prices and income. Furthermore, variables such as the incidence of corruption, stability, and effectiveness of the government and the status of violence have been essential to minimize the PoU (Lee et al., 2016; Moore et al., 2006). This study looks into the impact of food prices, regardless of their effects infused due to income (i.e., determining the price sensitivity of undernutrition prevalence). Additionally, the most often asked topic from previous research is revisited, namely, what impacts (if any) economic development and contraction may be predicted to have on aggregate undernutrition (i.e. assessing the elasticity of income in terms of the the incidence of undernourishment).

9.2.1 Prices for Food and the PoU

Food costs fell dramatically over the world in the late twentieth century. However, after reaching the lowest levels in the early 2000s, the trend has shifted (FAO, 2011). Food prices have risen steadily since 2006, reaching a maximum in 2008 and sparking a global food crisis that has never been overcome (Lee et al., 2016). A hike in food prices exacerbates concerns about food security by raising the likelihood of famine and reducing family intake of healthful foods. The recession worldwide in 2007, which combined food, fuel, and financial crises, shows a drop in the intake of food score (determining the frequency and variety of food intakes) due to food price increases in Haiti, Nepal, and Niger (Christian, 2010[1]). This made families to face vulnerability as a result of a logical compromise in terms of both food quality and quantity, leading to more malnutrition. According to Darnton-Hill and Cogill (2010), these vulnerabilities have an influence on mother and child health by reducing micronutrients and increasing infectious illnesses, which may lead to death in severe circumstances. Similarly, rising food costs have had a significant and detrimental impact on nutrition, resulting in greater

[1] Further read Darnton-Hill and Cogill (2010) and Brinkman et al. (2010).

rates of newborn and infant mortality in both developing nations as well as least developed nations (LDCs) between 2001 and 2011 (Lee et al., 2016). These findings compliment the notion that nations having lower income have a greater percentage of household spending on food, particularly in LDCs, and therefore rely substantially on the import of food from outside. A review of the previous literature especially across the poorer countries shows an incidence of increasing hunger, malnutrition, and death fuelled by the rising food prices among newborns and children (ADB, 2012; FAO, 2008). In this connection, the program named 'Alliances to Improve the Situation of Children, Food Security and Nutrition' have improved nutritional outcome especially among the children through providing access to adequate financial resources and education (UNSDF, n.d.). The relatively wealthier part of the society (measured in monetary terms) is further reported to experience massive cutbacks in private expenditure and adopting dramatic measures such as increased usage of food banks (Butler, 2012; Selke, 2011), attempting suicide due to poverty, and dying from extended famine (Gentleman, 2014). When such techniques are implemented, micronutrient and calorie shortages occur, exacerbating the incidence of malnutrition (Ortiz et al., 2011). Furthermore, such hikes in the food prices are evident to substantially reverse the achievements so far as the Sustainable Development Goals (SDGs) are concerned. The rise in food costs is expected to increase both child and newborn death rates (Goal 3), which were reduced by a quarter between 1990 and the year 2015 (Brinkman et al., 2010). A spike in global food commodity prices (as assessed by the FAO Food Price Index) occurs with a one-period lag (Lee & Park, 2013). Furthermore, increased food prices do not result in increased advantages for the marginal farmers (Wise & Harvie, 2009). To summarize, increasing food commodity prices increase food insecurity and, as a result, the incidence of undernourishment, with negative consequences for other development indicators.

9.2.2 Per Capita Income and the PoU

Income becomes yet another crucial aspect in determining food security. The availability of a consistent and enough income may help purchase a healthy meal, so reducing the prevalence of undernourishment (PoU) and consequently malnutrition. The sufficiency of income requirement cannot be justified. However, expanding financial services may help to promote economic stability by providing a chance to save. Recent research indicates that the income elasticity of individual energy consumption in emerging nations is lower than previously believed (Behrman et al., 1988; Bouis & Haddad, 1988). Policies aimed to raise the income especially for the marginalized either through accelerating the growth or addressing inequality. Raising the poor's income (via policies aimed at boosting growth and decreasing income disparity) has helped to overcome hunger, which is said to be one of the most important drivers of development (Ravallion, 1990). The income sensitivity of calories may range from 0 to infinite. Empirical data reveals a size ranging from 0.08 to 0.14 in Philippines (Bouis & Haddad, 1992). Further, this value arrives at 0.55 across the poorest families of Maharastra (India) (Subramanian & Deaton, 1996). Research findings remain statistically significant at 0.37 for India in Alderman (1993) and Strauss and Thomas (1995) (Behrman & Deolalikar, 1987). However, the substitution effect between nutritional values and other dietary qualities within food-staple aggregates might overwhelm the sensitivity of PoU to income (Behrman & Deolalikar, 1987). When the non-linear nature of the Engel Curve for calories is taken into account, the impoverished exhibit more sensitivity than the rich (Ravallion, 1990; Reutlinger & Selowsky, 1976; Strauss & Thomas, 1990). Especially across the families with lower income, it was observed that demand with respect to price of food was more sensitive (Green et al., 2013). Furthermore, the sensitivity of food quality to income provides inadequate explanation for such results (Alderman, 1993). Bouis and Haddad (1988) present an alternative interpretation fearing erroneous variables.

Economic downturns often exacerbate infant deaths as assessed by per capita GDP (Baird et al., 2011). Falling incomes as a result of the economic downturn are communicated to households through lower food and health costs (FAO, 2012), whereas improved earnings is likely to provide a variety of improved prospects, including better housing, adequate nutrition with a supply of hygienic water and sanitation, and, most importantly, access to health-care services. However, panel studies considering a number of economies together demonstrate that economic progress may not be resulting into improved health conditions unless purposeful governmental interventions are taken (Cutler et al., 2006; Dreze & Sen, 2002). Even with a long-term view, the findings remain equivocal (Lee et al., 2016).

9.2.3 Food Production and the PoU

Despite being a critical economic sector in emerging nations, employing a significant percentage of the labor population, agricultural productivity froze between 1960 and 1980 and has been steadily increasing since then. Review of Warr (2014) shows that increase in agricultural productivity, while assuming the prices to stay constant, improves the incidence of undernourishment. Policies that aim for agricultural protection and decrease substantially the reliance on the imports often inadvertently raises food prices through achieving food self-sufficiency, which ultimately aggravates malnutrition (Tiwari & Zaman, 2010; Warr, 2014). The pace of growth in this sector has been suspected slow most likely due to the unsupportive policies and limitation of the resources (Alston & Pardey, 2015; Mattoo et al., 2020). The proportion of Official Development Assistance (ODA) dedicated to agriculture slips down from 15 to 20% in 1970–1980 (Cabral & Howell, 2012) to less than 3% in 2007 (Webb, 2010) and has never surpassed 4% throughout 2012–2020. This negligent behavior, along with increased demand for food, has rendered the nationwide agricultural research programs unviable. Operations such as seed multiplication and provisioning of loans across the rural belts are on the brink of elimination (Webb, 2010). Furthermore, with the advent of COVID-19, especially across the poor and underdeveloped nations, a weak Food Supply Chain (FSC) leads to the danger of food insecurity (Béné, 2020). Food shortages and disruptions in the supply chain have a direct influence on the health and well-being of a third of the world's population (Khan et al., 2022). High fluctuation in transaction prices leads to more interruptions in FSC, limiting food supply (Hobbs, 2020; Reardon et al., 2020). However, the impact on under nutrition seems to be transient. Several challenges, including harvest losses, environmental disasters, insufficient food storage, and outdated FSC technologies, must be addressed to ensure food availability for billions of undernourished people (Baležentis et al., 2021; Tian et al., 2016; Webb, 2010).

9.3 Empirical Connections Between Financial Inclusion and Zero Hunger (SDG 2)

Sample 4 of this study aims to evaluate the influence of the changes in prices and production of food, individual's income, and the outreach of banking services on the prevalence of undernourishment (PoU) among

61 countries from 2004 to 2019. Furthermore, the study assesses the impacts of corruption, the performance of the government, and political equilibrium on PoU, with the absence of violence serving as a supplementary but substantial influence. The structure of the available data consists of 61 countries majorly drawn from the least developed and developing category covering a 16- year period and therefore the panel data econometrics appear as one of the prominent options leading to exploration of the existent relationship.

The data shown in Table 9.1 indicates that the average amount of people at risk of starvation is roughly 2.26 million, having a variability of 4.49 million across the nations as well as periods studied. The reliability of the mean seems to be very unlikely due to the fact that the variance is almost twice the mean value. The food index of prices currently stands around 81.7, compared to 100 in its founding year of 2015, and has a 1.45-point spread across the mean. The average value of the food production index (FPI) equals 89.93, which has a 1.2-point variance, indicating a reduction in the overall production of food all through the entire sample. Across the sample, the average per capita income annually stood at USD 2028.5, which further shows a fluctuation of USD 2.77 across the panel. Accessibility to banking services (mainly bank branches and the ATMs) on an average available to nearly 0.1 million customers stood at 9.55 (~10) and 18.90 (~20) respectively with a variation of 8.17 and 20.86 respectively. The estimated average of the corruption, government's efficiency and political stability with absence of violence remained negative as well as close to (−1). However, moment conditions for these variables have not been discussed as they fall within the ±2.5 range of a conventional normal curve.

In Panel A of Table 9.2, the correlation between the outcome variable and its explanatory factors are observed to be significant and negative. This gives a primitive idea that the variables are drifting apart in opposing directions. Nevertheless, differencing makes the linear association disappear for the relevant variables in Panel B of the same table (excluding the exogeneous section). Further the strength of relation between the explanatory variables and unobserved errors under both the parametric (t and F statistics) and non-parametric (Chi-square statistics) test alternatives are found to be weakly exogeneous for the variables supposed to have an auxiliary impact on PoU (see Table 9.2.1 in the Appendix).

Table 9.3 illustrates that all the factors considered at level contains unit-root [i.e. I (1)]. Thus, application of the OLS estimator will lead into

Table 9.1 Descriptive statistics

Variables	*Variable description*	*Average*	*Stdev*
lnPoU (Millions)	Natural Logarithm of the number of undernourished people in millions	0.816	1.502
lnFPI_C	Natural Logarithm of Consumer Food Price Index (Base Year 2015 = 100) [Using the Ceiling]	4.403	0.372
lnFPI	Natural Logarithm of the Food production index [It considers edible food containing nutrients exclusive of Coffee and Tea]	4.499	0.185
lnGNIpc	Natural Logarithm of per capita Gross National Income [Measured in US$]	7.615	1.020
Bran_D	Commercial Bank branches per 100,000 adults	9.548	8.170
ATM_D	Automated Teller Machines per 100,000 adults	18.902	20.855
CoC_estimate	An estimate of Corruption measured on a standard normal curve with a range of $\pm 2.5\ \sigma$	−0.596	0.498
GE_estimate	An estimate of Government Effectiveness measured on a standard normal curve with a range of $\pm 2.5\ \sigma$	−0.465	0.573
PSAV_estimate	An estimate of the Political Stability and Absence of Violence/Terrorism measured on a standard normal curve with a range of $\pm 2.5\ \sigma$	−0.628	0.784

Source Authors calculation

erroneous results - a phenomenon known as spurious regression. Furthermore, following the first differencing, all of the series attain stationarity, indicating that the factors under analysis may be co-integrated.

This book uses both the residual-based panel co-integration tests (i.e. Pedroni and Kao's test) to identify the linear arrangement [i.e. I (0)] across the variables that are not stationary in Table 9.4. With non-zero weights and involving separate specifications both the tests records an affirmative response to accept the presence of a linear trend while estimating the deterministic trend that use both individual and group estimations offered by Pedroni (unweighted and weighted) and substantiated further by the lagged leftovers with Kao's specification.

Table 9.5 exhibits the convergence in the estimates of long-run influence under parametric alternative (i.e. the DOLS) and non-parametric alternative (i.e. the FMOLS) with regard to the movement between food prices and the PoU. Furthermore, the production of food, per capita income, and the presence of bank branches and ATMs are shown to have a negative influence on the PoU. The results obtained under different

Table 9.2 Tests of linear associations

Panel A: At Level

	lnPoU (Millions)	*lnFPI_C*	*lnFPI*	*LnGNIpc*	*Bran_D*	*ATM_D*	*CoC_ estimate*	*GE_ estimate*	*PSAV_ estimate*
lnPoU (Millions)	1								
lnFPI_C	−0.143***	1							
lnFPI	−0.054	0.713***	1						
lnGNIpc	−0.398***	0.318***	0.415***	1					
Bran_D	−0.426***	0.168***	0.248***	0.712***	1				
ATM_D	−0.418***	0.359***	0.397***	0.834***	0.602***	1			
CoC_ estimate	−0.301***	0.067*	0.147***	0.402***	0.428***	0.519***	1		
GE_ estimate	−0.320***	0.079**	0.146***	0.567***	0.452***	0.718***	0.819***	1	
PSAV_ estimate	−0.611***	0.124***	0.012	0.290***	0.312***	0.312***	0.519***	0.502***	1

Panel B: At 1st differences

	ΔlnPoU (Millions)	*ΔlnFPI_C*	*ΔlnFPI*	*ΔlnGNIpc*	*ΔBran_D*	*ΔATM_D*	*ΔCoC_ estimate*	*ΔGE_ estimate*	*ΔPSAV_ estimate*
ΔlnPoU (Millions)	1								
ΔlnFPI_C	−0.002	1							
ΔlnFPI	−0.10***	0.052	1						
ΔlnGNIpc	−0.138***	0.168***	0.215***	1					

(continued)

Table 9.2 (continued)

Panel B: At 1st differences

	Δ*lnPoU* (*Millions*)	Δ*lnFPI_C*	Δ*lnFPI*	Δ*lnGNIpc*	Δ*Bran_D*	Δ*ATM_D*	Δ*CoC_ estimate*	Δ*GE_ estimate*	Δ*PSAV_ estimate*
ΔBran_D	−0.206***	0.113***	0.078*	0.067***	1				
ΔATM_D	−0.086**	0.029	−0.005	-0.018	0.159***	1			
ΔCoC_ estimate	0.002	−0.046	0.029	0.068	0.059	−0.079	1		
ΔGE_ estimate	−0.059	−0.010	0.043	0.128***	0.028	0.003	0.312***	1	
ΔPSAV_ estimate	−0.056	−0.089	0.047	0.127***	0.039	0.012	0.148***	0.212***	1

Note ***, ** and * shows 1%, 5% and 10% level of significance
Source Authors calculation

Table 9.3 Panel data unit-root tests

	At Level				*At 1st difference*			
	LLC	*ADF*	*PP*	*Order of integration*	*LLC*	*ADF*	*PP*	*Order of integration*
lnPoU (Millions)	2.198	110.132	35.992	I (1) (T)	−7.983***	226.479***	239.978***	I (0)
lnFPI_C	22.178	6.945	0.785	I (1)	−9.897***	279.786***	359.486***	I (0)
lnFPI	19.156	8.131	7.021	I (1)	−17.824***	579.656***	719.013***	I (0)
lnGNIpc	18.021	11.256	3.982	I (1)	−17.625***	471.818***	458.185***	I (0)
Bran_D	6.386	49.495	39.581	I (1)	−12.986***	421.456***	456.789***	I (0)
ATM_D	16.446	35.875	7.982	I (1)	−6.517***	241.501***	319.712***	I (0)
CoC_estimate	−6.986***	167.856***	147.485**	I (0) (T)	−20.069***	402.402***	534.795***	I (0)
GE_estimate	−7.836***	196.837***	190.215***	I (0) (T)	−21.573***	459.415***	675.294***	I (0)
PSAV_estimate	−7.631***	187.478***	212.158***	I (0) (T)	−16.892***	421.122***	580.016***	I (0)

Notes
(i) ***, ** and * shows 1%, 5% and 10% level of significance
(ii) Schwarz Information Criterion is used as the Lag selection criteria; Include in test equation: no intercept
(iii) I (1) and I (0) indicate the order of integration [i.e. Non-stationary for I (1) and Stationary for I (0) respectively]
Source Authors calculation

Table 9.4 Review of panel co-integration

pedroni's residual co-integration test					*kao's residual co-integration test*	
Specification		*No intercept*	*Intercept only*	*Intercept & trend*	*Specification*	*Intercept only*
Group ADF		−8.510***	−5.680***	−6.423***	ADF	−4.384***
Group PP		−2.946***	−6.808***	−12.382***		
Non-weighted	Panel ADF-statistics	−6.214***	−4.591***	−7.576***	$Residual_{t-1}$	−0.302***
	Panel pp-statistic	−2.992***	−6.652***	−7.856***	$\Delta Residual_{t-1}$	0.198***
Weighted	Panel ADF-statistics	−8.215***	−6.813***	−5.127***	$\Delta Residual_{t-2}$	0.213***
	Panel pp-statistic	−2.892***	−6.213***	−4.492***	$\Delta Residual_{t-3}$	0.141***
					Summary	R-squared = 0.186 DW-stat = 1.78

Note ***, ** and * represents significance at 1%, 5% and 10% level
Source Authors calculation

scenarios are all significant, indicating that the null hypotheses explored in this research should be rejected. The findings are also in agreement with the hypothesis that rising prices for food will contribute to the prevalence of undernourishment (Anríquez et al., 2013; Lee et al., 2016; Warr, 2014; Webb, 2010). Additionally, a fall in the production of food, individual's income, and the outreach of banking services will raise the incidence of inadequate nutrition (Bryce et al., 2008; Tiwari & Zaman, 2010).

The transient influence on PoU by the explanatory variables along with the estimated Error Correction (EC) term (lagged one) of the specification is provided in Table 9.6. The EC term represents the divergence from a long-run equilibrium that tend to influence the short-run dynamics. It is the adjustment factor that shows the speed of adjustment of the dependent variable with respect to its own lagged values (in this instance PoU). The correction factor to equilibrium (i.e. the EC term) is reasonably slow at 0.24% but still significant (at 1% level) and negative. The food production in the previous year resulted in a significant reduction of the prevalence of undernourishment (PoU) in the short term, at a confidence level of 1%. The prevalence of undernourishment was further reduced by the reduction in food prices and an increase in individual income

Table 9.5 Assessment of long-run associations

Response variable: lnPoU (Millions)		
	Panel DOLS	*Panel FMOLS*
lnFPI_C	0.278***	0.324***
	(0.073)	(0.062)
lnFPI	−0.396***	−0.475***
	(0.131)	(0.110)
lnGNIpc	−0.171**	−0.162***
	(0.070)	(0.061)
Bran_D	−0.023***	−0.021***
	(0.059)	(0.006)
ATM_D	−0.009***	−0.010***
	(0.003)	(0.003)

Notes
(i) ***, ** and * represents significance at 1%, 5% and 10% level
(ii) Values in the parenthesis represents the standard errors
Source Authors calculation

(both at 10% level) while enhanced presence of commercial bank branches (at 1% level). In conjunction with the other external components in the mathematical framework, researchers observe that at the 5% level, the estimated level of Corruption significantly raises PoU. Further estimations of government effectiveness, political stability, and lack of violence significantly diminish the PoU at 5% and 1%, respectively. Lee et al. (2016) discovered a comparable impact of external influences on the PoU.

The causal connections across the panel structure between the variables were estimated using the Granger Causality Test (Granger, 1981). The findings from Table 9.7 indicate that alterations in food production, together with individual wealth, significantly influence the shifts in PoU at a 10% level of significance. Further, the change in the availability of bank branches per 0.1 million customers is also observed to significantly influence PoU at 5% level. The combined influence of all the explanatory variables taken together is observed to influence PoU significantly at 1% level. Although the connections observed are mostly unidirectional except for the prices of food and the availability of the ATMs which were found to have no causal linkages with PoU. Thus, with observed statistically significant influences the majority of the differenced explanatory variables are found useful to estimate the changes in PoU.

Further, it is reasonable to believe that an increase in the prices of food may lead to compromising the quantity as well as quality of food

Table 9.6 Panel error correction model

Response Variable: ΔlnPoU (Millions)	
ΔlnPoU (Millions)$_{t-1}$	0.098**
	(2.532)
ΔlnFPI_C$_{t-1}$	−0.02
	(−0.176)
ΔlnFPI$_{t-1}$	−0.143***
	(−2.615)
ΔlnGNIpc$_{t-1}$	−0.043
	(−0.816)
ΔBRAN_D$_{t-1}$	0.001
	(0.070)
ΔATM_D$_{t-1}$	0.001
	(0.700)
CoC_estimate	0.023**
	(2.346)
GE_estimate	−0.032**
	(−2.317)
PSAV_estimate	−0.023**
	(−2.698)
ECM$_{t-1}$	−0.0024***
	(−2.623)
Constant	0.019*
	(1.933)
R^2	0.162
F-Statistics	6.032

Notes
(i) *t*-statistics are presented within brackets
(ii) *, ** and *** indicates significance at 10%, 5% and 1% level respectively
Source Authors calculation

Table 9.7 Granger causality test

Response Variable: ΔlnPoU (Millions)	
	Chi-square statistic
ΔlnFPI_C	3.78
ΔlnFPI	6.73*
ΔlnGNIpc	7.69*
ΔBran_D	10.49**
ΔATM_D	3.52
Combined influence	38.72***

Note ***, ** and * represents 1%, 5% and 10% significance levels respectively
Source Authors calculation

consumption and may thus increase the prevalence of undernourishment under the conventional ceteris paribus assumption of keeping all other things constant. People often tend to skip their meals or compromise with the quantity and quality of food when the prices of food are on the surge. Further, it is not the nutritional requirements or the dietary intakes, rather the status of nourishment (under or over nourishment) that may present the impact in somewhat a promising way. Additionally, while assuming the prices of the food to be a constant, an improvement in the production of food will certainly increase the availability of food and thus may reduce the status of undernourishment. This is quite obvious, as the availability of food increases people those have been deprived from food or surviving starvation may have access to food and thus the status of undernourishment might fall (Warr, 2014).

Income of an individual is an important factor that helps to decide the status of his/ her nourishment. It is reasonable to believe that as income of an individual increases the consumption surges and thus status of nourishment goes better. However, empirical evidence exists where increment of income has a detrimental effect on the food consumption pattern (Farrell et al., 2018). It has been observed that as income of an individual increases the person proportionately reduces the expenses on his/ her food intakes. However, income spent on food depends on the household's behavior and the price of other goods that are in demand by that particular household (Andersen, 2009).

Among other factors accessibility to financial services helps to stabilize income in a way that meets certain contingencies when raised. Especially when individuals are faced with loss of income by way of unemployment or any other reason (e.g. the outbreak of pandemic) they might turn to their savings to take control of the situation. In this way, availability of financial services puts an additional cushion to refrain from undernourishment.

9.4 Conclusions

One of the most important problems that many countries are currently dealing with is undernourishment. In light of the COVID-19 pandemic, achieving the Sustainable Development Goal of ending hunger by 2030 is now inevitable. The number of persons impacted by the epidemic is expected to rise even in the absence of the negative consequences that hunger is predicted to have. The five international agencies that work together to combat malnutrition—FAO, IFAD, UNICEF, WFP, and WHO - have unanimously concluded that present efforts will not be sufficient to eradicate it in the coming ten years (FAO, 2020).

It is preferable to guarantee year-round access to wholesome food that promotes a healthy diet for everyone rather than just obtaining food. But, the affordability of healthy diets is linked to their accessibility. The majority of people, particularly in developing and least developed nations, cannot afford nutritious diets because of their high cost (high food prices) and inadequate income (FAO, 2020). This chapter focuses on two main topics and examines through empirical research if there is a notable distinction connecting the unit elasticity of food price and income elastic properties of inadequate nutrition.

Additionally, a serious worldwide condition is the rising need for food commodities over time as a result of population expansion. Food insecurity arises from insufficient food supply, especially if the agricultural component of Official Development Assistance (ODA) is stagnant (Cabral & Howell, 2012; Webb, 2010). Additionally, inadequate government expenditure on nutrition and health contributes to this issue (Chen et al., 2023; Nandy et al., 2016). Agricultural productivity is decreased when farmers cannot obtain better seeds, fertilizers, or extension guidance (Webb, 2010). The study aims to assess whether the food production elasticity of undernourishment varies substantially from the unit elasticity, considering the negative impact of increased uncertainty in FSC costs for transactions (Khan et al., 2022) and the reduced benefits to farmers despite rising food prices (Wise & Harvie, 2009).

Based on the results, all the components considered are non-stationary at the initial level and become stationary when their first difference is taken. Additionally, the non-stationary series' linear I (0) combination indicates that the variables should be co-integrated. According to long-term impact estimates using the parametric (DOLS) and non-parametric (FMOLS) approaches, rising food prices greatly raise PoU, while increases in food production, per capita income, and the availability of ATMs and commercial bank branches dramatically lower PoU. The VECM yielded a significant and negative EC term for PoU, indicating a small but significant annual adjustment of approximately 0.24% to reach equilibrium. In addition, an upsurge in agricultural output from last year considerably reduces the PoU in the near run, with an average loss of 1%. Aside from the exogenous variables considered in the study, corruption has a substantial positive impact on the level of poverty (at a 5% level of significance). On the other hand, government effectiveness, political stability, and the absence of violence have significant negative effects on the level of poverty, with significance levels of 5% and 1%, respectively. The results of this investigation refute each of the three unit-elastic theories that were previously spelled out in theoretical terms. The immediate-run dynamics-focused Granger causality test results validate the one-way significant causal relationship between FPI and PoU, GNIpc and PoU, and Bran_D and PoU. Functional form to review causal connections between ATM_D to PoU and FPI_C to PoU, however, are still unclear. Growing food costs can have a volatile short-term effect on PoU. According to the studies we have evaluated thus far, higher food prices lead to greater food insecurity and, hence, PoU. In the long run, this study finds significant insights; but, in the short run, it is not accepted. Additionally, the VECM yields equivocal results that indicate a favorable impact in the second lag but a negative though negligible impact in the first and third lags. Similar findings were noted in the Irz et al. (2013) investigation. Such ambiguous short-term results could be the consequence of an inadequate time series. This limitation may be taken into account for next research. We suggest that characteristics such as unemployment rates and income inequality should be taken into consideration in order to make insightful generalizations.

Since it analyzes the influence of the theoretically important PoU factors across the 61 developing and least developed nations chosen between 2004 and 2019, the study makes a substantial contribution from a policy standpoint. Understanding the range of results in cross-sectional and temporal dimensions might help create a comprehensive representation of the connection. Undernutrition has hidden costs both to health (SDG 3) and the environment (SDG 13) as a result of food consumption behaviors. By 2030, it is projected that insufficient dietary intake (both in quantity and quality) will lead to higher levels of mortality and illnesses that are not transmissible. This will result in annual health expenses exceeding USD 1.3 trillion. Additionally, there will be an estimated annual social cost of USD 1.7 trillion due to greenhouse gas emissions. Evidence from China can be quite helpful in this area. The numbers for China that date back to 2000 have been revised, and the result is a significant decline in the global undernutrition rate. Furthermore, research indicates that China's agricultural production has significantly increased as a result of a spectacular resuscitation of the sector (Bryce et al., 2008), providing nutrient-dense food for the nation with the largest population.

In addition to revival techniques, this study suggests raising agricultural output in the event that food commodity prices stay the same (Ravallion, 1990). Food price increases appear to be primarily driven by the per-unit cost of agricultural inputs, the cost of the energy consumed, and the wage index, which measures labor costs. These factors require calibration in order to maximize productivity and manage the PoU (Irz et al., 2013). Among the key elements required to lower the PoU are adequate income and accessible savings choices (Warr, 2014). Therefore, the government must implement measures that provide a sufficient and steady income by lowering unemployment and fostering financial inclusion. To ensure that the availability of food supplies keeps up with the growing demand, adjustments have been proposed in the spending plan for agribusiness (Nandy et al., 2016) and the agricultural component of the ODA needs to increase over time (Webb, 2010). In addition, it becomes essential that the government establishes and implements targeted nutritional programs that treat undernourishment concurrently, as noted by Bryce et al. (2008). When contrasted with the United States and the United Kingdom, the current state of conquering the PoU in this sense can be described as impeding growth throughout the cohort (Bishai & Nalubola, 2002; De Benoist et al., 2006).

References

Alderman, H. (1993). New research on poverty and malnutrition: What are the implications for policy? In *Including the poor* (pp. 115–132). World Bank.

Alston, J. M., & Pardey, P. G. (2015). Agricultural R&D, food prices, poverty, and malnutrition redux. In *The fight against hunger and malnutrition: The role of food, agriculture, and targeted policies* (pp. 208–239). Oxford University Press.

Andersen, P. (2009). Food security: Definition and measurement. *Food Security, 1*(1), 5–7. https://doi.org/10.1007/s12571-008-0002-y

Anríquez, G., Daidone, S., & Mane, E. (2013). Rising food prices and undernourishment: A cross-country inquiry. *Food Policy, 38*, 190–202.

Asian Development Bank. (2012). *Food security and poverty in Asia and the Pacific: Key challenges and policy issues*. Manila. https://www.adb.org/publications/food-security-and-poverty-asia-and-pacific-key-challenges-and-policy-issues

Baird, S., Friedman, J., & Schady, N. (2011). Aggregate income shocks and infant mortality in the developing world. *Review of Economics and statistics, 93*(3), 847–856. https://www.jstor.org/stable/23016081

Baležentis, T., Li, T., & Chen, X. (2021). Has agricultural labor restructuring improved agricultural labor productivity in China? A Decomposition Approach. *Socio-Economic Planning Sciences, 76*, Article 100967. https://doi.org/10.1016/j.seps.2020.100967

Behrman, J. R. (1988). *Nutrition and incomes: Tightly wedded or loosely meshed?* Pew/Cornell Lecture Series on Food and Nutrition Policy-Cornell University (EUA).

Behrman, J. R., & Deolalikar, A. B. (1987). Will developing country nutrition improve with income? A case study for rural South India. *Journal of political Economy, 95*(3), 492–507. https://doi.org/10.1086/261469

Behrman, J. R., Deolalikar, A. B., & Wolfe, B. L. (1988). Nutrients: impacts and determinants. *The World Bank Economic Review, 2*(3), 299–320. https://www.jstor.org/stable/3989840

Bishai, D., & Nalubola, R. (2002). The history of food fortification in the United States: Its relevance for current fortification efforts in developing countries. *Economic Development and Cultural Change, 51*(1), 37–53.

Bouis, H. E., & Haddad, L. J. (1988). *Comparing calorie-income elasticities using calories derived from reported food purchases and a twenty-four hour recall of food intakes: An application using Philippine data*. University of Warwick.

Bouis, H. E., & Haddad, L. J. (1992). Are estimates of calorie-income elasticities too high? A recalibration of the plausible range. *Journal of Development Economics, 39*(2), 333–364. https://doi.org/10.1016/0304-3878(92)90043-9

Brinkman, H. J., De Pee, S., Sanogo, I., Subran, L., & Bloem, M. W. (2010). High food prices and the global financial crisis have reduced access to nutritious food and worsened nutritional status and health. *The Journal of Nutrition, 140*(1), 153S-161S. https://doi.org/10.3945/jn.109.110767

Bryce, J., Coitinho, D., Darnton-Hill, I., Pelletier, D., Pinstrup-Andersen, P., & Maternal and Child Undernutrition Study Group. (2008). Maternal and child undernutrition: Effective action at national level. *The Lancet, 371*(9611), 510–526.

Butler, P. (2012, June 25). Demand for food parcels explodes as welfare cuts and falling pay hit home—early-warning indicator should set alarm bells ringing about poverty levels, government told. *The Guardian*. https://www.theguardian.com/society/2012/jun/25/breadline-britain-growth-food-parcels

Béné, C. (2020). Resilience of local food systems and links to food security—A review of some important concepts in the context of COVID-19 and other shocks. *Food Security, 12*(4), 805–822. https://doi.org/10.1007/s12571-020-01076-1

Cabral, L., & Howell, J. (2012). *Measuring aid to agriculture and food security* (ODI Briefing Paper, 72).

Chen, X., Shuai, C., & Wu, Y. (2023). Global food stability and its socio-economic determinants towards sustainable development goal 2 (Zero Hunger). *Sustainable Development, 31*(3), 1768–1780. https://doi.org/10.1002/sd.2482

Christian, P. (2010). Impact of the economic crisis and increase in food prices on child mortality: Exploring nutritional pathways. *The Journal of Nutrition, 140*(1), 177S-181S. https://doi.org/10.3945/jn.109.111708

Cutler, D., Deaton, A., & Lleras-Muney, A. (2006). The determinants of mortality. *Journal of Economic Perspectives, 20*(3), 97–120.

Darnton-Hill, I., & Cogill, B. (2010). Maternal and young child nutrition adversely affected by external shocks such as increasing global food prices. *The Journal of Nutrition, 140*(1), 162S-169S. https://doi.org/10.3945/jn.109.111682

De Benoist, B., Dary, O., & Hurrell, R. (2006). *Guidelines on food fortification with micronutrients* (Vol. 126; L. Allen, Ed.). World Health Organization.

Dreze, J., & Sen, A. (2002). *India: Development and participation.* Oxford University Press on Demand.

Farrell, P., Thow, A. M., Abimbola, S., Faruqui, N., & Negin, J. (2018). How food insecurity could lead to obesity in LMICs: When not enough is too much: A realist review of how food insecurity could lead to obesity in low-and middle-income countries. *Health Promotion International, 33*(5), 812–826. https://doi.org/10.1093/heapro/dax026

Food and Agriculture Organization. (2008). *The state of food insecurity in the world 2008: High food prices and food security—Threats and opportunities.* Rome. https://www.fao.org/publications/card/en/c/3d431db5-34c8-5e01-9ad9-b2b0a2b9b918/

Food and Agriculture Organization. (2011). *The state of food insecurity in the world 2011: How does international price volatility affect domestic economics and food security?* Rome. https://www.fao.org/publications/card/fr/c/b9dff790-3555-55c3-a8c6-c5efc0e5c304/

Food and Agriculture Organization. (2012). *The state of food insecurity in the world 2012: Economic growth is necessary but not sufficient to accelerate reduction of hunger and malnutrition.* Rome.

Food and Agriculture Organization. (2020). The state of food security and nutrition in the world 2020: Transforming food systems for affordable healthy diets. *Rome.* https://doi.org/10.4060/ca9692en

Gentleman, A. (2014, February 28). Vulnerable man starved to death after benefits were cut—44-year-old died months after sickness and housing benefits were stopped following Atos fitness-for-work assessment.

The Guardian. https://www.theguardian.com/society/2014/feb/28/man-starved-to-death-after-benefits-cut.
Ghosh S., & Sahu T. N. (2020). How financially inclusive the Asian's are–an empirical analysis. *Rajagiri Management Journal, 14*(1). https://doi.org/10.1108/RAMJ-01-2020-0003
Granger, C. W. J. (1981). Some properties of time series data and their use in econometric model specification. *Journal of Econometrics, 16*, 121–130.
Green, R., Cornelsen, L., Dangour, A. D., Turner, R., Shankar, B., Mazzocchi, M., & Smith, R. D. (2013). The effect of rising food prices on food consumption: systematic review with meta-regression. *BMJ*, *346*. https://doi.org/10.1136/bmj.f3703
Hobbs, J. E. (2020). Food supply chains during the COVID-19 pandemic. *Canadian Journal of Agricultural Economics/revue Canadienne D'agroeconomie, 68*(2), 171–176. https://doi.org/10.1111/cjag.12237
Irz, X., Niemi, J., & Liu, X. (2013). Determinants of food price inflation in Finland—The role of energy. *Energy Policy, 63*, 656–663. https://doi.org/10.1016/j.enpol.2013.09.023
Khan, S. A. R., Razzaq, A., Yu, Z., Shah, A., Sharif, A., & Janjua, L. (2022). Disruption in food supply chain and undernourishment challenges: An empirical study in the context of Asian countries. *Socio-Economic Planning Sciences, 82*, Article 101033. https://doi.org/10.1016/j.seps.2021.101033
Lee, H. H., Lee, S. A., Lim, J. Y., & Park, C. Y. (2016). Effects of food price inflation on infant and child mortality in developing countries. *The European Journal of Health Economics, 17*(5), 535–551.
Lee, H. H., & Park, C. Y. (2013). International transmission of food prices and volatilities: A panel analysis. *Asian Development Bank Economics Working Paper Series, 373.*
Mattoo, A., Rocha, N., & Ruta, M. (2020). *The evolution of deep trade agreements* (Policy Research Working Paper No. 9283). World Bank.
Moore, S., Teixeira, A. C., & Shiell, A. (2006). The health of nations in a global context: Trade, global stratification, and infant mortality rates. *Social Science & Medicine, 63*(1), 165–178. https://doi.org/10.1016/j.socscimed.2005.12.009
Nandy, S., Daoud, A., & Gordon, D. (2016). Examining the changing profile of undernutrition in the context of food price rises and greater inequality. *Social Science & Medicine, 149*, 153–163. https://doi.org/10.1016/j.socscimed.2015.11.036
Ortiz, I., Chai, J., & Cummins, M. (2011). *Escalating food prices: The threat to poor households and policies to safeguard a Recovery for All* (UNICEF Social and Economic Policy Working Paper). UNICEF.
Ravallion, M. (1987). *Markets and famines*. Oxford University Press.

Ravallion, M. (1990). Income effects on undernutrition. *Economic development and cultural change, 38*(3), 489–515. https://doi.org/10.1086/451812
Reardon, T., Mishra, A., Nuthalapati, C. S., Bellemare, M. F., & Zilberman, D. (2020). COVID-19's disruption of India's transformed food supply chains. *Economic and Political Weekly, 55*(18), 18–22.
Reutlinger, S., & Selowsky, M. (1976). *Malnutrition and poverty. Magnitude and policy options*. Published for the World Bank by the Johns Hopkins University Press.
Sachs, J. (2002). *Macroeconomics and health*. WHO.
Sahn, D. (1987). *Causes and implications of seasonal variability in household food security*. Food Policy Research Institute.
Selke, S. (Ed.). (2011). *Food banks in Germany: Aspects of a social movement between food redistribution and poverty intervention*. Springer Science & Business Media.
Sen, A. (1982). *Poverty and famines: An essay on entitlement and deprivation*. Oxford University Press.
Sen, A. (1995). *Inequality reexamined*. Harvard University Press.
Srinivasan, T. (1981). Malnutrition: Some measurement and policy issues. *Journal of Development Economics, 8*(1), 3–19. https://doi.org/10.1016/0304-3878(81)90043-2
Strauss, J., & Thomas, D. (1990). *The shape of the calorie-expenditure curve* (Center Discussion Paper, No. 595).
Strauss, J., & Thomas, D. (1995). Human resources: Empirical modeling of household and family decisions. *Handbook of Development Economics, 3*, 1883–2023. https://doi.org/10.1016/s1573-4471(05)80006-3
Subramanian, S., & Deaton, A. (1996). The demand for food and calories. *Journal of Political Economy, 104*(1), 133–162. https://doi.org/10.1086/262020
Sukhatme, P. V. (1982). *Newer concepts in nutrition and their implications for policy*. Maharashtra Association for the Cultivation of Science Research Institute.
Tian, J., Bryksa, B. C., & Yada, R. Y. (2016). Feeding the world into the future-food and nutrition security: The role of food science and technology. *Frontiers in Life Science, 9*(3), 155–166. https://doi.org/10.1080/21553769.2016.1174958
Tiwari, S., & Zaman, H. (2010). *The impact of economic shocks on global undernourishment* (World Bank Policy Research Working Paper, 5215).
United Nations Sustainable Development Goals Fund. (n.d.). *Partnerships to combat malnutrition in Guatemala*. https://www.sdgfund.org/sites/default/files/Case%20Study%20-%20Guatemala%20Nutricion%20-%20EN.pdf

Warr, P. (2014). Food insecurity and its determinants. *Australian Journal of Agricultural and Resource Economics, 58*(4), 519–537. https://doi.org/10.1111/1467-8489.12073

Webb, P. (2010). Medium-to long-run implications of high food prices for global nutrition. *The Journal of Nutrition, 140*(1), 143S-147S. https://doi.org/10.3945/jn.109.110536

Wise, T. A., & Harvie, A. (2009). Boom for whom? Family farmers saw lower on-farm income despite high prices. *GDAE Policy Brief, 09-02.*

CHAPTER 10

Conclusion and Policy Recommendations

10.1 Conclusion and Policy Recommendations

10.1.1 Summary of the Study

The study emphasizes on the significance of financial inclusion in promoting economic development and achieving sustainable development goals (SDGs) by 2030. Initially it aims to empirically characterize financial inclusion and further analyze the relationship between financial inclusion and various development indicators, specifically focusing on poverty reduction (SDG 1), hunger alleviation (SDG 2), and economic growth (SDG 8).

By construction, the study is structured into nine chapters, starting with an introduction to financial inclusion and its relevance to sustainable development. Subsequent chapters provide a historical perspective on financial inclusion in India, assess current achievements, and explore the impact of various initiatives, such as the Mahatma Gandhi National Rural Employment Guarantee Scheme and the Pradhan Mantri Jan Dhan Yojana especially across the Indian territories. Further, it emphasizes the need for a comprehensive understanding of the factors influencing financial inclusion and its effects on development indicators. It also highlights the lack of consistent measures for evaluating financial inclusion across different regions and countries. The final chapters focus on comparative analyses, the relationship between financial inclusion and economic

S. Ghosh et al., *Financial Inclusion*, Sustainable Development Goals Series, https://doi.org/10.1007/978-981-95-3385-5_10

development and policy recommendations for enhancing financial inclusion efforts globally. Overall, the study aims to provide valuable insights for policymakers and contribute to the ongoing discourse on financial inclusion and its role in sustainable development.

The present study offers extensive reviews on cross-country experiences, focusing on aspects such as understanding financial concepts, budgeting, and accessibility to formal financial products. Further, it highlights that a significant portion of adults lack understanding of financial concepts like diversification and compounding, and many are unaware of their financial security after retirement. There is also a notable gender gap in financial knowledge, with men generally having higher financial knowledge compared to women. The present study emphasizes the importance of proper budgeting and the need for individuals to consider their ability to afford before making purchase decisions. Additionally, it addresses the accessibility and usage of formal financial products, revealing disparities across different nations. The World Bank Group and IMF have committed to enabling access to basic financial services for 1 billion people under the Universal Financial Access (UFA) 2020 Plan, with a focus on private sector engagement in financial inclusion. The document also discusses the challenges faced by specific countries, such as limited coverage of financial services, conservative lending policies and the reliance on informal finance. It provides insights into the financial inclusion status of countries like Mozambique and Cote d'Ivoire, highlighting the percentage of active account holders, electronic transaction usage, and the proportion of unbanked population.

Additionally, the study discusses the global efforts towards achieving the Sustainable Development Goals (SDGs) as outlined by the United Nations. It emphasizes the importance of evaluating and surveilling the developmental goals at both regional and global levels. A Global Indicator Framework comprising 243 indicators has been designed to measure SDG progress. The High-Level Political Forum (HLPF) on the Sustainable Development Goals plays a crucial role in the global follow-up and assessment of the SDGs. The document also highlights India's journey towards achieving the SDGs, including its emphasis on poverty elimination, access to education, healthcare, and renewable energy. India's strong political commitment to the SDGs, reduction in poverty, improved access to education and healthcare, and progress in renewable energy are also discussed. Further, the study mentions specific initiatives undertaken by India to advance the SDGs, such as direct purchases of agricultural commodities from farmers and the encouragement of e-trading.

So far the results in Chapter 4 is concerned, we classify the available data into four categories (samples) to empirically characterize the phenomena of financial inclusion. Since, financial participation is a broad concept to explore. Presentation of the concept by a single number involves consolidation of a number of variables. However, the variables that build up the index or are significant contributors are numerous as well as diverse in nature. In this connection, Sect. 4.3 focuses to estimate the achievements in financial inclusion across the 28 states and 7 union territories of India for 2003–22 applying the Max–min. approach for normalization of the three variables and the inverse of the Euclidean distance to consolidate the different dimensions. Section 4.4 aims to estimate the financial inclusion achievements across the 26 Asian countries for 2013–22 using Factor Analysis along with the weighted arithmetic mean in the first stage and the inverse of Euclidian distance solely in the second stage. Factor analysis has been used to find the common structure and derive the relative weight of the theoretically relevant variables. Section 4.5 considers measuring financial inclusion across the 12 Asian and 12 European nations using Max–min. approach for normalization and applying the inverse of Euclidean distance for consolidation. This sample uses the principal component analysis as weighing strategies to allocate relevant weights to the 20 variables those are found to be theoretically relevant. Further, Sect. 4.6 provides a global perspective through measuring the financial inclusion achievement scores across the 46 countries taken out of 5 continents using 14 variables for 2012–21.

So far, the results from Sect. 4.3 are concerned; the union territory of Chandigarh drives ahead in regards to average IFI scores, obtaining an average score of 0.715, followed by the state of Goa (0.702) and Kerala (0.539). Furthermore, the North-East states [Manipur (0.041) and Nagaland (0.026)] occupy the bottom of the ranking (Chotia & Rao, 2014). Apart from volatility (i.e. variance in IFI achievements), Andaman and Nicobar (0.088) appears to be the most stable, whereas the state of West Bengal (0.94) seems to be in the most volatile position. However, these scores are based on three indicators that represent the dimension of service outreach (i.e. per capita number of bank branches, savings, and credit accounts) and therefore might suffer the limited representation of the concerned phenomena. Results for the top three contenders (i.e. Chandigarh, Goa, and Kerala) in financial inclusion achievements are due to the fact that these states/UT's have a relatively high standard of living backed by relatively high per capita income and educational facilities (Kumar & Mishra, 2011). Further the states like Manipur and Nagaland have relatively poor performance due to their geographical position, unemployment and low-level of education.

Review of the results from Sect. 4.4 shows that using an average of the degree of inclusion attained through the investigation across the 26 Asian nations, the study discovers that South Korea leads with a score of 0.641, ahead of countries from the whole continent's eastern and northern regions. Further, the South Korea is acclaimed to have the highest number of bank account ownership (Bae, 2022). This sample takes into account 14 distinct metrics that include critical characteristics such as Stability, Service Potential, Accessibility, Availability, and Efficiency, which characterize financial inclusion comprehensively. In parallel, a review of the country experiences across the time, based on the reported variations, we observe South Korea has maintained a steadier presence than its rivals in the overall order of the IFI rankings. However, when it comes to IFI accomplishments, Myanmar's performance is quite weak. It displays substantial changes over time, with an average accomplishment value of 0.170 contrasted with South Korea (the leader). India, for example, holds the third position in overall mean IFI values obtained for the study period, and holds the top position in the South Asian Association for Regional Cooperation (SAARC) category to pass the 0.307 (2nd quartile) threshold. The study found that the coefficient of variation, or CV, for the collection of countries according to average IFI scores from 2013 to 2022 shows that a large number of nations in the Southeast as well as the East are ahead. As far as mean IFI values during 2013–22 are concerned, Japan (ranked first), Malaysia (ranked second), and South Korea (ranked third) are among the most stable economies. Nevertheless, the inclusive atmosphere in countries such as Turkey (23rd), Azerbaijan (24th), Jordan (25th), and Mongolia (26th) is determined to be particularly volatile (see Table 4.4). Year on year financial inclusion achievements along with the dimensional contributions throughout the study period (i.e. 2013–22) can be studied from Table 4.3.1–4.3.10 provided in the appendix of the study.

Further, a strong positive relationship between income and mean IFI values across countries categorized on GNP per capita (PPP terms on constant international $) in Table 4.4 is observed. As demonstrated by the statistical conclusion in the next section of the study, shifting from a low-income to a high-income nation is suspected to boost the mean IFI scores.

Among the stated income groups, South Korea, Georgia, India, and Nepal had the strongest IFI average scores for the concerned period of study. When average IFI scores across the research period are taken into account, it is clear that only South Korea (3rd), Georgia (15th), India (20th), and Nepal (5th) have an acceptable degree of consistency.

While elaborating on its performance, the Deputy Governor, Bank of Korea, states that the economy is sufficiently stable and secure to manage its financial transactions (Bae, 2022). The account ownership remains as high as 95% of its population and is barely achieved by any of the advanced economies. Further, the penetration of digitization allows a real-time transfer over a 24 × 7 window from 2001 onwards, when Korea became the first country to introduce retail based fast payment system under the aegis of 'Electronic Banking System'. With a high smart phone penetration rate (i.e. 95%) offering a steady growth in the number of internet and mobile banking users, usage have shown leapfrog in the numbers especially after the COVID-19 pandemic. Faster authentication (e.g. passwords, fingerprints etc.), pre-stored information in the mobile application and payment gateways at the terminals are some of the initiatives those are accelerating the numbers to deliver hassle free payment and transfers. The whole system was part of the strategic move of the Bank of Korea and the government in action back in 2019 that allows the financial institutions to compete in open market and customize their services that fits the consumer needs and preferences. However, concerns are still there on the fairness of credit provisioning. Individuals with no or less collateral are devoid of getting loans from institutional arrangements and are exposed to predatory money lenders charging high rate of interest. The reliance on credit card seems to be exceptionally high (almost 53.8%) due to the promotion of credit cards following the 1997's financial crisis. Certain policies that promote to increase consumption and allow refunds of income tax based on credit card purchases somehow infuses instability in the individual's asset portfolio due to enhanced debt burdens.

The document further discusses the financial landscape along with the savings behavior in Japan (2nd Rank Holder) and Myanmar (Last Position), highlighting the relevant factors behind their standing. In Japan, the postal savings system is a significant financial institution, with a vast network of post offices and a high level of public trust (Yoshino et al., 2018). As of 1999, accumulated savings reached ¥262 trillion, nearly

50% of the country's GDP. Factors influencing savings behavior include income growth, household size, and the social welfare system (Horioka et al., 2007). The government plays a crucial role in channeling savings into loans through government banks, which have shifted their focus over the years from improving living standards to other areas.

In contrast, Myanmar is characterized with the status of being a struggling economy, with a large portion of its population living on very low incomes (Tambunlertchai, 2018). The country has low levels of education and employment, along with minimal internet and mobile connectivity. Access to formal financial services is limited, with only 30% of adults having access, while a significant portion relies on informal finance. The financial sector is dominated by state-owned banks (GIZ, 2013; Turnell, 2014), and there are ongoing efforts on transition to private ownership and expand services to underserved areas (Duflos et al., 2013).

Review of the results from Sect. 4.5 shows that the mean financial inclusion achievements over the time, Netherlands dominates having a score of 0.968, followed by Italy (0.962) and Japan (0.959) from the Asian continent. Despite being near the bottom, the Pakistan has a value that is higher than the average (i.e. 0.527). India holds the 11th rank considering all the nations taken together and 6th against the other Asian countries with an inclusion index score of 0.787.

Evaluating the components of stability and viability from the appendix shows that a favorable atmosphere continually increased financial inclusion in countries such as the Maldives, Turkey, and the Netherlands during the study period. Except in 2016, when India outperforms Pakistan, data reveal that both Pakistan and Bhutan do better among India's neighbors. Furthermore, assessing the scale of efficiency indicates mix evidence where countries like Saudi Arabia, Spain from Europe and Japan, Pakistan and Maldives from Asia have been among the most consistent countries across the period of study, while Bhutan and Greece were among the least consistent ones. Additionally, with regard to depth, Switzerland and Japan have constantly been at the top during the period of study except in 2022 where Malaysia takes the lead, whereas Pakistan has continually been at the bottom. Likewise, rankings in the banking and financial services outreach component remain stable, having Japan and Switzerland continuing to lead and Pakistan with Latvia (in 2014) at the bottom.

Among the high-income OECD countries 83% of the adult's report to have an account in a financial institution of which 95% as of 2014 agrees that they use their Debit-cum ATM card for making payments directly (Demirgüç-Kunt et al., 2015). Netherlands being one of the member countries of the OECD high-income group exhibits to have 99% of its adults to hold an account with a formal financial intermediary. Further, across the group of adults belonging to the poorest 40%, the households reported to have accessibility to formal financial accounts stood at 99% (World Bank, 2015). Among the Asian countries, Japan shows an account penetration rate of 97% of which 95% belongs to the poorest 40% of households. While going down at the bottom, Bhutan is reported to have an account penetration rate of 34% where the marginalized 40% in terms of their income shows to have account ownership at the rate of 25% only. Therefore, the findings of our study are in consonance with the results as provided by the World Bank (2015).

Considering the review of Sect. 4.6 covering 46 countries from the 5 continents over 2012–21 it is observed that Namibia, Bangladesh, Turkey, Costa Rica and Ecuador seems to be the most stable nations among the African, Asian, European, North and South American continents so far as the values of coefficient of variation in financial inclusion scores are concerned. Continental average of achievements in financial inclusion during the study period across the sample nations show Europe (0.417) to take the lead followed by Asia (0.407), South America (0.360), North America (0.339) and Africa (0.263). Further, Singapore (0.579), Mauritius (0.528), Italy (0.473), Brazil (0.463) and Costa Rica (0.393) from the continents of Asia, Africa, Europe, South America and North America respectively appears as the top contenders in regards to the average IFI scores throughout the study period.

Review of the dimensional achievements show that on an average throughout the period of study (i.e. 2012–21) considering the dimension of Capital Formation, China (0.760) takes the lead followed by Brazil (0.567), Mauritius (0.521), and Singapore (0.504). This dimension particularly aims to emphasize on the basic finance function and thus represents the intrinsic capability of an economy to create surplus out of the intermediation between the surplus and deficit units (Beck et al., 2007). However, review of the positions shows Burundi (0.163) to be at the bottom. Further, review of the dimension of Efficiency shows China (0.656) at the top followed by Malaysia (0.636), Singapore (0.627), and Malta (0.605), whereas Cameroon (0.202) to be placed at the bottom. This dimension particularly focuses to represent the effectiveness and efficiency of service providers to extend financial services to their intended clients. Finally, across the dimension of Outreach this study finds Seychelles (0.696) at the top followed by Croatia (0.615), Singapore (0.596), and Estonia (0.543). Across the dimension of Outreach, Burundi (0.014) finds place at the bottom. This dimension particularly gives an overview of the attainments that the service providers have achieved while extending the contours of the formal financial system within an economy.

Results show that China seems to be the top contender in terms of Capital Formation and the dimension of Efficiency. This result is substantiated by the fact that the savings rate in China is extremely high (Yang, 2012). Further, with 66% individuals reported to have access to a formal financial account in China, the country is positioned way beyond the achievements than its peers in the BRICS group. Among other members, the proportion of account ownership stood at 55% for Brazil, 44% in Russia, 37% in India, and 57% in South Africa. In fact, the proportion of population having a bank account in China is way above the average across the world (Demirgüc-Kunt & Klapper, 2013). These cross-sectional deviations can be explained by the country's economic development status, which is replicated by the per capita output (i.e. GDP per capita) (Demirgüc-Kunt & Klapper, 2013). However, per capita GDP for China is not the highest among the BRICS members showing that the usage of formal financial accounts is certainly higher than compared to the other members in the group (Fungáčová & Weill, 2015). Following the survey results of Fungáčová and Weill (2015), 82% of the respondents who saved in the past 12 months reported to save their earnings in a financial institution, which is far better than the BRICS results lying within the range of 50–72% and well above the world average of 36% and is corroborated further by the high aggregate deposits to GDP ratio for China among the G20 nations (Iorgova & Liu, 2013). The well positioning in terms of account openings with financial intermediaries those are having a formal recognition is highly motivated by the extremity of high savings rate across the country (Zhang et al., 2012). Further, it is observed that this high savings rate leads to higher propensity to park savings with formal financial intermediary rather going into the traps of informal channels unlike Russia where a majority of the depositors fall in the trap of shadow banking. The leading factors behind the high savings rate being the lesser availability of the social as well as private insurance alongside the gender imbalance with which Chinese households are associated with (Wei & Zhang, 2011).

However, following the survey of Fungáčová and Weill (2015), the extent of formal credit accessibility for China seems to be very low at 7% for the past year. This is far below the average of the high-income countries (i.e. 14%) as well as compared to BRICS average. The status of other BRICS members show that use of formal credits remained at 7% for Brazil, 9% for Russia, 8% for India and 11% for South Africa. Notwithstanding the Banks' private credit to GDP ratio remained at 121% following the Global Financial Development Database which classifies China in the high-income category (Čihak et al., 2012). A relatively high concentration of credit towards the large state-owned firms as compared to individual loan applicants is one among the prominent reasons behind the divergence (Hale & Long, 2011).

Further, in the dimension of Capital Formation and Outreach, Burundi is found to be at the bottom from among the sample countries considered in the study. Burundi, as classified under the category of Least Developed Countries (LDC's) is landlocked from its geographical positioning and is characterized with low and volatile growth rates (Nkurunziza et al., 2012). Faced with political and economic turmoil the country is constrained with flexibility in utilizing the public and private resources. Thus, it survives with regard to drag investment, create entrepreneurship, and accelerate the growth rates in the economy. High rates of poverty and unemployment forces the stagnation of the income and thus put severe constraints on the mechanism of capital formation as well as restricts people to access the available financial services (Nkurunziza et al., 2012).

Subsequently, Chapter 5 emphasizes on comprehensive review of the comparisons across mean FI achievements across four different samples as detailed in the previous section.

Section 5.2 showcases the performances across the 28 states and 7 union territories of India during 2003–22 and attempts to figure out the significance of change in mean FI scores across the different regions and income groups classified. Following the statistical chronology, we apply both the Analysis of Variance and Independent Samples t- test to figure out significant changes across the regions and income groups respectively. Section 5.3 considers 26 Asian countries across a period of 2013–22 and attempts to outline significant improvements in mean FI scores over the time applying Dependent samples t- test as well as Wilcoxon-Signed Rank test as a non-parametric alternative depending upon the fulfillment of the conditions prescribed. It further classifies the Asian countries into high-income and low-income categories and through application of the

Independent samples t- test attempts to observe whether high-income countries are significantly better than the lower ones in terms of mean FI achievements. Section 5.4 is in no way different, except that it further outlines the significance of the continental differences between Asia and Europe during the study period of 2013–22 applying the Independent samples t- test. Finally, Sect. 5.5 focuses to appreciate the differences across mean FI scores during 2012–21 considering an extended cross-sectional band of 46 countries selected from 5 different continents. This sample capitalizes on the magnitude of differences in mean FI scores across the 5 continents applying ANOVA (parametric alternative) as well as the Kruskal–Wallis test (non-parametric alternative).

Inquiry into data from India's 28 states and 7 union territories (i.e. Sample 1) for 2003–22 reveals a noticeable disparity in the mean IFI scores across regions for both income levels. Difference is also evident across the average IFI values between both the income groups belonging to a specific region. Results in Table 5.2 shows that both the independent categorical variables (i.e. groups based on income and regions) and their interaction factor significantly influences financial achievement scores. Hence, values of financial inclusion achievement are expected to differ significantly across the income groups and regions. The effect size as measured by the eta-squared shows the dependence to be more vibrant across the regions (0.928). However, robustness of such findings largely depends on the assumptions that underlie behind the functions of the model. According to Table 5.3, the mean variation in financial inclusion ratings between both low- and high-income groups is little (at 5%). The results are robust to the intrinsic assumption of applying the parametric independent samples *t*- test at an acceptable confidence level (i.e. normality, homogeneity of variance and heteroscedasticity). Therefore, financial inclusion achievements for high-income groups are better prima facie but not significantly in comparison to the low-income groups. Comparison across the regions in Table 5.4 shows a significant difference in mean FI score achievements after satisfying the intrinsic assumptions (Stoline, 1981). Tukey's Honestly Significant Difference (HSD) and the Ryan-Einot-Gabriel-Welsch (R-E–G-W-Q) test are used in Table 5.5 to identify particular locations that vary substantially from each other (Shingala & Rajyaguru, 2015). Table 5.5 displays the agreement of each of these post-hoc analyses at the five percent threshold of significance, indicating that each site seems to be statistically different so far as mean achievements in IFI scores are concerned during the period of study.

Findings of this analysis converge with the results of Laha and Kuri (2014). Such convergence was mainly due to the similarity in the mechanism of constructing the index as well as consideration of both the demand and supply side variables taken in to account. However, the results are somehow different apart from ranking of the areas based on their accomplishments in financial inclusion following Gupte et al. (2012). The source of contradiction results from the non-parametric assignment of weights rather employing a data-driven approach that the present study overcomes from (Ghosh & Sahu, 2020).

Regional disparity in the achievement scores of financial inclusions implies disproportional provisioning of financial assets (Chakravarty & Pal, 2013). A part from analyzing the regional achievements, the North-Eastern part of the country is most vulnerable so far as the dimension of stability is concerned. The average achievement scores in financial inclusion are relatively lesser due to fact of geographical inconvenience that most of the states of this region suffer from (Gupte et al., 2012). Additionally, economic disadvantage due to geographical positioning further suppresses the achievements.

The divergence in financial inclusion achievements across the productivity-based groups is statistically weak and further reinforces the states agenda of ensuring finance for all (Cámara & Tuesta, 2014). However, this might have an adverse consequence through increasing the existing levels of non-performing assets that the financial intermediaries may suffer from (Sarma & Pais, 2011).

Comparisons across the 26 Asian countries over a period of 2013–22 show significant improvements over the time in Financial Inclusion attainments at acceptable levels of significance. However, the trajectory of the growth is not so smooth, showing convergence with the World Bank's Report of 2020. Economic slowdown prior to the COVID-19 pandemic has reduced the pace of efforts targeting to financially include the excluded sections within its periphery. Thus, the erratic move in financial inclusion achievements looks somewhat inconsistent with the approach to attain an all-inclusive financial system, highlighting the inherited instability in the targeting of policy initiatives by countries from the Asian continent. Similar predictions have been made as part of suggestions in the studies of Rahman (2012) in the Malaysian context and of Amidžić et al. (2014) considering a broader cross-sectional band. Further, the High-income countries are observed to surpass the Low-income countries substantially at 5 percent level of significance in financial inclusion achievement scores. Similar evidences were observed in the analysis of Global Findex Database, which had coverage of 150,000 nationally representative adults above 15 years of age selected at random. This survey covering almost 148 economies by Gallop Inc. revealed a substantial variation in account penetration between the groups classified on the basis of income. Among the countries classified under the high-income group, 90% of the respondents reported to have an account with a formal financial intermediary whereas the percentage of account ownership for the Sub-Saharan Africa reported to be less than 18%. However, the developing Asia has fared better than compared to the Sub-Saharan African region with a score of 26.7%. Further, the variation within the Asian continent is also substantially high [see Fig. 5 at page number 5 of Ayyagari and Beck (2015)].

So far, comparison across the continents of Asia and Europe is concerned, significant differences are observed (at 10% level) though not substantial in financial inclusion achievements. Apart from analyses, we may conclude that nations throughout Europe and Asia appear to be more or less similar when it comes to evaluating the success in average financial inclusion, at least for the sample we examined. The sample countries considered from the continent of Europe are majorly from the middle-income group, whereas the Asian countries taken as sample more or less represent every possible group. This basically leads to non-significant differences (at 5 percent level) in financial inclusion achievements so far as the achievement scores are concerned.

However, a significant difference is observed between the countries from high-income and low-income group (with $p < 0.01$) during 2013–22. This certainly indicates that wealthier countries have a substantially better position in financial inclusion achievements than compared to the deprived ones and was part of previous literary evidence given by Demirgüç-Kunt (2006) and Ghosh and Sahu (2020). Additionally, countries with more resources tend to exhibit higher growth rates thus showing greater dependability on increasing the outreach of financial services. Thus, the drive of financial inclusion follows the growth in the real sector (Lucas Jr., 1988).

Considering as much as continents (i.e. 5 continents) across a large cross-sectional band covering 46 countries over a period of 2012–21 in Sample 4, it is observed that the countries within the continents are distributed approximately normally and share a constant variance across the groups. The differences between Africa (0.263) and Asia (0.407), between Africa (0.263) and Europe (0.417) and between Africa (0.263) and South America (0.3604) are significant. However, after the Bonferroni correction in the p- values only differences between Africa and Asia and between Africa and Europe appears significant. The F-statistic values for equality of the means across the five continents are consistently significant according to Wilks' lamda, Pillai's trace, Lawley-Hotelling trace and Roy's largest root showing the overall model fitness. The difference in the FI achievements across the continent of Europe and Africa and further between Asia and Africa emanates from the fact that countries within these continents are different in terms of their regulatory framework, economic structures, and many other factors those lead to differences in their financial inclusion achievement scores. Further, most of the countries from the continent of Africa are categorized as the Least Developed Countries (LDC's) based on their economic capacity (World Bank, 2015), whereas countries from the continent of Europe are mostly classified under the high or middle-income group. Thus, the difference in their economic capacity creates a difference in financial inclusion achievement scores which is found significant as a result in this study. Similar analogy is valid while comparing the continent of Asia with Africa. Although the economies across the continent of Asia in terms of their economic capacity are quite symmetric and normally distributed but their average capacity is substantially higher than compared to an average African nation (ADB, 2017). This basically leads to the conclusion that with higher economic capacity countries from the continent of Asia surpasses an average African country in terms of the achievement scores in financial inclusion.

Further, this sixth chapter attempts to characterize the relationship between financial inclusion and economic development (SDG 8). Since, the literature shows mixed evidence on the causative connections, this study has considered both the probable propositions (1) Whether, Financial Inclusion can accelerate the pace of Economic Development and subsequently (2) Whether, Economic Development improves the extent of Financial Inclusion (The case of Reverse Causality).

This study attempts to review the 1st proposition across a sample of 26 states and 4 union territories of India for 2002–19 to analyze the impact of financial inclusion (measured as natural logarithm of savings, credit, and branches) and Gross Fixed Capital Formation (indicating the investment) on the output growth (measured as natural logarithm of RNSDP, RealIP, and per capita RNSDP) after considering a dummy for the financial crisis of 2007.

A preliminary investigation of the growth in production and indicators of financial inclusion shows a co-movement. The production during the period of study across the states and UTs grew on an average by 6–8% and the growth across the indicators of financial inclusion on an average was observed roughly near 5–11%. This was further supported by a significant correlation between the variables at level. The long-run association as observed between growth in output and indicators of financial inclusion under both the parametric and non-parametric alternatives are valid as well as complimentary to each other. However, short-run dynamics are inconclusive due to the non-availability of data for a longer period of time (Pierse & Snell, 1995). Implications to the present study converges with earlier research of Apergis et al. (2007) which was an extension of the Christopoulos and Tsionas (2004) considering a cross-sectional portfolio of 10 developing nations across 1970–2000. However, literary evidence in support of finance-led growth (i.e. the supply leading hypothesis) is mixed. Watchel and Rousseau (1994) and Rousseau (1998) finds evidence in favor of this hypothesis whereas Arestis and Demetriades (1997), Luintel and Khan (1999) and Arestis et al. (2001) finds evidence of bi-directional causality.

The present study provides a microscopic view of the hypothesis by framing the heterogeneity across the states and UTs of India. Insights from the data show that in long-run financial inclusion promote economic growth. However, in the short-run the results are inconclusive and are quite reasonable due to the fact that improvement in outreach of financial services takes certain time to materialize in to output growth. It is quite reasonable to believe that opening of savings account or provision of credit does not influence output instantaneously especially across the provinces of India where it is unlikely to get a loan just after applying for the same (Ghosh & Sahu, 2024). Further, using a single lag to represent the equation reduces the degrees of freedom and thus in short-run results became inconclusive. Thus, in the long-run it is quite certain that increase in banking outlets and usage of savings and credit will raise the output levels. The Government of India in this connection assumes a proactive role by implementing a nation-wide strategy to improve the status of financial inclusion [through National Strategy for Financial Inclusion (NSFI) covering 2019–24]. Under such strategy, initiatives such as the Pradhan Mantri Jan Dhan Yojana (PMJDY) which focuses to open no-cost bank accounts, Pradhan Mantri Suraksha Bima Yojana (PMSBY) covering a one-year death cum disability insurance at a premium of Rs.

12/ year and Pradhan Mantri Jeevan Jyoti Beema Yojana (PMJJBY) to cover a life at a premium of Rs. 330/ year are worth mentioning.

While introspecting the 2nd proposition that whether Economic Development (measured in per capita state-wise Net Domestic Product) improves the extent of Financial Inclusion (measured in per capita number of savings accounts) after controlling for population, inflation, and government spending on health and education across the 27 Indian states and 5 territories of the union from 2002 to 2019, the study observes per capita state-wise Net Domestic Product to increase the income and thus the number of savings accounts in the economy. The impact (though less important) of other variables in the model on increasing the number of savings accounts are consistent with theoretical constructs. Analysis of the controlling factors further shows that an increase of 14.49 points of HEL and 100 points of CPI with respect to its base year includes an additional individual in the framework of the formal finance. Expenditure on health accounts almost 15 percent of the annual family income and turns "financially catastrophic" as they exceed the relative ability to pay (Wyszewianski, 1986). Especially considering the case of India where a significant share of population lives below the international poverty lines it appears obvious that they are unable to protect themselves not only from illness involving high costs/catastrophic illnesses (e.g. Cancer and Acquired Immune Deficiency Syndrome (AIDS) but also illnesses from acute urinary tract infection and a strep throat in absence of an adequate health insurance plan. Therefore, increased investment from government's part either by way of building up new hospitals or modifying the old ones or even by way of extending social health insurance schemes (e.g. Rashtriya Swasthya Bima Yojana, Ayushman Bharat, etc.) for the poor offloads to some extent the burden of catastrophic health care expenses on them resulting in to a positive association with their propensity to save (Berki, 1986) and hence promote financial inclusion.

Addressing poverty and inequality is one of the key policy goals pursued via increased access to different financial services, particularly for the most vulnerable segments of society (Sahoo et al., 2017). The majority of research found a negative relationship between poverty and inequality as well as accessibility to legally authorized banking services (Burgess & Pande, 2005; Galor & Zeira, 1993; Honohan, 2004; Jalilian & Kirkpatrick, 2002). Studies in this connection are motivated by empirical evidences of strong correlations between the distribution of income and the aggregate economic activity (measured by income per capita) (Kravis, 1960). The distribution of income is more concentrated for countries which are financially sounder (Lydall, 1968). Consistent evidence was reported in the findings of Torsten and Tabellini (1991) which exhibits that equity increases both the level of income as well as the growth in the economy. However, the extent of influence of both poverty and inequality on the access to financial services varies largely across the contexts being studied (Buckland et al., 2005; Clarke et al., 2006; Beck et al., 2007; Jeanneney & Kpodar, 2011; Inoue & Hamori, 2012 and Boukhatem, 2016). The growth of financial sector promotes economic development albeit qualifications exist on the preferential influence that the sectoral growth may have on the wealthier part of the population (Clarke et al., 2006). Financial inclusion as a significant contributor of the inclusive development mechanism can be viewed as a relevant and complementary measure for poverty eradication (Chibba, 2009). In this respect, using Sample 3, this study in the seventh chapter, aims to determine the relationship between financial inclusion and output growth, as well as to examine whether inclusion-led output development reduces poverty. Table 7.1.2 in the appendix provides a primary study of the link between the variables using Pearsons' correlation. The results show ΔlnGDP to negatively (although not significant) correlate with the measure of poverty. Further, the growth in the measure of financial inclusion is also found to share a significant positive correlation with ΔlnGDP. Such characterization gives us reasonable evidence that financial inclusion may improve the output growth and may also reduce poverty through improving the output growth. Using the two-stage least squares method the study finds financial inclusion to improve the output growth and reduce poverty significantly across the 67 countries from 2004 to 2018.

Capitalizing on the eighth chapter, the present study observes that proper nutrition is a key factor for assessing the effectiveness of the strategies emphasizing on development. Development programs usually emphasizes on the relevance of important economic factors, notably income and pricing of food, in addressing undernutrition. Surprisingly, adequate nutrition is not found to improve certain indicators of health instead increasing the productivity (Behrman et al., 1988). Recent techniques have relied on econometric estimations that exhibit the connections between energy intakes and income, prices and other factors based on household level data. However, it is the appropriateness of these intakes relative to each person's requirements that is important, not the intakes themselves (Sahn, 1987).

The eighth chapter of this study aims to evaluate the influence of the changes in prices and production of food, individual's income, and the availability of financial service on the prevalence of undernourishment (PoU) among 61 countries from 2004 to 2019. Furthermore, the study assesses the impacts of corruption, the performance of the government, and political equilibrium on PoU, with the absence of violence serving as a supplementary but substantial influence. Using the Panel Co-integration and error correction techniques across the 61 countries mostly drawn from the Developing and Least-developed group of countries, the study observes that rise in food prices and corruption to increase undernourishment significantly in the long-run using both the parametric and non-parametric alternatives. Further, food production, individual's income, availability of bank branches and ATMs significantly reduces undernourishment in the long-run.

The transient influence on PoU by the explanatory variables along with the estimated Error Correction (EC) term (lagged one) of the specification is provided in Table 8.6. The correction factor to equilibrium (i.e. the EC term) is reasonably slow at 0.24 percent but still significant (at 1% level) and negative. The food production in the previous year resulted in a significant reduction of the prevalence of undernourishment (PoU) in the short term, at a confidence level of one percent. The prevalence of undernourishment was further reduced by the reduction in food prices (at 10% level), increase in individual income (at 10% level), and enhanced presence of commercial bank branches (at 1% level). In conjunction with the other external components in the mathematical framework, researchers observe that at the 5% level, the estimated level of Corruption significantly raises PoU, whereas estimations of government effectiveness, political stability, and lack of violence significantly diminish the PoU at 5% and 1%, respectively.

The combined influence of all the explanatory variables taken together is observed to influence PoU significantly at 1% level. Although the connections observed are mostly unidirectional except for the prices of food and the availability of the ATMs which were found to have no causal linkages with PoU. Thus, with observed statistically significant influences the majority of the differenced explanatory variables are found useful to estimate the changes in PoU.

10.2 Conclusion

Measuring the different contributors lending themselves towards building up of an all-inclusive financial system and consolidating their contribution towards computation of the overall achievement of a region is a work of rigor. The financial inclusion attainments help achieving the targeted wide range of developmental characteristics. While capturing the attainments in financial inclusion across the constructs, this study takes into account 4 distinct samples spread-over diverse study periods. Applying the max–min technique of normalization along with parametric assignment of weights (both FA and PCA) we determine the achievement scores and the contribution of theoretically relevant variables and dimensions that constitute the concept of financial inclusion. Identification of the theoretically relevant variables and dimensions through FA and PCA helps to find the latent structure along with their impact on the overall FI achievements. After computing the achievement scores this study focuses on the comparisons across the regions within a country, income groups within the country and outside, comparison across the continents and over the study period. This comparison gives us the idea of the effectiveness of the policies that focuses to enlarge the contour of formal finance.

Extending financial supplies is one strategy to stimulate growth within an economy, particularly in areas active signs of vulnerability are present (Beck et al., 2007). In this connection, this study makes an effort to estimate the immediate and long-run implications that these FI achievements hold on certain indicators of development. The study considers SDG1, SDG2, and SDG8 as the relevant indicators of development across 3 different samples to analyze the impact. We further consider a separate sample to check the presence of reverse causality theoretically prominent in SDG 8. We consider studying the impact both in Indian as well as in a cross-country setup exploring the properties of both cross-sectional and time-series elements together. The study's conclusions demonstrate that long-term FI successes, regardless of how the cross-sectional units are categorized, can greatly improve the developmental goals. Short-run impact is however in some specific samples are subject to certain qualifications.

10.3 Policy Recommendations

This section of the study offers evidence-based guidance to policymakers that will help to embark upon informed decisions, leading to critical customizations in the policies floated. It aims to highlight pressing social, economic, or environmental challenges, ensuring that policies are aligned with public needs and priorities. So far the discussion of this study is concerned, we categorize the policy prescription into two broad classifications. The first category aims to prescribe recommendations for the Indian authorities and the later focuses on the rest of the world. The Government of India through implementation of the National Strategy for Financial Inclusion (NSFI) took aggressive policy initiatives to promote financial inclusion especially across the rural settlements. NSFI through different policies like the PMJDY, PMSBY, and PMJJBY over 2029–24 aims to strengthen the rural households which remain the backbone of the Indian economy. Further, the relative position of India as compared to the other nations is praiseworthy. It stands 3rd among the 26 Asian countries that our study considers in Sample 2 and remains the top contender among the SAARC group in financial inclusion achievements. A consistent growth (except in the dimension of availability) from 2016 onwards is also evident across the samples we have considered. However, review of the driving forces behind the achievement in financial inclusion reveals the per capita income, geographical position of the areas, status of unemployment and availability of educational facilities as the significant factors that needs attention. Further states and UTs those have relatively better industrialization are on the heavier side of credit consumption compared to those having less industrialization. Thus, policies those aim to promote financial inclusion must initially consider improving the per capita income through reducing unemployment, so that the demand for financial services gets improved. In this connection, promoting industrialization can have significant impact as it offers better employment and thus improved income for individuals. Promoting industrialization further helps to increase the consumption of credit which ensures consistency in usage of the available funds. Subsequently, awareness campaigns on increasing financial literacy along with better educational facilities would help increasing the need for access to financial services leading to growth of the economy. So far comparisons are made across the regions, regional disparity are vibrant across the Indian states and UTs. Adverse geographical positioning hinders the economic growth of a region. In this connection, the north-east part of India due to the hilly terrain is devoid of economic advantages leading to vulnerable attainments in financial inclusion (Gupte et al., 2012).

While capitalizing on the finance-led growth hypothesis, achievements in financial inclusion are observed to significantly drive the growth of output especially in the long-run. It is quite reasonable to believe that provision of financial services help individuals and the economy as a large to grow. However, the empirical evidence in the short-run remains inconclusive which is due to the fact that it takes time to reflect the changes especially in India, where it is unlikely to get a loan instantaneously. Thus, a simplified loan application procedure that doesn't compromise with the integrity of evaluating credit worthiness of the applicant is of worth recommending.

While emphasizing on the case of reverse causality, we observe the ratio of health expenditure to the total GDP to improve achievements in financial inclusion. This highlights that enhanced expenditure by the Indian government in health sector would boost the attainments in financial inclusion. In this connection, building up of new hospitals or modifications to the existing ones and improving the coverage of the social health insurance schemes (e.g. Rastriya Swasthya Bima Yojana, Ayushman Bharat, etc.) would be worth recommending.

At the global level, experiences from the leaders would help to frame effective strategies on improving financial inclusion. From the continent of Asia, South Korea is observed to be the leader with consistently higher achievement scores in financial inclusion throughout the period of study. The success story behind its positioning emanates from the fact that this country enjoys the top position in term of accessibility and availability of financial services. With a 95% of its population having access to a formal financial account and with real-time transfer over a 24 × 7 window from 2001 onwards, the country has a high smartphone penetration rate (i.e. 95%). This offers a steady growth in the number of internet and mobile banking users, showing leapfrog in the numbers especially after the COVID-19 pandemic. Faster authentication (e.g. passwords, fingerprints etc.), pre-stored information in the mobile application and payment gateways at the terminals are some of the initiatives those are accelerating the numbers to deliver hassle free payment and transfers. The whole system was part of the strategic move of the Bank of Korea and the government in action back in 2019 that allows the financial institutions to compete in open market and customize their services that fits the consumer needs and preferences. In contrast, the financial landscape of Myanmar is underdeveloped and dominated by banks [mainly state-owned banks (GIZ, 2013)] holding nearly 99% of the formal financial sectors' assets (Turnell, 2014). Thus, adequate infrastructure with regulated competition among the providers of financial services is one among the key takeaways that is part of recommendations from these experiences.

Further, the Japanese experience in this connection would be worth mentioning. While identifying the driving forces behind the intensity to save, Horioka et al. (2007) find three major factors. Firstly, growth in income or the economy per se is followed by the savings behavior of an individual. That means when an individual observes his/her income is growing they start to save more. Secondly, the amount and intensity to save is driven by the number of children a particular household contains. People tend to save more when they have more children. Finally, the condition of the social welfare system determines the individual's drive to save for his/her future or retirement per se.

Among the latest buzz going around the adoption of revised Payment Services Directive (PSD2) and Open Banking (OB) across the European Union, very ample evidence shows up the implications on the underbanked or even the unbanked population (Preziuso et al., 2023). Although across the EU this proportion of unbanked is merely close to 4%, but the far-flung consequences of adopting PSD2 and OB are yet to emphasize. The conventional financial intermediaries have become incompetent to innovate especially after the financial crisis of 2008 (Arner et al., 2015). The digital wave as rippled by the introduction of the FinTechs have captivated the customers confidence by offering tailored services to their clients making the transactions more agile, transparent, and affordable (Gomber et al., 2017). Through offering personalized and data driven solutions, the FinTechs are able to conquer the 1.4 billion of unbanked worldwide (Findex, 2022). Through democratizing the services, FinTechs aim to ease out the economic, social, and geographical barriers (Philippon, 2019). By improving transparency, the reliance on intermediation can be mitigated resulting into crowdfunding to collect funds in somewhat unconventional way though being under the regulations (Nguyen et al., 2021) and by aborting blockchain platforms (Scott et al., 2017). This could indeed help the vulnerable especially the disadvantaged groups (e.g. lower-income groups, women and immigrants) to survive and eventually earn a living (Dorfleitner et al., 2021). The disruptions in the conventional banking habits are already in place through rising the income and consumption levels (Lashitew et al., 2019) alongside the enhanced usage of the mobile phones to transact money (Ouma et al., 2017). As a downside, this can also contribute to increase indebtedness and infuse instability which might shake the overall stability of an economy (Bateman et al., 2019). Therefore, the need of a suitable balance becomes an obvious trade-off between the conventional banking and the FinTechs to have consumer protection as well as financial stability associated to innovative products and healthy competition (Van Loo, 2018). In this regard, the open banking (OB) regulations can be instrumental to make the system flexible to adaptations to the changing needs of the clients (Zachariadis & Ozcan, 2017). Adoption of OB can increase competition offering affordable rates of financial products while maintaining confidentiality of consumer rights simultaneously (European Commission, 2015). Penetration of technology can improve productivity and profitability while offering customized solutions to clients at competitive rates (Berger, 2003; Romanova et al., 2018). Thus, joining hands between FinTechs and the conventional banks could offer benefits and safely avoid the disruptive consequence that interventions of technology usually have (Schumpeter, 1942).

10.4 Scope for Further Studies

In order to present a comprehensive view of the situation, the study makes an effort to quantify the level of financial inclusion while taking into account a wide range of factors. As the concept of financial inclusion is wide enough in spread it requires a lot of variables to be perfectly portrayed. Works of literature so far have adequately captured the essence of inclusion considering various combinations of variables. Yet they had limitations in the projection of the achievement value due to a number of reasons like data availability, its consistency, its sufficiency, and like. As a result, a number of significant variables are forced to drop while constructing an IFI. The problem amplifies when an attempt is made to consider the construction of an index for a number of years. Due to this, relatively few researches have created an index of financial inclusion that spans over a long time period and has multiple dimensions. The problems further augment when the index is supposed to be calculated for cross-country comparisons. The present study addresses these problems to a large extent but fails to provide a complete remedy. The amount of financial inclusion can be represented by a greater variety of indicators, which would improve the presentation's comprehensiveness in further research.

Further, while assessing the influence of financial inclusion on the sustainable development indicators, we refrain from analysis beyond 2019 due to insufficient availability of the data for a longer period especially after the structural break. The impact of COVID 19 has been instrumental to shape different category of population differently. Although it has prompted the use of digital finance, yet the large chunk of population were unable to pivot towards the use of digitally available financial products. Particularly across the women, rural dwellers and poor population, access to internet coupled with ownership of a smart phone and adequate skills to operate the same have been among the prominent restrictions people were faced with. Women across the developing nations are 20 percent less likely than men to have a smart phone or using mobile internet (GSMA, 2020a). Review across the China and Mexico shows that nearly one-third of the women owning a mobile do not use internet stating that they lack the necessary skills, whereas men are reported to use only 15 percent of the total category (Andrieu et al., 2019). Further data shows that among the rural consumers 600 million are still devoid of getting a mobile coverage (GSMA, 2020b). Affordability is yet another an important factor that impacts the ownership of mobile phone and access to internet. Data availability for a sufficiently longer period after 2019 especially across the usage of digital means of finance can bring new insights induced by the structural break and be considered as part of further research of this present study.

References

ADB. (2017). *Asian development outlook 2017 update*. Asian Development Bank, Manila, Philippines.

Amidžić, G., Massara, M. A., & Mialou, A. (2014). *Assessing countries' financial inclusion standing—A new composite index* (IMF Working Paper WP/14/36). https://doi.org/10.5089/9781475569681.001

Andrieu, E., Borgonovi, F., & Liebender, A. S. (2019). *The role of education and skills in bridging the digital gender divide: Evidence from APEC economies.*

Apergis, N., Filippidis, I., & Economidou, C. (2007). Financial deepening and economic growth linkages: A panel data analysis. *Review of World Economics*, *143*, 179–198. https://www.jstor.org/stable/40441131

Arestis, P., & Demetriades, P. (1997). Financial development and economic growth: Assessing the evidence. *The Economic Journal, 107*(442), 783–799. https://www.jstor.org/stable/2957802

Arestis, P., Demetriades, P. O., & Luintel, K. B. (2001). Financial development and economic growth: The role of stock markets. *Journal of Money, Credit and Banking*, 16–41. https://www.jstor.org/stable/2673870

Arner, D. W., Barberis, J., & Buckley, R. P. (2015). The evolution of Fintech: A new post-crisis paradigm. *Geo. J. Int'l l., 47*, 1271.

Ayyagari, M., & Beck, T. (2015). Financial inclusion in Asia: An overview. *Asian Development Bank Economics Working Paper Series*, (449). https://hdl.handle.net/11540/5142

Bae, J. (2022). The Bank of Korea's CBDC research: Current status and key considerations. *BIS Papers*, 107.

Bateman, M., Duvendack, M., & Loubere, N. (2019). Is fin-tech the new panacea for poverty alleviation and local development? Contesting Suri and Jack's M-Pesa findings published in Science. *Review of African Political Economy, 46*(161), 480–495. https://doi.org/10.1080/03056244.2019.1614552

Beck, T., Demirguc-Kunt, A., & Levine, R. (2007). Finance, inequality, and the poor. *Journal of Economic Growth, 12*(1), 27–49. https://doi.org/ https://doi.org/10.1007/s10887-007-9010-6

Behrman, J. R., Deolalikar, A. B., & Wolfe, B. L. (1988). Nutrients: Impacts and determinants. *The World Bank Economic Review, 2*(3), 299–320. https://www.jstor.org/stable/3989840

Berger, A. N. (2003). The economic effects of technological progress: Evidence from the banking industry. *Journal of Money, Credit and Banking*, 141–176. https://www.jstor.org/stable/3649852

Berki, S. E. (1986). A look at catastrophic medical expenses and the poor. *Health Affairs, 5*(4), 138–145. https://doi.org/10.1377/hlthaff.5.4.138

Boukhatem, J. (2016). Assessing the direct effect of financial development on poverty reduction in a panel of low-and middle-income countries. *Research in International Business and Finance, 37*, 214–230. https://doi.org/10.1016/j.ribaf.2015.11.008

Buckland, J., Guenther, B., Boichev, G., Geddie, H., & Mutch, M. (2005). There are no banks here: Financial & insurance exclusion in Winnipeg's north end. *Winnipeg Inter- City Research Alliance (WIRA).*

Burgess, R., & Pande, R. (2005). Do rural banks matter? Evidence from the Indian Social Banking Experiment. *American Economic Review, 95*(3), 780–795.: https://www.jstor.org/stable/4132740

Čihák, M., Demirgüç-Kunt, A., Feyen, E., & Levine, R. (2012). *Benchmarking financial systems around the world* (World Bank policy research working paper (6175)). https://ssrn.com/abstract=2152254

Cámara, N., & Tuesta, D. (2014). *Measuring financial inclusion: A multidimensional index* (BBVA Research Working Paper (WP No 14/26)).

Chakravarty, S. R., & Pal, R. (2013). Financial inclusion in India: An axiomatic approach. *Journal of Policy Modeling, 35*, 813–837. https://doi.org/10.1016/j.jpolmod.2012.12.007

Chibba, M. (2009). Financial inclusion, poverty reduction and the millennium development goals. *The European Journal of Development Research, 21*(2), 213–230. https://doi.org/10.1057/ejdr.2008.17

Christopoulos, D. K., & Tsionas, E. G. (2004). Financial development and economic growth: Evidence from panel unit root and cointegration tests. *Journal of Development Economics, 73*(1), 55–74. https://doi.org/10.1016/j.jdeveco.2003.03.002

Chotia, V., & Rao, N. M. (2014). Financial inclusion and human development: A state-wise analysis from India. *International Journal of Economics, Commerce and Management, II, 5*, 1–24.

Clarke, G., Xu, L., & Zou, H. (2006). Finance and income inequality: What do the data tell us? *Southern Economic Journal, 72*(3), 578–596.

Demirgüç-Kunt, A. (2006). Finance and economic development: Policy choices for developing countries. *The World Bank*. https://doi.org/10.1596/1813-9450-3955

Demirgüç-Kunt, A., & Klapper, L. (2013). Measuring financial inclusion: Explaining variation in use of financial services across and within countries (The Global Findex Database). *Brookings papers on economic activity*. Brookings Institution.

Demirgüç-Kunt, A., Klapper, L. F., Singer, D., & Van Oudheusden, P. (2015). *The global findex database 2014: Measuring financial inclusion around the world* (World Bank Policy Research Working Paper, (7255)).

Dorfleitner, G., Oswald, E. M., & Zhang, R. (2021). From credit risk to social impact: On the funding determinants in interest-free peer-to-peer lending. *Journal of Business Ethics, 170*, 375–400. https://doi.org/10.1007/s10551-019-04311-8

Duflos, E., Luchtenburg, P., Ren, L., & Chen, L. Y. (2013). *Microfinance in Myanmar sector assessment*. CGAP, IFC.

European Commission. (2015). Payment services. *The European Commission*. https://ec.europa.eu/info/law/paymentservices-psd-2-directive-eu-2015-2366_en

Findex. (2022). *The global findex database 2021*. World Bank Group.

Fungáčová, Z., & Weill, L. (2015). Understanding financial inclusion in China. *China Economic Review, 34*, 196–206. https://doi.org/10.1016/j.chieco.2014.12.004

Galor, O., & Zeira, J. (1993). Income distribution and macroeconomics. *Review of Economic Studies, 60*(1), 35–52. https://www.jstor.org/stable/2297811

Ghosh S., & Sahu T. N. (2020). How financially inclusive the Asian's are–an empirical analysis. *Rajagiri Management Journal, 14*(1). https://doi.org/10.1108/RAMJ-01-2020-0003.

Ghosh, S., Sahu, T. N., & Mallik, G. (2024). Does financial inclusion stimulate India's economy: A panel data analysis. *Journal of Public Affairs, 24*(1), Article e2891. https://doi.org/10.1002/pa.2891

GIZ. (2013). *Myanmar's Financial sector: A challenging environment for banks.* Deutsche Gesellschaft für Internationale Zusammenarbeit (GIZ) GmbH.

Gomber, P., Koch, J. A., & Siering, M. (2017). Digital finance and FinTech: Current research and future research directions. *Journal of Business Economics, 87*, 537–580. https://doi.org/10.1007/s11573-017-0852-x

GSMA. (2020a). *Connected women.* The Mobile Gender Gap Report 2020.

GSMA. (2020b). Driving the digital revolution with improved mobile coverage. https://www.gsma.com/connectivity-for-good/spectrum/wp-content/uploads/2020/11/Rural-Mobile-Coverage-Positions.pdf

Gupte, R., Venkataramani, B., & Gupta, D. (2012). Computation of financial inclusion index for India. *Procedia—Social and Behavioral Sciences, 37*, 133–149. https://doi.org/10.1016/j.sbspro.2012.03.281

Hale, G., & Long, C. (2011). What are the sources of financing for Chinese firms?. In *The evolving role of Asia in global finance* (Vol. 9, pp. 313–339). Emerald Group Publishing Limited. https://doi.org/10.1108/S1574-8715(2011)0000009018

Horioka, C. Y., Suzuki, W., & Hatta, T. (2007). Aging, savings, and public pensions in Japan. *Asian Economic Policy Review, 2*(2), 303–319.

Honohan, P. (2004). *Financial development, growth and poverty: How close are the links?* (World Bank Policy Research Working Paper 3203). http://documents.worldbank.org/curated/en/518641468763230741/Financial-development-growth-and-poverty-how-close-are-the-links

Inoue, T., & Hamori, S. (2012). How has financial deepening affected poverty reduction in India? Empirical analysis using state-level panel data. *Applied Financial Economics, 22*(5), 395–408. https://doi.org/10.1080/09603107.2011.613764

Iorgova, S., & Liu, Y. (2013). Structure of the banking sector and implications for financial stability. *China's road to greater financial stability: Some policy perspectives* (U. Das, J. Fiechter, & T. Sun, Ed., pp. 123–139). IMF.

Jalilian, H., & Kirkpatrick, C. (2002). Financial development and poverty reduction in developing countries. *International Journal of Finance and Economics, 7*(2), 97–108. https://doi.org/10.1002/ijfe.179

Jeanneney, S., & Kpodar, K. (2011). Financial development and poverty reduction: Can there be a benefit without a cost? *Journal of Development Studies, 47*(1), 143–163. https://doi.org/ https://doi.org/10.1080/00220388.2010.506918

Kravis, I. B. (1960). International differences in the distribution of income. *The Review of Economics and Statistics,* 408–416. https://doi.org/10.2307/1925690

Kumar, C., & Mishra, S. (2011). Banking outreach and household level access: Analyzing financial inclusion in India. *13th Annual Conference on Money and Finance in the Indian Economy* (pp. 25–26).

Laha, A., & Kuri, P. K. (2014). Demand for and supply of financial inclusion in India: An inter-state analysis. *Elk Asia Pacific Journal of Finance & Risk Management, 5*(1), 103–116.

Lashitew, A. A., van Tulder, R., & Liasse, Y. (2019). Mobile phones for financial inclusion: What explains the diffusion of mobile money innovations? *Research Policy, 48*(5), 1201–1215. https://doi.org/10.1016/j.respol.2018.12.010

Lucas, R. E., Jr. (1988). On the mechanics of economic development. *Journal of Monetary Economics, 22*(1), 3–42. https://doi.org/10.1016/0304-3932(88)90168-7

Luintel, K. B., & Khan, M. (1999). A quantitative reassessment of the finance–growth nexus: Evidence from a multivariate VAR. *Journal of Development Economics, 60*(2), 381–405. https://doi.org/10.1016/S0304-3878(99)00045-0

Lydall, H. F. (1968). Technical progress in Australian manufacturing. *The Economic Journal, 78*(312), 807–826. https://doi.org/10.2307/2229179

Nguyen, L. T., Hoang, T. G., Do, L. H., Ngo, X. T., Nguyen, P. H., Nguyen, G. D., & Nguyen, G. N. (2021). The role of blockchain technology-based social crowdfunding in advancing social value creation. *Technological Forecasting and Social Change, 170*, Article 120898.

Nkurunziza, J. D., Ndikumana, L., & Nyamoya, P. (2012). *The financial sector in Burundi* (No. w18289). National Bureau of Economic Research. https://doi.org/ https://doi.org/10.3386/w18289

Ouma, S. A., Odongo, T. M., & Were, M. (2017). Mobile financial services and financial inclusion: Is it a boon for savings mobilization? *Review of Development Finance, 7*(1), 29–35.

Philippon, T. (2019). On fntech and fnancial inclusion. (BIS Working Papers 84). Bank for International Settlements. https://www.bis.org/events/conf190628/philippon.pdf

Pierse, R. G., & Snell, A. J. (1995). Temporal aggregation and the power of tests for a unit root. *Journal of Econometrics, 65*(2), 333–345. https://doi.org/10.1016/0304-4076(93)01589-E

Preziuso, M., Koefer, F., & Ehrenhard, M. (2023). Open banking and inclusive finance in the European Union: Perspectives from the Dutch stakeholder ecosystem. *Financial Innovation, 9*(1), 111. https://doi.org/10.1186/s40854-023-00522-1

Rahman, Z. A. (2012). Financial inclusion in Malaysia: Tracking progress using Index. *IFC Workshop on Financial Inclusion Indicators Co-hosted by Bank Negara Malaysia and Central Bank of Malaysia.*

Romanova, I., Grima, S., Spiteri, J., & Kudinska, M. (2018). The payment services Directive II and competitiveness: The perspective of European fintech companies. *European Research Studies, 21*(2), 3–22.

Rousseau, P. L. (1998). The permanent effects of innovation on financial depth: Theory and US historical evidence from unobservable components models. *Journal of Monetary Economics, 42*(2), 387–425. https://doi.org/10.1016/S0304-3932(98)00027-0

Sahn, D. (1987). *Causes and implications of seasonal variability in household food security*. Food Policy Research Institute.

Sahoo, A. K., Pradhan, B. B., & Sahu, N. C. (2017). Determinants of financial inclusion in Tribal Districts of Odisha: An empirical investigation. *SAGE Publications, 47*(1), 45–64.

Sarma, M., & Pais, J. (2011). Financial inclusion and development. *Journal of International Development,* 613–628. https://doi.org/10.1002/jid.1698

Schumpeter, J. A. (1942). *Capitalism, socialism, and democracy*. Harper & Bros

Scott, B., Loonam, J., & Kumar, V. (2017). Exploring the rise of blockchain technology: Towards distributed collaborative organizations. *Strategic Change, 26*(5), 423–428. https://doi.org/10.1002/jsc.2142

Shingala, M. C., & Rajyaguru, A. (2015). Comparison of post hoc tests for unequal variance. *International Journal of New Technologies in Science and Engineering, 2*(5), 22–33.

Stoline, M. R. (1981). The status of multiple comparisons: Simultaneous estimation of all pair-wise comparisons in one-way ANOVA designs. *The American Statistician, 35*(3), 134–141.

Tambunlertchai, K. (2018). Determinants and barriers to financial inclusion in Myanmar: What determines access to financial services and what hinders it? *The Singapore Economic Review, 63*(01), 9–26.

Turnell, S. (2014). Banking and financial regulation and reform in Myanmar. *Journal of Southeast Asian Economies, 31*, 225–240.

Torsten, P., & Tabellini, G. (1991). *'Is Inequality Harmful for Growth? Theory and Evidence* (No. 91–155). Working Paper.

Van Loo, R. (2018). Making innovation more competitive: The case of fintech. *UCLA Law Review 65*, 232.

Wachtel, P., & Rousseau, P. (1994). *Financial intermediation and economic growth: A historical comparison of the US, UK and Canada* (No. 94–04).

Wei, S. J., & Zhang, X. (2011). The competitive saving motive: Evidence from rising sex ratios and savings rates in China. *Journal of Political Economy, 119*(3), 511–564.

World Bank. (2015). *World development indicators.*

Wyszewianski, L. (1986). Financially catastrophic and high-cost cases: Definitions, distinctions, and their implications for policy formulation. *Inquiry*, 382–394. https://www.jstor.org/stable/29771827

Yang, D. T. (2012). Aggregate savings and external imbalances in China. *Journal of Economic Perspectives, 26*(4), 125–146. https://doi.org/10.1257/jep.26.4.125

Yoshino, N., Helble, M., Takeda, A., & Shimazaki, S. (2018). Postal savings and financial inclusion in Japan. *Postal savings: Reaching everyone in Asia*, 268–285.

Zachariadis, M., & Ozcan, P. (2017). *The API economy and digital transformation in financial services: The case of open banking*. https://doi.org/10.2139/ssrn.2975199

Zhang, J., Wang, L., & Wang, S. (2012). Financial development and economic growth: Recent evidence from China. *Journal of Comparative Economics, 40*(3), 393–412. https://doi.org/10.1016/j.jce.2012.01.001

Appendix

S. Ghosh et al., *Financial Inclusion*, Sustainable Development Goals Series, https://doi.org/10.1007/978-981-95-3385-5

Table 4.3.1 Dimensional achievements and financial inclusion scores for 2022

Country name	*Stability*	*Service potential*	*Availability*	*Accessibility*	*Efficiency*	*IFI*
Afghanistan	1.00	0.35	0	0	0.39	0.205
Armenia	0.68	0.57	0.10	1	0.46	0.635
Azerbaijan	0.23	0.46	0.05	0.74	0.60	0.469
Bangladesh	0.56	0.61	0.32	0.02	0.68	0.234
Bhutan	0.45	0.50	0.02	0.85	0.61	0.431
Brunei Darussalam	1.00	0.61	0.07	0.91	0.60	0.457
Cambodia	1.00	0.73	0.01	0.69	0.49	0.458
Georgia	1.00	0.66	0.07	1	0.44	0.363
India	0.90	0.60	0.23	0.83	0.44	0.653
Indonesia	0.68	0.55	0.10	0.88	0.44	0.276
Japan	0.56	1.00	0.70	0.96	0.75	0.625
Jordan	0.00	0.50	0.05	0.72	0.60	0.357
Korea, Rep.	0.56	0.50	0.62	1	0.64	0.686
Lebanon	0.23	0.98	0.53	0.74	0.64	0.318
Malaysia	0.78	0.86	0.05	0.97	0.77	0.474
Maldives	0.23	0.50	0.88	0.61	0.69	0.347
Mongolia	1.00	0.53	0.00	0.97	0.36	0.676
Myanmar	0.23	0.57	0.01	0.01	0.51	0.163
Nepal	1.00	0.56	0.06	0.02	0.69	0.259
Pakistan	0.23	0.51	0.06	0.79	0.60	0.358
Philippines	0.10	0.56	0.15	0.81	0.52	0.353
Saudi Arabia	0.10	0.52	0.01	1.00	0.72	0.327
Thailand	0.69	1.00	0.17	0.98	0.62	0.534
Turkey	0.45	0.69	0.10	0.91	0.53	0.615
United Arab Emirates	0.23	0.69	0.10	1.00	0.72	0.424
Vietnam	0.90	0.83	0.07	0.81	0.55	0.507

Source Authors Calculation

Table 4.3.2 Dimensional achievements and financial inclusion scores for 2021

Country name	*Stability*	*Service potential*	*Availability*	*Accessibility*	*Efficiency*	*IFI*
Afghanistan	0.96	0.34	0	0	0.37	0.21
Armenia	0.65	0.54	0.09	0.96	0.44	0.57
Azerbaijan	0.22	0.44	0.05	0.71	0.57	0.42
Bangladesh	0.53	0.58	0.30	0.02	0.65	0.25
Bhutan	0.43	0.47	0.02	0.81	0.58	0.41
Brunei Darussalam	0.38	0.58	0.07	0.87	0.57	0.45
Cambodia	1.00	0.69	0.01	0.66	0.46	0.45
Georgia	0.96	0.63	0.07	0.98	0.42	0.38
India	0.86	0.57	0.22	0.79	0.42	0.59
Indonesia	0.65	0.52	0.09	0.83	0.42	0.31
Japan	0.53	1	0.67	0.91	0.72	0.62
Jordan	0.00	0.47	0.05	0.68	0.57	0.34
Korea, Rep.	0.53	0.48	0.59	1.00	0.61	0.68
Lebanon	0.22	0.94	0.51	0.71	0.61	0.36
Malaysia	0.74	0.82	0.05	0.93	0.73	0.47
Maldives	0.22	0.47	0.83	0.58	0.66	0.36
Mongolia	0.96	0.51	0.01	0.93	0.35	0.59
Myanmar	0.22	0.54	0.00	0.01	0.49	0.16
Nepal	1.00	0.53	0.06	0.02	0.66	0.28
Pakistan	0.22	0.49	0.06	0.75	0.57	0.35
Philippines	0.09	0.53	0.14	0.78	0.50	0.34
Saudi Arabia	0.09	0.50	0.02	0.96	0.68	0.32
Thailand	0.74	1.00	0.16	0.94	0.56	0.52
Turkey	0.43	0.66	0.09	0.87	0.51	0.55
United Arab Emirates	0.22	0.66	0.09	0.96	0.68	0.41
Vietnam	0.86	0.79	0.07	0.78	0.52	0.49

Source Authors Calculation

Table 4.3.3 Dimensional achievements and financial inclusion scores for 2020

Country name	*Stability*	*Service potential*	*Availability*	*Accessibility*	*Efficiency*	*IFI*
Afghanistan	0.92	0.32	0	0	0.35	0.21
Armenia	0.62	0.52	0.09	0.92	0.42	0.51
Azerbaijan	0.21	0.42	0.03	0.67	0.54	0.37
Bangladesh	0.51	0.55	0.29	0.02	0.62	0.27
Bhutan	0.41	0.45	0.02	0.77	0.55	0.39
Brunei Darussalam	0.21	0.55	0.07	0.83	0.54	0.39
Cambodia	1.00	0.66	0.01	0.63	0.44	0.44
Georgia	0.92	0.60	0.07	0.94	0.40	0.40
India	0.82	0.54	0.21	0.75	0.40	0.54
Indonesia	0.62	0.50	0.09	0.79	0.40	0.33
Japan	0.51	0.99	0.64	0.87	0.68	0.61
Jordan	0.00	0.45	0.05	0.65	0.54	0.31
Korea, Rep.	0.51	0.95	0.96	0.99	0.58	0.67
Lebanon	0.21	0.89	0.49	0.67	0.58	0.40
Malaysia	0.71	0.78	0.06	0.88	0.69	0.47
Maldives	0.21	0.45	0.79	0.55	0.63	0.38
Mongolia	0.92	0.49	0.01	0.88	0.33	0.52
Myanmar	0.21	0.52	0.02	0.01	0.46	0.16
Nepal	1.00	0.51	0.06	0.02	0.63	0.30
Pakistan	0.21	0.46	0.06	0.72	0.54	0.33
Philippines	0.09	0.51	0.13	0.74	0.47	0.32
Saudi Arabia	0.09	0.47	0.03	0.92	0.65	0.31
Thailand	0.71	0.95	0.15	0.89	0.53	0.51
Turkey	0.41	0.63	0.09	0.83	0.49	0.49
United Arab Emirates	0.21	0.63	0.09	0.92	0.65	0.40
Vietnam	0.82	0.75	0.07	0.74	0.50	0.47

Source Authors Calculation

Table 4.3.4 Dimensional achievements and financial inclusion scores for 2019

Country name	*Stability*	*Service potential*	*Availability*	*Accessibility*	*Efficiency*	*IFI*
Afghanistan	0.87	0.30	0	0	0.34	0.21
Armenia	0.59	0.49	0.07	0.87	0.40	0.47
Azerbaijan	0.20	0.40	0.05	0.64	0.51	0.33
Bangladesh	0.48	0.53	0.27	0.02	0.59	0.28
Bhutan	0.39	0.43	0.02	0.74	0.53	0.37
Brunei Darussalam	0.16	0.53	0.08	0.79	0.51	0.39
Cambodia	0.96	0.63	0.03	0.60	0.42	0.43
Georgia	0.87	0.57	0.07	0.89	0.38	0.42
India	0.78	0.51	0.20	0.71	0.38	0.49
Indonesia	0.59	0.47	0.09	0.76	0.38	0.37
Japan	0.48	1.00	0.61	0.83	0.65	0.61
Jordan	0.00	0.43	0.05	0.62	0.51	0.29
Korea, Rep.	0.48	0.90	0.91	0.99	0.56	0.66
Lebanon	0.20	0.85	0.46	0.64	0.56	0.38
Malaysia	0.67	0.75	0.06	0.84	0.66	0.51
Maldives	0.20	0.43	0.76	0.53	0.60	0.40
Mongolia	0.87	0.46	0.01	0.84	0.32	0.46
Myanmar	0.20	0.49	0.00	0.01	0.44	0.17
Nepal	0.96	0.48	0.06	0.02	0.60	0.31
Pakistan	0.20	0.44	0.04	0.68	0.51	0.32
Philippines	0.08	0.48	0.13	0.70	0.45	0.31
Saudi Arabia	0.08	0.45	0.02	0.87	0.62	0.31
Thailand	0.67	0.90	0.15	0.85	0.50	0.50
Turkey	0.39	0.60	0.09	0.79	0.46	0.44
United Arab Emirates	0.20	0.60	0.09	0.87	0.62	0.39
Vietnam	0.78	0.71	0.07	0.70	0.47	0.46

Source Authors Calculation

Table 4.3.5 Dimensional achievements and financial inclusion scores for 2018

Country name	*Stability*	*Service potential*	*Availability*	*Accessibility*	*Efficiency*	*IFI*
Afghanistan	0.83	0.29	0	0	0.32	0.22
Armenia	0.56	0.47	0.07	0.83	0.38	0.43
Azerbaijan	0.19	0.38	0.03	0.61	0.49	0.30
Bangladesh	0.51	0.50	0.26	0.02	0.56	0.30
Bhutan	0.41	0.41	0.02	0.70	0.50	0.36
Brunei Darussalam	0.12	0.50	0.08	0.75	0.49	0.43
Cambodia	0.91	0.60	0.03	0.57	0.40	0.42
Georgia	0.83	0.54	0.05	0.85	0.36	0.43
India	0.74	0.49	0.19	0.68	0.36	0.46
Indonesia	0.56	0.45	0.09	0.72	0.36	0.37
Japan	0.46	0.95	0.58	0.79	0.62	0.61
Jordan	0.00	0.41	0.05	0.59	0.49	0.27
Korea, Rep.	0.46	0.86	0.87	0.99	0.53	0.65
Lebanon	0.19	0.81	0.44	0.61	0.53	0.45
Malaysia	0.64	0.71	0.05	0.80	0.63	0.51
Maldives	0.19	0.41	0.72	0.50	0.57	0.39
Mongolia	0.83	0.44	0.01	0.80	0.30	0.41
Myanmar	0.19	0.47	0.00	0.01	0.42	0.17
Nepal	0.91	0.46	0.04	0.02	0.57	0.32
Pakistan	0.19	0.42	0.04	0.65	0.49	0.31
Philippines	0.08	0.46	0.12	0.67	0.43	0.31
Saudi Arabia	0.08	0.43	0.02	0.83	0.59	0.28
Thailand	0.64	0.86	0.14	0.81	0.48	0.48
Turkey	0.37	0.57	0.09	0.75	0.44	0.41
United Arab Emirates	0.19	0.57	0.07	0.83	0.59	0.38
Vietnam	0.74	0.68	0.05	0.67	0.45	0.44

Source Authors Calculation

Table 4.3.6 Dimensional achievements and financial inclusion scores for 2017

Country name	*Stability*	*Service potential*	*Availability*	*Accessibility*	*Efficiency*	*IFI*
Afghanistan	0.79	0.28	0.01	0	0.30	0.22
Armenia	0.53	0.45	0.07	0.79	0.36	0.40
Azerbaijan	0.18	0.36	0.05	0.58	0.47	0.30
Bangladesh	0.44	0.48	0.25	0.02	0.53	0.32
Bhutan	0.35	0.39	0.02	0.67	0.48	0.35
Brunei Darussalam	0.95	0.48	0.07	0.71	0.47	0.46
Cambodia	0.87	0.57	0.03	0.54	0.38	0.41
Georgia	0.79	0.51	0.07	0.81	0.34	0.43
India	0.70	0.47	0.18	0.65	0.34	0.44
Indonesia	0.53	0.43	0.09	0.69	0.34	0.38
Japan	0.44	0.90	0.55	0.75	0.59	0.61
Jordan	0.00	0.39	0.05	0.56	0.47	0.26
Korea, Rep.	0.44	0.82	0.83	1.03	0.50	0.66
Lebanon	0.18	0.77	0.42	0.58	0.50	0.46
Malaysia	0.61	0.68	0.05	0.76	0.60	0.48
Maldives	0.18	0.39	0.69	0.48	0.54	0.43
Mongolia	0.79	0.42	0.01	0.76	0.29	0.38
Myanmar	0.18	0.45	0.00	0.01	0.40	0.19
Nepal	0.87	0.44	0.06	0.02	0.54	0.31
Pakistan	0.18	0.40	0.04	0.62	0.47	0.32
Philippines	0.08	0.44	0.11	0.64	0.41	0.31
Saudi Arabia	0.08	0.41	0.02	0.79	0.56	0.31
Thailand	0.61	0.82	0.13	0.77	0.46	0.50
Turkey	0.35	0.54	0.09	0.71	0.42	0.39
United Arab Emirates	0.18	0.54	0.10	0.79	0.56	0.38
Vietnam	0.70	0.65	0.07	0.64	0.43	0.45

Source Authors Calculation

Table 4.3.7 Dimensional achievements and financial inclusion scores for 2016

Country name	*Stability*	*Service potential*	*Availability*	*Accessibility*	*Efficiency*	*IFI*
Afghanistan	0.90	0.11	0	0	0.58	0.23
Armenia	0.63	0.34	0.14	0.59	0.47	0.41
Azerbaijan	0.21	0.27	0.07	0.40	0.24	0.23
Bangladesh	0.53	0.36	0.38	0.09	0.35	0.33
Bhutan	0.42	0.32	0.01	0.42	0.73	0.34
Brunei Darussalam	0.53	0.41	0.11	0.57	0.44	0.39
Cambodia	0.95	0.49	0.03	0.27	0.70	0.40
Georgia	0.93	0.38	0.09	0.70	0.86	0.49
India	0.63	0.40	0.26	0.43	0.33	0.40
Indonesia	0.63	0.29	0.14	0.45	0.69	0.41
Japan	0.53	0.90	0.82	0.83	0.18	0.57
Jordan	0.00	0.50	0.06	0.16	0.46	0.21
Korea, Rep.	0.53	0.89	0.81	0.84	0.27	0.60
Lebanon	0.21	0.85	0.64	0.45	0.40	0.47
Malaysia	0.74	0.79	0.07	0.50	0.34	0.43
Maldives	0.21	0.27	0.95	0.35	1	0.45
Mongolia	0.53	0.36	0.01	0.75	0.27	0.34
Myanmar	0.21	0.27	0.00	0.02	0.28	0.15
Nepal	0.63	0.48	0.07	0.10	0.59	0.33
Pakistan	0.21	0.25	0.08	0.32	0.53	0.26
Philippines	0.11	0.35	0.16	0.11	0.43	0.22
Saudi Arabia	0.11	0.38	0.02	0.48	0.45	0.26
Thailand	0.32	0.77	0.22	0.56	0.41	0.42
Turkey	0.21	0.44	0.13	0.63	0.51	0.36
United Arab Emirates	0.21	0.59	0.15	0.47	0.44	0.35
Vietnam	0.74	0.63	0.11	0.32	0.36	0.39

Source Authors Calculation

Table 4.3.8 Dimensional achievements and financial inclusion scores for 2015

Country name	*Stability*	*Service potential*	*Availability*	*Accessibility*	*Efficiency*	*IFI*
Afghanistan	0.95	0.16	0	0	0.37	0.22
Armenia	0.42	0.36	0.16	0.57	0.41	0.37
Azerbaijan	0.21	0.34	0.09	0.38	0.28	0.25
Bangladesh	0.53	0.39	0.42	0.08	0.38	0.34
Bhutan	0.42	0.30	0.01	0.38	0.76	0.33
Brunei Darussalam	0.42	0.40	0.12	0.54	0.51	0.38
Cambodia	1	0.46	0.03	0.21	0.70	0.38
Georgia	0.95	0.39	0.11	0.66	0.71	0.48
India	0.63	0.42	0.28	0.38	0.40	0.41
Indonesia	0.53	0.33	0.16	0.43	0.58	0.39
Japan	0.53	0.88	0.89	0.86	0.35	0.64
Jordan	0.00	0.45	0.07	0.32	0.48	0.24
Korea, Rep.	0.53	0.86	0.85	0.83	0.32	0.62
Lebanon	0.21	0.82	0.73	0.44	0.44	0.48
Malaysia	0.74	0.78	0.08	0.44	0.43	0.44
Maldives	0.21	0.31	0.96	0.30	0.92	0.44
Mongolia	0.53	0.37	0.01	0.70	0.40	0.36
Myanmar	0.21	0.31	0.00	0.01	0.33	0.16
Nepal	0.63	0.42	0.07	0.09	0.60	0.32
Pakistan	0.21	0.27	0.09	0.17	0.56	0.25
Philippines	0.11	0.36	0.18	0.11	0.45	0.23
Saudi Arabia	0.11	0.38	0.03	0.45	0.51	0.27
Thailand	0.32	0.78	0.25	0.51	0.45	0.43
Turkey	0.21	0.46	0.15	0.60	0.48	0.36
United Arab Emirates	0.21	0.52	0.17	0.45	0.55	0.41
Vietnam	0.74	0.56	0.12	0.29	0.42	0.39

Source Authors Calculation

Table 4.3.9 Dimensional achievements and financial inclusion scores for 2014

Country name	*Stability*	*Service potential*	*Availability*	*Accessibility*	*Efficiency*	*IFI*
Afghanistan	0.95	0.44	0	0	0.27	0.25
Armenia	0.42	0.55	0.16	0.57	0.36	0.39
Azerbaijan	0.21	0.50	0.09	0.42	0.45	0.32
Bangladesh	0.53	0.59	0.40	0.06	0.41	0.37
Bhutan	0.42	0.29	0.02	0.39	0.74	0.29
Brunei Darussalam	0.42	0.57	0.13	0.51	0.51	0.41
Cambodia	1.00	0.54	0.03	0.23	0.56	0.38
Georgia	0.95	0.58	0.11	0.65	0.60	0.51
India	0.63	0.50	0.27	0.41	0.32	0.39
Indonesia	0.42	0.58	0.16	0.43	0.53	0.41
Japan	0.53	0.79	0.91	0.80	0.24	0.59
Jordan	0.00	0.25	0.07	0.13	0.38	0.15
Korea, Rep.	0.53	0.79	0.86	0.90	0.25	0.60
Lebanon	0.21	0.56	0.73	0.44	0.33	0.44
Malaysia	0.74	0.67	0.09	0.51	0.40	0.44
Maldives	0.21	0.53	0.95	0.33	0.92	0.50
Mongolia	0.53	0.44	0.01	0.61	0.37	0.37
Myanmar	0.21	0.61	0.00	0.01	0.30	0.20
Nepal	0.63	0.43	0.07	0.07	0.48	0.29
Pakistan	0.21	0.48	0.09	0.17	0.45	0.26
Philippines	0.11	0.47	0.17	0.09	0.35	0.22
Saudi Arabia	0.11	0.44	0.02	0.49	0.45	0.27
Thailand	0.32	0.81	0.25	0.53	0.39	0.43
Turkey	0.21	0.61	0.15	0.64	0.41	0.38
United Arab Emirates	0.21	0.41	0.17	0.47	0.47	0.33
Vietnam	0.74	0.55	0.12	0.30	0.29	0.37

Source Authors Calculation

Table 4.3.10 Dimensional achievements and financial inclusion scores for 2013

Country name	*Stability*	*Service potential*	*Availability*	*Accessibility*	*Efficiency*	*IFI*
Afghanistan	0.95	0.36	0	0	0.39	0.26
Armenia	0.42	0.49	0.16	0.54	0.49	0.41
Azerbaijan	0.21	0.44	0.09	0.16	0.44	0.25
Bangladesh	0.53	0.54	0.40	0.07	0.46	0.38
Bhutan	0.42	0.30	0.02	0.37	0.80	0.29
Brunei Darussalam	0.42	0.50	0.14	0.56	0.53	0.41
Cambodia	1.00	0.48	0.01	0.21	0.61	0.37
Georgia	0.95	0.52	0.12	0.64	0.67	0.51
India	0.63	0.50	0.24	0.38	0.40	0.42
Indonesia	0.42	0.53	0.16	0.42	0.55	0.40
Japan	0.53	0.87	0.91	0.82	0.27	0.61
Jordan	0.00	0.32	0.07	0.14	0.46	0.18
Korea, Rep.	0.53	0.75	0.88	0.90	0.30	0.61
Lebanon	0.21	0.64	0.75	0.45	0.36	0.45
Malaysia	0.74	0.74	0.09	0.52	0.44	0.51
Maldives	0.21	0.49	0.92	0.33	0.89	0.49
Mongolia	0.53	0.44	0.01	0.60	0.45	0.37
Myanmar	0.21	0.51	0.00	0.00	0.27	0.21
Nepal	0.63	0.45	0.07	0.07	0.52	0.31
Pakistan	0.21	0.45	0.08	0.17	0.42	0.25
Philippines	0.11	0.44	0.17	0.09	0.40	0.23
Saudi Arabia	0.11	0.41	0.02	0.48	0.46	0.27
Thailand	0.32	0.81	0.25	0.53	0.42	0.43
Turkey	0.21	0.58	0.16	0.61	0.46	0.38
United Arab Emirates	0.21	0.43	0.18	0.42	0.48	0.33
Vietnam	0.74	0.56	0.12	0.29	0.35	0.38

Source Authors Calculation

Table 4.5.1 Dimensional achievements and FI scores for 2022

Countries/ indicators	*Stability &viability*	*Efficiency*	*Depth*	*Service outreach*	*IFI*
Latvia	0.57	0.83	0.35	0.18	0.60
Italy	0.96	0.95	0.75	0.66	0.98
Estonia	0.70	0.83	0.56	0.28	0.69
Croatia	0.84	0.96	0.61	0.44	0.84
Czech Republic	0.30	0.88	0.43	0.39	0.22
Greece	0.96	0.52	0.86	0.44	0.70
Netherlands	0.97	0.82	0.89	0.70	0.99
Spain	0.73	0.96	0.91	0.61	0.91
Switzerland	0.70	0.59	0.97	0.90	0.99
Bosnia and Herzegovina	0.55	0.97	0.45	0.25	0.69
Macedonia, FYR	0.49	0.89	0.33	0.33	0.58
Georgia	0.61	0.95	0.33	0.23	0.68
Saudi Arabia	0.95	0.95	0.54	0.15	0.80
Israel	0.59	0.96	0.53	0.73	0.95
Japan	0.64	0.92	0.95	1.00	0.97
Malaysia	0.48	0.96	0.98	0.37	0.67
Maldives	0.97	0.97	0.21	0.68	0.79
Turkey	0.97	0.96	0.44	0.30	0.30
Armenia	0.81	0.98	0.27	0.26	0.68
Pakistan	0.61	0.97	0.09	0.07	0.44
Indonesia	0.73	0.96	0.16	0.25	0.55
Philippines	0.18	0.97	0.37	0.29	0.74
India	0.61	0.97	0.44	0.41	0.81
Bhutan	0.63	0.59	0.38	0.18	0.53

Source Calculated by Authors

Table 4.5.2 Dimensional achievements and FI scores for 2021

Countries/ indicators	*Stability &viability*	*Efficiency*	*Depth*	*Service outreach*	*IFI*
Latvia	0.56	0.81	0.34	0.18	0.61
Italy	0.99	0.98	0.74	0.65	0.97
Estonia	0.69	0.81	0.55	0.28	0.69
Croatia	0.82	0.99	0.60	0.43	0.85
Czech Republic	0.29	0.86	0.42	0.39	0.33
Greece	0.99	0.51	0.84	0.43	0.73
Netherlands	1.00	0.80	0.87	0.68	0.99
Spain	0.71	0.99	0.89	0.60	0.91
Switzerland	0.69	0.58	1.00	0.88	0.98
Bosnia and Herzegovina	0.53	1.00	0.44	0.24	0.69
Macedonia, FYR	0.48	0.87	0.32	0.32	0.60
Georgia	0.60	0.98	0.32	0.22	0.68
Saudi Arabia	0.94	0.98	0.53	0.14	0.79
Israel	0.58	0.99	0.52	0.72	0.94
Japan	0.62	0.95	0.98	0.98	0.97
Malaysia	0.47	0.99	0.96	0.36	0.72
Maldives	1.00	1.00	0.21	0.66	0.79
Turkey	1.00	0.99	0.43	0.30	0.43
Armenia	0.79	0.96	0.26	0.25	0.69
Pakistan	0.60	1.00	0.09	0.07	0.46
Indonesia	0.71	0.99	0.15	0.24	0.57
Philippines	0.18	1.00	0.36	0.29	0.73
India	0.60	1.00	0.43	0.40	0.81
Bhutan	0.61	0.58	0.38	0.18	0.55

Source Calculated by Authors

Table 4.5.3 Dimensional achievements and FI scores for 2020

Countries/ indicators	*Stability & viability*	*Efficiency*	*Depth*	*Service outreach*	*IFI*
Latvia	0.55	0.81	0.34	0.17	0.61
Italy	0.98	0.97	0.73	0.64	0.97
Estonia	0.68	0.81	0.55	0.27	0.70
Croatia	0.82	0.98	0.59	0.43	0.85
Czech Republic	0.29	0.85	0.42	0.38	0.43
Greece	0.98	0.51	0.83	0.43	0.75
Netherlands	0.99	0.79	0.86	0.68	0.98
Spain	0.71	0.98	0.88	0.59	0.91
Switzerland	0.68	0.57	0.99	0.87	0.97
Bosnia and Herzegovina	0.53	0.99	0.44	0.24	0.69
Macedonia, FYR	0.47	0.86	0.32	0.32	0.62
Georgia	0.60	0.97	0.32	0.22	0.68
Saudi Arabia	0.93	0.97	0.52	0.14	0.79
Israel	0.57	0.98	0.51	0.71	0.92
Japan	0.62	0.94	0.97	0.97	0.97
Malaysia	0.46	0.98	0.95	0.36	0.77
Maldives	0.99	0.99	0.21	0.66	0.80
Turkey	0.99	0.98	0.43	0.29	0.53
Armenia	0.78	0.95	0.26	0.25	0.69
Pakistan	0.60	0.99	0.09	0.07	0.48
Indonesia	0.71	0.98	0.15	0.24	0.59
Philippines	0.18	0.99	0.36	0.28	0.72
India	0.60	0.99	0.43	0.39	0.80
Bhutan	0.61	0.57	0.37	0.17	0.57

Source Calculated by Authors

Table 4.5.4 Dimensional achievements and FI scores for 2019

Countries/ indicators	*Stability & viability*	*Efficiency*	*Depth*	*Service outreach*	*IFI*
Latvia	0.55	0.80	0.34	0.17	0.62
Italy	0.97	1.00	0.72	0.64	0.97
Estonia	0.68	0.80	0.54	0.27	0.70
Croatia	0.81	0.97	0.58	0.42	0.85
Czech Republic	0.28	0.84	0.41	0.38	0.51
Greece	0.97	0.50	0.82	0.42	0.77
Netherlands	0.98	0.79	0.85	0.67	0.98
Spain	0.70	0.97	0.88	0.58	0.91
Switzerland	0.68	0.57	0.98	0.87	0.96
Bosnia and Herzegovina	0.52	0.98	0.43	0.24	0.70
Macedonia, FYR	0.47	0.85	0.31	0.31	0.64
Georgia	0.59	1.00	0.31	0.22	0.68
Saudi Arabia	0.92	1.00	0.52	0.14	0.79
Israel	0.57	0.97	0.51	0.70	0.90
Japan	0.61	0.93	0.96	1.00	0.97
Malaysia	0.46	0.97	0.94	0.36	0.80
Maldives	0.98	0.98	0.21	0.65	0.80
Turkey	0.98	0.97	0.42	0.29	0.62
Armenia	0.78	0.94	0.26	0.25	0.70
Pakistan	0.59	0.98	0.09	0.06	0.50
Indonesia	0.70	0.97	0.15	0.24	0.60
Philippines	0.17	0.98	0.36	0.28	0.70
India	0.59	0.98	0.42	0.39	0.80
Bhutan	0.60	0.57	0.37	0.17	0.58

Source Calculated by Authors

Table 4.5.5 Dimensional achievements and FI scores for 2018

Countries/ indicators	*Stability & viability*	*Efficiency*	*Depth*	*Service outreach*	*IFI*
Latvia	0.54	0.78	0.33	0.17	0.62
Italy	0.95	0.98	0.72	0.63	0.96
Estonia	0.66	0.78	0.54	0.27	0.71
Croatia	0.79	0.99	0.58	0.42	0.85
Czech Republic	0.28	0.82	0.41	0.37	0.58
Greece	0.98	0.49	0.81	0.42	0.80
Netherlands	0.99	0.77	0.85	0.66	0.98
Spain	0.69	0.99	0.87	0.58	0.92
Switzerland	0.66	0.56	1.00	0.86	0.95
Bosnia and Herzegovina	0.51	1.00	0.43	0.24	0.70
Macedonia, FYR	0.46	0.84	0.31	0.31	0.66
Georgia	0.58	0.98	0.31	0.21	0.68
Saudi Arabia	0.90	0.98	0.51	0.14	0.78
Israel	0.56	0.95	0.50	0.70	0.88
Japan	0.60	0.91	0.95	0.99	0.96
Malaysia	0.45	0.98	0.93	0.35	0.82
Maldives	0.99	0.99	0.20	0.64	0.80
Turkey	0.99	0.95	0.42	0.29	0.70
Armenia	0.76	0.92	0.26	0.25	0.70
Pakistan	0.58	0.99	0.09	0.06	0.51
Indonesia	0.69	1.00	0.15	0.24	0.62
Philippines	0.17	0.99	0.35	0.28	0.69
India	0.58	0.99	0.42	0.39	0.80
Bhutan	0.59	0.56	0.36	0.17	0.60

Source Calculated by Authors

Table 4.5.6 Dimensional achievements and FI scores for 2017

Countries/ indicators	*Stability & viability*	*Efficiency*	*Depth*	*Service outreach*	*IFI*
Latvia	0.53	0.77	0.33	0.17	0.63
Italy	0.93	0.99	0.70	0.62	0.96
Estonia	0.65	0.77	0.53	0.26	0.72
Croatia	0.78	1.00	0.57	0.41	0.86
Czech Republic	0.27	0.81	0.40	0.37	0.64
Greece	0.99	0.48	0.80	0.41	0.82
Netherlands	1.00	0.76	0.83	0.65	0.97
Spain	0.67	1.00	0.85	0.57	0.92
Switzerland	0.65	0.55	0.98	0.84	0.94
Bosnia and Herzegovina	0.50	0.98	0.42	0.23	0.71
Macedonia, FYR	0.45	0.82	0.30	0.30	0.68
Georgia	0.57	1.00	0.30	0.21	0.68
Saudi Arabia	0.88	1.00	0.50	0.14	0.78
Israel	0.55	0.93	0.49	0.68	0.87
Japan	0.59	0.89	0.93	1	0.96
Malaysia	0.44	0.99	0.91	0.35	0.83
Maldives	1	0.97	0.20	0.63	0.80
Turkey	1	0.93	0.41	0.28	0.75
Armenia	0.75	0.90	0.25	0.24	0.71
Pakistan	0.57	1	0.08	0.06	0.53
Indonesia	0.67	0.98	0.15	0.23	0.64
Philippines	0.17	1	0.35	0.27	0.67
India	0.57	1	0.41	0.38	0.79
Bhutan	0.58	0.55	0.36	0.17	0.62

Source Calculated by Authors

Table 4.5.7 Dimensional achievements and FI scores for 2016

Countries/ indicators	*Stability & viability*	*Efficiency*	*Depth*	*Service outreach*	*IFI*
Latvia	0.50	0.73	0.31	0.16	0.63
Italy	0.89	0.94	0.67	0.59	0.96
Estonia	0.62	0.73	0.50	0.25	0.76
Croatia	0.74	0.95	0.54	0.39	0.86
Czech Republic	0.26	0.77	0.38	0.35	0.66
Greece	0.94	0.46	0.76	0.39	0.84
Netherlands	0.97	0.72	0.79	0.62	0.96
Spain	0.64	0.96	0.81	0.54	0.93
Switzerland	0.62	0.52	0.96	0.80	0.92
Bosnia and Herzegovina	0.48	0.93	0.40	0.22	0.7
Macedonia, FYR	0.43	0.78	0.29	0.29	0.67
Georgia	0.54	0.98	0.29	0.20	0.68
Saudi Arabia	0.84	0.98	0.48	0.13	0.76
Israel	0.52	0.89	0.47	0.65	0.86
Japan	0.56	0.85	0.89	1.00	0.98
Malaysia	0.42	0.94	0.87	0.33	0.82
Maldives	1.00	0.92	0.19	0.60	0.82
Turkey	1.00	0.89	0.39	0.27	0.80
Armenia	0.71	0.86	0.24	0.23	0.70
Pakistan	0.54	1.00	0.08	0.06	0.53
Indonesia	0.64	0.96	0.14	0.22	0.64
Philippines	0.16	1.00	0.33	0.26	0.59
India	0.55	1.00	0.39	0.36	0.77
Bhutan	0.55	0.52	0.34	0.16	0.61

Source Calculated by Authors

Table 4.5.8 Dimensional achievements and FI scores for 2015

Countries/ indicators	*Stability & viability*	*Efficiency*	*Depth*	*Service outreach*	*IFI*
Latvia	0.51	0.74	0.35	0.17	0.66
Italy	1.00	0.95	0.69	0.62	0.97
Estonia	0.33	0.90	0.51	0.25	0.7
Croatia	0.79	1.00	0.59	0.41	0.88
Czech Republic	0.58	0.96	0.40	0.37	0.79
Greece	0.96	0.50	0.81	0.44	0.87
Netherlands	1.00	0.78	0.79	0.67	0.98
Spain	0.67	1.00	0.83	0.58	0.95
Switzerland	0.70	0.52	0.95	0.84	0.94
Bosnia and Herzegovina	0.59	0.99	0.43	0.24	0.75
Macedonia, FYR	0.68	0.87	0.34	0.32	0.76
Georgia	0.46	0.95	0.28	0.20	0.66
Saudi Arabia	1.00	0.97	0.47	0.13	0.77
Israel	0.38	1.00	0.49	0.67	0.83
Japan	0.66	1.00	0.88	1.00	0.95
Malaysia	0.59	1.00	0.88	0.35	0.88
Maldives	0.93	0.74	0.21	0.59	0.8
Turkey	1.00	0.96	0.41	0.28	0.81
Armenia	0.84	0.81	0.26	0.24	0.72
Pakistan	1.00	0.92	0.09	0.07	0.6
Indonesia	0.76	0.99	0.17	0.23	0.68
Philippines	0.41	1.00	0.35	0.26	0.69
India	0.55	1.00	0.41	0.37	0.79
Bhutan	0.73	0.50	0.38	0.15	0.66

Source Calculated by Authors

Table 4.5.9 Dimensional achievements and FI scores for 2014

Countries/ indicators	*Stability & viability*	*Efficiency*	*Depth*	*Service outreach*	*IFI*
Latvia	0.58	0.60	0.37	0.17	0.66
Italy	0.99	0.93	0.68	0.42	0.92
Estonia	0.54	0.82	0.51	0.26	0.76
Croatia	0.75	1.00	0.63	0.33	0.86
Czech Republic	0.20	1.00	0.41	0.36	0.67
Greece	0.75	0.62	0.80	0.46	0.89
Netherlands	0.98	0.64	0.77	0.49	0.92
Spain	0.56	1.00	0.84	0.52	0.92
Switzerland	0.49	0.51	0.92	0.70	0.87
Bosnia and Herzegovina	0.33	0.93	0.45	0.28	0.7
Macedonia, FYR	0.54	0.81	0.34	0.33	0.73
Georgia	0.52	1.00	0.28	0.24	0.69
Saudi Arabia	0.84	1.00	0.37	0.19	0.75
Israel	0.37	0.99	0.48	0.46	0.78
Japan	0.35	1.00	0.86	0.82	0.91
Malaysia	0.22	0.94	0.85	0.39	0.77
Maldives	1.00	0.79	0.23	0.42	0.78
Turkey	1.00	1.00	0.42	0.33	0.83
Armenia	0.85	0.87	0.26	0.37	0.77
Pakistan	0.54	1.00	0.12	0.27	0.64
Indonesia	0.79	0.98	0.18	0.41	0.75
Philippines	0.01	0.81	0.32	0.45	0.57
India	0.35	0.96	0.41	0.55	0.77
Bhutan	0.75	0.43	0.39	0.45	0.75

Source Calculated by Authors

Table 4.5.10 Dimensional achievements and FI scores for 2013

Countries/ indicators	*Stability & viability*	*Efficiency*	*Depth*	*Service outreach*	*IFI*
Latvia	0.59	0.61	0.43	0.21	0.69
Italy	0.88	1.00	0.68	0.61	0.97
Estonia	0.59	0.90	0.54	0.29	0.8
Croatia	0.75	0.99	0.66	0.42	0.9
Czech Republic	0.44	0.88	0.44	0.41	0.77
Greece	0.94	0.61	0.80	0.50	0.91
Netherlands	1.00	0.55	0.75	0.70	0.94
Spain	0.76	1.00	0.85	0.63	0.98
Switzerland	0.57	0.51	0.91	0.85	0.91
Bosnia and Herzegovina	0.58	0.92	0.47	0.25	0.76
Macedonia, FYR	0.56	0.85	0.37	0.32	0.75
Georgia	0.52	0.93	0.26	0.20	0.66
Saudi Arabia	0.91	1.00	0.35	0.12	0.72
Israel	0.35	0.99	0.49	0.65	0.82
Japan	0.46	1.00	0.86	1.00	0.96
Malaysia	0.33	0.85	0.86	0.37	0.8
Maldives	1.00	0.92	0.27	0.53	0.84
Turkey	1.00	0.99	0.40	0.28	0.81
Armenia	0.85	0.68	0.27	0.22	0.7
Pakistan	0.65	1.00	0.14	0.07	0.59
Indonesia	0.74	0.92	0.21	0.22	0.69
Philippines	0.04	0.66	0.31	0.23	0.48
India	0.47	0.98	0.42	0.31	0.74
Bhutan	0.68	0.49	0.43	0.14	0.65

Source Calculated by Authors

Table 4.5.11 Variable-specific weights applying principal component analysis (1st stage)

	2022	*2021*	*2020*	*2019*	*2018*	*2017*	*2016*	*2015*	*2014*	*2013*
LRI	0.509	0.485	0.516	0.473	0.488	0.456	0.480	0.470	0.460	0.490
CII	0.479	0.504	0.476	0.523	0.508	0.546	0.520	0.530	0.540	0.510
BZ	0.321	0.305	0.276	0.280	0.284	0.283	0.261	0.328	0.304	0.285
CRWA	0.288	0.291	0.272	0.271	0.277	0.289	0.267	0.312	0.293	0.272
ROA	0.224	0.197	0.180	0.170	0.154	0.138	0.232	0.191	0.208	0.227
CDR	0.167	0.207	0.270	0.276	0.283	0.289	0.240	0.185	0.203	0.218
OCTA	0.222	0.147	0.168	0.147	0.151	0.157	0.085	0.079	0.098	0.102
NIM	0.218	0.173	0.147	0.188	0.176	0.179	0.167	0.191	0.150	0.149
NPLG	0.126	0.218	0.250	0.246	0.240	0.250	0.233	0.108	0.146	0.167
CR	0.216	0.214	0.260	0.250	0.245	0.244	0.269	0.440	0.376	0.345
LOAN	0.217	0.248	0.18	0.17	0.19	0.175	0.25	0.18	0.23	0.24
PRIVY	0.276	0.289	0.312	0.267	0.304	0.272	0.289	0.284	0.280	0.276
DMBAG	0.280	0.283	0.328	0.261	0.293	0.285	0.283	0.277	0.271	0.270
LLG	0.170	0.138	0.185	0.232	0.208	0.218	0.289	0.283	0.276	0.272
CTA	0.274	0.289	0.191	0.240	0.203	0.227	0.138	0.154	0.170	0.180
DEPO	0.147	0.157	0.085	0.079	0.098	0.102	0.168	0.172	0.176	0.179
BDG	0.173	0.179	0.167	0.191	0.150	0.149	0.250	0.246	0.245	0.244
BRAN_A	0.248	0.250	0.233	0.108	0.146	0.167	0.178	0.188	0.186	0.175
ATM_P	0.218	0.244	0.269	0.440	0.376	0.345	0.147	0.147	0.151	0.157
ATM_A	0.214	0.175	0.25	0.18	0.23	0.24	0.26	0.25	0.24	0.25

Source Authors Calculation
The dimension of 'Stability and Viability' is decomposed into two sub-dimensions namely 'Internal' and 'External' where they share equal weightage, so far 2nd of PCA is concerned

Table 4.7.1 Multivariate test of the covariance matrix (Of the variables)

Year	*Null hypothesis*	*LR chi-square*	*Degree of freedom*	*Level of significance*
2012	Covariance matrix is diagonal	627.38***	91	0.0000
	Covariance matrix is spherical	691.56***	104	0.0000
	Covariance matrix is compound symmetric	662.59***	103	0.0000
2013	Covariance matrix is diagonal	634.35***	91	0.0000
	Covariance matrix is spherical	704.55***	104	0.0000
	Covariance matrix is compound symmetric	661.26***	103	0.0000
2014	Covariance matrix is diagonal	591.05***	91	0.0000
	Covariance matrix is spherical	668.85***	104	0.0000
	Covariance matrix is compound symmetric	625.75***	103	0.0000
2015	Covariance matrix is diagonal	587.77***	91	0.0000
	Covariance matrix is spherical	658.38***	104	0.0000
	Covariance matrix is compound symmetric	612.00***	103	0.0000
2016	Covariance matrix is diagonal	566.77***	91	0.0000
	Covariance matrix is spherical	636.68***	104	0.0000
	Covariance matrix is compound symmetric	593.26***	103	0.0000
2017	Covariance matrix is diagonal	564.78***	91	0.0000
	Covariance matrix is spherical	603.54***	104	0.0000

(continued)

Table 4.7.1 (continued)

Year	*Null hypothesis*	*LR chi-square*	*Degree of freedom*	*Level of significance*
	Covariance matrix is compound symmetric	577.68***	103	0.0000
2018	Covariance matrix is diagonal	578.69***	91	0.0000
	Covariance matrix is spherical	618.10***	104	0.0000
	Covariance matrix is compound symmetric	595.65***	103	0.0000
2019	Covariance matrix is diagonal	553.20***	91	0.0000
	Covariance matrix is spherical	596.42***	104	0.0000
	Covariance matrix is compound symmetric	579.07***	103	0.0000
2020	Covariance matrix is diagonal	552.02***	91	0.0000
	Covariance matrix is spherical	588.60***	104	0.0000
	Covariance matrix is compound symmetric	570.23***	103	0.0000
2021	Covariance matrix is diagonal	562.35***	91	0.0000
	Covariance matrix is spherical	609.47***	104	0.0000
	Covariance matrix is compound symmetric	593.13***	103	0.0000

Source Authors Calculation
Note ***Refers to rejection of the null hypothesis at 1% level of significance

Table 4.7.2 Multivariate test of the covariance matrix (of the dimensions)

Year	*Null*	*LR chi-square*	*Degree of freedom*	*Level of significance*
2012	Covariance matrix is diagonal	33.10***	3	0.0000
	Covariance matrix is spherical	40.76***	5	0.0000
	Covariance matrix is compound symmetric	10.12**	4	0.0385
2013	Covariance matrix is diagonal	31.93***	3	0.0000
	Covariance matrix is spherical	45.27***	5	0.0000
	Covariance matrix is compound symmetric	18.32***	4	0.0011
2014	Covariance matrix is diagonal	33.12***	3	0.0000
	Covariance matrix is spherical	42.05***	5	0.0000
	Covariance matrix is compound symmetric	12.23**	4	0.0157
2015	Covariance matrix is diagonal	35.50***	3	0.0000
	Covariance matrix is spherical	46.79***	5	0.0000
	Covariance matrix is compound symmetric	15.77***	4	0.0033
2016	Covariance matrix is diagonal	31.11***	3	0.0000
	Covariance matrix is spherical	46.51***	5	0.0000
	Covariance matrix is compound symmetric	20.19***	4	0.0005
2017	Covariance matrix is diagonal	25.24***	3	0.0000
	Covariance matrix is spherical	50.11***	5	0.0000

(continued)

Table 4.7.2 (continued)

Year	*Null*	*LR chi-square*	*Degree of freedom*	*Level of significance*
	Covariance matrix is compound symmetric	31.19***	4	0.0000
2018	Covariance matrix is diagonal	23.14***	3	0.0000
	Covariance matrix is spherical	44.27***	5	0.0000
	Covariance matrix is compound symmetric	27.48***	4	0.0000
2019	Covariance matrix is diagonal	21.08***	3	0.0001
	Covariance matrix is spherical	38.87***	5	0.0000
	Covariance matrix is compound symmetric	25.96***	4	0.0000
2020	Covariance matrix is diagonal	19.72***	3	0.0002
	Covariance matrix is spherical	28.00***	5	0.0000
	Covariance matrix is compound symmetric	12.93**	4	0.0116
2021	Covariance matrix is diagonal	17.45***	3	0.0006
	Covariance matrix is spherical	25.06***	5	0.0001
	Covariance matrix is compound symmetric	13.97***	4	0.0074

Source Authors Calculation
Note *** and **Refers to rejection of the null hypothesis at 1% and 5% level of significance

Table 4.7.3 Dimensional achievements

Country name	*2012*			*2013*			*2014*		
	Capital formation	*Efficiency*	*Outreach*	*Capital formation*	*Efficiency*	*Outreach*	*Capital formation*	*Efficiency*	*Outreach*
Bangladesh	0.1769	0.4132	0.0893	0.1632	0.3942	0.1322	0.1663	0.3961	0.1331
Botswana	0.3347	0.3893	0.1758	0.3286	0.3641	0.2228	0.3457	0.3296	0.2259
Brazil	0.6209	0.4299	0.5115	0.5933	0.4042	0.5000	0.5830	0.4059	0.4889
Burundi	0.2047	0.2313	0.0074	0.2134	0.2312	0.0108	0.1624	0.2612	0.0112
Cameroon	0.1923	0.1985	0.0119	0.1680	0.2066	0.0165	0.1407	0.1884	0.0187
China	0.7306	0.6907	0.1439	0.7139	0.6779	0.1475	0.7424	0.6969	0.1679
Colombia	0.4293	0.2985	0.2850	0.4273	0.2908	0.3817	0.4554	0.2959	0.3899
Costa Rica	0.4953	0.3160	0.3477	0.4939	0.3024	0.4489	0.5090	0.3175	0.4641
Croatia	0.4445	0.4563	0.6300	0.4261	0.4316	0.6864	0.4248	0.4400	0.6744
Djibouti	0.2908	0.3765	0.0301	0.2830	0.3558	0.0415	0.2311	0.3578	0.0492
Dominican Republic	0.3907	0.2870	0.2050	0.3961	0.2665	0.2472	0.4107	0.2601	0.2519
Ecuador	0.3917	0.2857	0.1869	0.3982	0.2722	0.2274	0.3748	0.2745	0.2466
El Salvador	0.3746	0.4066	0.1963	0.3854	0.3935	0.2390	0.4164	0.4699	0.2970
Estonia	0.4885	0.4541	0.5566	0.4462	0.4639	0.6793	0.4600	0.4521	0.6607
Eswatini	0.3355	0.2726	0.1757	0.3221	0.2352	0.1950	0.3116	0.2272	0.2167
Georgia	0.4285	0.3287	0.3770	0.4294	0.3593	0.4399	0.4480	0.3712	0.4483
Ghana	0.2999	0.3419	0.0753	0.2926	0.3353	0.1162	0.3002	0.3136	0.1218
Guinea	0.2257	0.1261	0.0085	0.1762	0.1784	0.0132	0.1443	0.2006	0.0175
Israel	0.4164	0.4512	0.5786	0.4041	0.4408	0.5833	0.4050	0.4520	0.5642
Italy	0.5593	0.4860	0.5600	0.5353	0.4661	0.5494	0.5459	0.4883	0.5318
Latvia	0.5468	0.3518	0.4546	0.5020	0.3489	0.5102	0.4573	0.3703	0.4734
Lesotho	0.2123	0.3288	0.0636	0.2089	0.3257	0.0913	0.1784	0.3188	0.0918

(continued)

Table 4.7.3 (continued)

Country name	*2012*			*2013*			*2014*		
	Capital formation	*Efficiency*	*Outreach*	*Capital formation*	*Efficiency*	*Outreach*	*Capital formation*	*Efficiency*	*Outreach*
Madagascar	0.1880	0.2368	0.0084	0.1480	0.2301	0.0118	0.1239	0.2471	0.0131
Malaysia	0.5094	0.7347	0.2941	0.5052	0.7206	0.3229	0.5018	0.7056	0.3169
Malta	0.3134	0.7149	0.4238	0.2628	0.6782	0.5064	0.2280	0.7126	0.5006
Mauritius	0.5657	0.5721	0.4290	0.5847	0.5758	0.5661	0.5518	0.5628	0.5480
Moldova	0.1684	0.3002	0.4415	0.1865	0.3030	0.6024	0.1578	0.2988	0.5823
Namibia	0.3368	0.4147	0.2701	0.3323	0.3995	0.3214	0.3407	0.3873	0.3480
Nigeria	0.1146	0.2833	0.1185	0.0905	0.2525	0.1700	0.0777	0.2529	0.1723
North Macedonia	0.4518	0.3772	0.3353	0.4308	0.3748	0.4028	0.4336	0.3917	0.4017
Pakistan	0.2844	0.3480	0.0637	0.2821	0.2925	0.0910	0.1535	0.3048	0.0948
Paraguay	0.5074	0.2890	0.1088	0.4679	0.3409	0.1254	0.4788	0.3521	0.1232
Peru	0.4042	0.3452	0.2067	0.4088	0.3442	0.2488	0.4004	0.3542	0.2988
Philippines	0.3556	0.3511	0.1268	0.3629	0.3788	0.1606	0.3371	0.3932	0.1647
Poland	0.4064	0.3892	0.3881	0.4042	0.3838	0.4518	0.4175	0.3984	0.4614
Rwanda	0.3877	0.2419	0.0479	0.3661	0.1897	0.0662	0.3378	0.2232	0.0566
Seychelles	0.2555	0.3929	0.4173	0.2188	0.3720	0.6576	0.1683	0.4122	0.6587
Singapore	0.5081	0.6919	0.5112	0.5335	0.6857	0.6535	0.5520	0.6852	0.6416
Thailand	0.4483	0.6368	0.4706	0.4454	0.6373	0.5292	0.4562	0.6402	0.5529
Turkey	0.3610	0.4014	0.3704	0.3678	0.3931	0.4464	0.3844	0.4000	0.4459
Uganda	0.1292	0.3088	0.0369	0.1406	0.2680	0.0495	0.1064	0.2361	0.0494
Ukraine	0.4611	0.3511	0.6817	0.4488	0.3475	0.6413	0.4562	0.3204	0.6033

Country name	*2012*			*2013*			*2014*		
	Capital formation	*Efficiency*	*Outreach*	*Capital formation*	*Efficiency*	*Outreach*	*Capital formation*	*Efficiency*	*Outreach*
United Arab Emirates	0.4195	0.4524	0.3058	0.3870	0.4705	0.3515	0.3996	0.4900	0.3593
Uruguay	0.4189	0.2980	0.2463	0.4075	0.3094	0.2958	0.3866	0.3077	0.3294
Uzbekistan	0.4486	0.2084	0.1916	0.4557	0.1879	0.2689	0.4467	0.1948	0.2664
Zambia	0.3183	0.2354	0.0536	0.3025	0.2318	0.0800	0.2971	0.2526	0.0874

Country name	*2015*			*2016*			*2017*		
	Capital formation	*Efficiency*	*Outreach*	*Capital formation*	*Efficiency*	*Outreach*	*Capital formation*	*Efficiency*	*Outreach*
Bangladesh	0.1603	0.3533	0.1688	0.1673	0.3387	0.1784	0.1735	0.3452	0.1830
Botswana	0.3411	0.3064	0.2344	0.3100	0.2747	0.2266	0.3450	0.3166	0.2328
Brazil	0.5704	0.3737	0.4606	0.5579	0.3748	0.4448	0.5686	0.3986	0.4271
Burundi	0.1445	0.2528	0.0136	0.1749	0.2476	0.0140	0.1790	0.3520	0.0127
Cameroon	0.1560	0.1932	0.0244	0.1966	0.1876	0.0255	0.2238	0.2131	0.0226
China	0.7428	0.6846	0.2136	0.6975	0.6961	0.2224	0.7392	0.6411	0.2152
Colombia	0.4667	0.3070	0.4181	0.4717	0.2935	0.4180	0.5083	0.3375	0.4257
Costa Rica	0.5094	0.2734	0.4341	0.5184	0.2624	0.4517	0.5449	0.2818	0.4488
Croatia	0.4063	0.3836	0.6944	0.3742	0.4044	0.6984	0.3916	0.3917	0.6966
Djibouti	0.2370	0.3275	0.0587	0.2850	0.3215	0.0578	0.2901	0.3509	0.0724
Dominican Republic	0.3994	0.2343	0.2710	0.3425	0.2027	0.2760	0.3823	0.2778	0.2795

(continued)

Table 4.7.3 (continued)

Country name	*2015*			*2016*			*2017*		
	Capital formation	*Efficiency*	*Outreach*	*Capital formation*	*Efficiency*	*Outreach*	*Capital formation*	*Efficiency*	*Outreach*
Ecuador	0.3667	0.2488	0.2600	0.3251	0.2342	0.2592	0.3669	0.2989	0.2538
El Salvador	0.4015	0.3630	0.3204	0.3880	0.3409	0.3318	0.4257	0.3556	0.2772
Estonia	0.4528	0.4483	0.6091	0.4310	0.4282	0.6010	0.4090	0.4041	0.5925
Eswatini	0.2967	0.2136	0.2261	0.2815	0.1850	0.2148	0.2651	0.2849	0.2136
Georgia	0.4589	0.3589	0.4603	0.4204	0.3469	0.4875	0.4494	0.4272	0.4874
Ghana	0.3100	0.2871	0.1547	0.3100	0.2589	0.1479	0.3170	0.3726	0.1644
Guinea	0.1470	0.2045	0.0212	0.2163	0.2154	0.0212	0.2550	0.3050	0.0216
Israel	0.4059	0.4311	0.5108	0.3773	0.4241	0.5187	0.4074	0.4182	0.5446
Italy	0.5339	0.4584	0.5562	0.4778	0.4576	0.5478	0.5306	0.4234	0.4928
Latvia	0.4257	0.3593	0.4678	0.3944	0.3390	0.4423	0.5023	0.3157	0.4561
Lesotho	0.1895	0.2627	0.1035	0.2244	0.2260	0.1091	0.3846	0.3347	0.1099
Madagascar	0.1590	0.2410	0.0185	0.2072	0.2153	0.0194	0.2353	0.3330	0.0201
Malaysia	0.4927	0.6557	0.3112	0.4492	0.6407	0.2942	0.4949	0.5886	0.2767
Malta	0.2030	0.6526	0.6619	0.2015	0.6634	0.6289	0.3243	0.5958	0.5950
Mauritius	0.5552	0.5417	0.5662	0.5303	0.5432	0.5296	0.5569	0.5557	0.5221
Moldova	0.3405	0.3223	0.4080	0.3042	0.2916	0.4415	0.3332	0.3073	0.4490
Namibia	0.3468	0.3602	0.3633	0.3577	0.3333	0.3842	0.3845	0.4071	0.3638
Nigeria	0.2618	0.2291	0.1775	0.2501	0.2291	0.2023	0.2982	0.2770	0.2197
North Macedonia	0.4205	0.3775	0.4268	0.3937	0.3615	0.4263	0.4212	0.3840	0.4166
Pakistan	0.1452	0.2977	0.1121	0.1311	0.2786	0.1177	0.2840	0.2878	0.1186
Paraguay	0.4956	0.2723	0.1381	0.4657	0.2806	0.1407	0.4276	0.3406	0.1538

Country name	*2015*			*2016*			*2017*		
	Capital formation	*Efficiency*	*Outreach*	*Capital formation*	*Efficiency*	*Outreach*	*Capital formation*	*Efficiency*	*Outreach*
Peru	0.4470	0.3419	0.4646	0.4146	0.3358	0.4376	0.4544	0.3807	0.4210
Philippines	0.2835	0.3913	0.1772	0.2519	0.3925	0.1831	0.3056	0.3861	0.1841
Poland	0.4015	0.3751	0.4863	0.3508	0.3778	0.4968	0.3734	0.3735	0.4660
Rwanda	0.3640	0.2262	0.0579	0.3840	0.2045	0.0672	0.4123	0.2697	0.0656
Seychelles	0.1622	0.3995	0.7063	0.2977	0.3956	0.7248	0.3255	0.4366	0.7202
Singapore	0.5292	0.6298	0.6608	0.4852	0.6386	0.6394	0.5190	0.5784	0.6603
Thailand	0.4507	0.6117	0.5499	0.4110	0.6009	0.5436	0.4715	0.5708	0.5385
Turkey	0.4486	0.3809	0.4630	0.4169	0.3839	0.4831	0.4667	0.3999	0.4944
Uganda	0.1350	0.2033	0.0543	0.3149	0.2102	0.0608	0.3757	0.3135	0.0790
Ukraine	0.4265	0.2583	0.5495	0.3926	0.2573	0.5005	0.3965	0.2393	0.5283
United Arab Emirates	0.4682	0.5161	0.3661	0.4544	0.5112	0.3848	0.4724	0.4787	0.3810
Uruguay	0.4043	0.3348	0.3354	0.3826	0.2358	0.3939	0.4935	0.3168	0.4017
Uzbekistan	0.3601	0.1833	0.2587	0.3658	0.1812	0.2943	0.4987	0.2214	0.3405
Zambia	0.3529	0.2161	0.0954	0.3726	0.1746	0.0972	0.3568	0.3065	0.0965

Country name	*2018*			*2019*			*2020*		
	Capital formation	*Efficiency*	*Outreach*	*Capital formation*	*Efficiency*	*Outreach*	*Capital formation*	*Efficiency*	*Outreach*
Bangladesh	0.1820	0.3416	0.1958	0.1616	0.3279	0.2043	0.2515	0.3088	0.1742
Botswana	0.2908	0.3107	0.2594	0.2880	0.3091	0.2200	0.3115	0.3022	0.1908

(continued)

Table 4.7.3 (continued)

Country name	*2018*			*2019*			*2020*		
	Capital formation	*Efficiency*	*Outreach*	*Capital formation*	*Efficiency*	*Outreach*	*Capital formation*	*Efficiency*	*Outreach*
Brazil	0.5190	0.3808	0.4050	0.5482	0.3730	0.2834	0.5728	0.4059	0.2605
Burundi	0.1090	0.3824	0.0130	0.1354	0.4096	0.0130	0.1615	0.4170	0.0189
Cameroon	0.2829	0.1903	0.0329	0.2893	0.1944	0.0285	0.2925	0.2153	0.0237
China	0.7923	0.6324	0.2251	0.8057	0.6137	0.1154	0.8226	0.6247	0.1199
Colombia	0.4698	0.3235	0.4317	0.4545	0.3155	0.4116	0.4299	0.2794	0.3452
Costa Rica	0.4891	0.2667	0.4743	0.4359	0.2603	0.4138	0.4619	0.2705	0.3402
Croatia	0.3712	0.4064	0.6811	0.3623	0.3814	0.5256	0.3624	0.3888	0.4549
Djibouti	0.2423	0.3261	0.0902	0.2305	0.3096	0.0875	0.1752	0.3243	0.0795
Dominican Republic	0.2986	0.2480	0.2818	0.2779	0.2356	0.2327	0.3217	0.2421	0.1955
Ecuador	0.2880	0.3031	0.2536	0.2898	0.2974	0.2479	0.3405	0.2763	0.2065
El Salvador	0.3132	0.3535	0.2721	0.2846	0.3602	0.2321	0.3270	0.3664	0.2050
Estonia	0.3433	0.4164	0.5766	0.3211	0.3729	0.4807	0.3801	0.3942	0.3677
Eswatini	0.2122	0.2898	0.1894	0.2435	0.2851	0.0830	0.2810	0.3449	0.0864
Georgia	0.4043	0.4324	0.4969	0.3998	0.4063	0.3752	0.4429	0.3841	0.3599
Ghana	0.2643	0.3604	0.1860	0.2562	0.3364	0.1849	0.2809	0.3688	0.1385
Guinea	0.2711	0.3254	0.0216	0.2610	0.3416	0.0218	0.2372	0.3329	0.0215
Israel	0.3260	0.4166	0.5338	0.3128	0.4085	0.3914	0.3838	0.4336	0.3451
Italy	0.4406	0.4155	0.4734	0.4107	0.4051	0.3630	0.4381	0.4172	0.3733
Latvia	0.3689	0.3090	0.4292	0.3626	0.2846	0.3272	0.3982	0.2880	0.2502
Lesotho	0.3146	0.3181	0.1125	0.3110	0.3087	0.0949	0.3001	0.2428	0.0803
Madagascar	0.3166	0.3559	0.0209	0.3125	0.3182	0.0219	0.3369	0.2948	0.0224

Country name	2018			2019			2020		
	Capital formation	*Efficiency*	*Outreach*	*Capital formation*	*Efficiency*	*Outreach*	*Capital formation*	*Efficiency*	*Outreach*
Malaysia	0.4865	0.6157	0.2664	0.4845	0.5761	0.2193	0.5238	0.5696	0.1837
Malta	0.3647	0.4877	0.4692	0.3728	0.4537	0.4049	0.3344	0.5583	0.3833
Mauritius	0.4372	0.5561	0.5089	0.4580	0.5267	0.4766	0.4841	0.5638	0.3771
Moldova	0.2610	0.2784	0.4702	0.2603	0.2993	0.4298	0.2750	0.2692	0.3893
Namibia	0.3095	0.3838	0.3852	0.3104	0.3803	0.3252	0.3462	0.3793	0.2771
Nigeria	0.2901	0.2520	0.2360	0.2845	0.2402	0.2482	0.3112	0.2384	0.2078
North Macedonia	0.3346	0.3908	0.4033	0.3425	0.3692	0.3445	0.3543	0.3734	0.3117
Pakistan	0.1886	0.2804	0.1210	0.1819	0.2806	0.1155	0.2234	0.3055	0.1124
Paraguay	0.3796	0.3427	0.1960	0.3634	0.3350	0.1624	0.3976	0.3329	0.1433
Peru	0.3467	0.3880	0.4183	0.3407	0.3756	0.3188	0.4096	0.3298	0.2777
Philippines	0.2456	0.3837	0.1947	0.2525	0.3831	0.1835	0.3218	0.3934	0.1613
Poland	0.3689	0.3716	0.4635	0.3543	0.3559	0.3940	0.3844	0.3135	0.3520
Rwanda	0.3167	0.2866	0.0738	0.2949	0.3376	0.0676	0.3529	0.3597	0.0593
Seychelles	0.2714	0.4662	0.7184	0.2658	0.4507	0.7693	0.3297	0.4584	0.7854
Singapore	0.4773	0.5871	0.6648	0.4667	0.5726	0.6216	0.4843	0.5997	0.4755
Thailand	0.3948	0.5824	0.5192	0.3812	0.5613	0.3989	0.4695	0.5695	0.3315
Turkey	0.4145	0.4005	0.5060	0.4125	0.3817	0.4433	0.4574	0.3982	0.4005
Uganda	0.3423	0.2734	0.0640	0.3516	0.2864	0.0615	0.3345	0.2785	0.0515
Ukraine	0.3636	0.2707	0.4778	0.3536	0.3579	0.4043	0.3799	0.2958	0.3076
United Arab Emirates	0.4551	0.4821	0.3704	0.4438	0.4738	0.2958	0.4986	0.4741	0.2464

(continued)

Table 4.7.3 (continued)

Country name	*2018*			*2019*			*2020*		
	Capital formation	*Efficiency*	*Outreach*	*Capital formation*	*Efficiency*	*Outreach*	*Capital formation*	*Efficiency*	*Outreach*
Uruguay	0.3695	0.3194	0.4700	0.3362	0.3841	0.4662	0.3743	0.3732	0.4148
Uzbekistan	0.4683	0.2027	0.3450	0.4945	0.2651	0.3421	0.5506	0.2697	0.3805
Zambia	0.3197	0.2301	0.0849	0.3200	0.2298	0.0580	0.3683	0.2059	0.0466

Country name	*2021*		
	Capital formation	*Efficiency*	*Outreach*
Bangladesh	0.2340	0.2789	0.1659
Botswana	0.3241	0.3071	0.1783
Brazil	0.5394	0.4271	0.2222
Burundi	0.1428	0.4279	0.0211
Cameroon	0.2922	0.2326	0.0246
China	0.8097	0.5965	0.0981
Colombia	0.4099	0.2508	0.2854
Costa Rica	0.4404	0.3051	0.3100
Croatia	0.3442	0.3918	0.4115
Djibouti	0.1352	0.3289	0.0846
Dominican Republic	0.2990	0.2773	0.1753
Ecuador	0.3414	0.3069	0.1879
El Salvador	0.3283	0.3460	0.1887
Estonia	0.3342	0.3911	0.3050
Eswatini	0.2914	0.4203	0.0679

Country name	*2021*		
	Capital formation	*Efficiency*	*Outreach*
Georgia	0.4218	0.4440	0.3267
Ghana	0.2788	0.3960	0.1316
Guinea	0.2392	0.3352	0.0212
Israel	0.3893	0.4601	0.2952
Italy	0.4046	0.4334	0.3258
Latvia	0.3723	0.2944	0.2221
Lesotho	0.3181	0.2313	0.0757
Madagascar	0.3407	0.3315	0.0237
Malaysia	0.5086	0.5524	0.1629
Malta	0.3316	0.5334	0.3554
Mauritius	0.4806	0.5715	0.3441
Moldova	0.2796	0.2999	0.3610
Namibia	0.3461	0.3845	0.2524
Nigeria	0.2967	0.2262	0.1984
North Macedonia	0.3613	0.3358	0.2874
Pakistan	0.2331	0.2768	0.1118
Paraguay	0.3744	0.3285	0.1378
Peru	0.3707	0.3400	0.2438
Philippines	0.3146	0.3878	0.1532
Poland	0.3647	0.3279	0.3226
Rwanda	0.3612	0.3899	0.0542
Seychelles	0.3233	0.4421	0.7967

(continued)

Table 4.7.3 (continued)

Country name	*2021*		
	Capital formation	*Efficiency*	*Outreach*
Singapore	0.4803	0.6015	0.4332
Thailand	0.4717	0.5644	0.2888
Turkey	0.4462	0.3837	0.3832
Uganda	0.3342	0.3181	0.0487
Ukraine	0.3249	0.2579	0.2536
United Arab Emirates	0.4796	0.4945	0.2126
Uruguay	0.3608	0.3481	0.3790
Uzbekistan	0.4993	0.2318	0.4017
Zambia	0.3370	0.3433	0.0434

Source Authors Calculation

Table 4.7.4 Factor loadings with eigenvalues

	2012			*2013*			*2014*		
	Capital formation	*Efficiency*	*Outreach*	*Capital formation*	*Efficiency*	*Outreach*	*Capital formation*	*Efficiency*	*Outreach*
CII	0.5784			0.6006			0.5839		
CDR	0.6478			0.6501			0.6633		
CR	−0.6195			−0.6180			−0.5331		
DMBAG	0.5374			0.5442			0.5370		
ROA		−0.5734			−0.5526			−0.4766	
OCTA		0.8601			0.8921			0.8786	
NIM		−0.8169			−0.8619			−0.8277	
PRIVY		0.9286			0.9079			0.9055	
BDG		0.8660			0.8629			0.8500	
LLG		0.9007			0.8890			0.8991	
DEPO			0.7177			0.8025			0.7855
ATM_P			0.7579			0.7088			0.6876
ATM_A			0.2908			0.3199			0.3180
BRAN_P			0.4676			0.5232			0.5225
Eigenvalues	1.4268	4.1593	1.3927	1.4615	4.2034	1.5226	1.3535	4.0351	1.4640
	2015			*2016*			*2017*		
	Capital formation	*Efficiency*	*Outreach*	*Capital formation*	*Efficiency*	*Outreach*	*Capital formation*	*Efficiency*	*Outreach*
CII	0.5642			0.5020			0.5753		
CDR	0.6463			0.6482			0.6453		
CR	−0.5421			−0.6003			−0.6336		

(continued)

Table 4.7.4 (continued)

	2015			*2016*			*2017*		
	Capital formation	*Efficiency*	*Outreach*	*Capital formation*	*Efficiency*	*Outreach*	*Capital formation*	*Efficiency*	*Outreach*
DMBAG	0.5496			0.5405			0.5594		
ROA		−0.4069			−0.1072			−0.6423	
OCTA		0.8946			0.8847			0.8835	
NIM		−0.8874			−0.8719			−0.9075	
PRIVY		0.8913			0.8780			0.8497	
BDG		0.8529			0.8597			0.8296	
LLG		0.8988			0.9028			0.8807	
DEPO			0.7724			0.7764			0.774
ATM_P			0.6439			0.6536			0.6568
ATM_A			0.3343			0.3234			0.3245
BRAN_P			0.5616			0.5852			0.5336
Eigenvalues	1.3319	4.0830	1.4384	1.3247	3.8795	1.4770	1.46175	4.20255	1.42045

	2018			*2019*			*2020*			*2021*		
	Capital formation	*Efficiency*	*Outreach*	*Capital formation*	*Efficiency*	*Outreach*	*Capital formation*	*Efficiency*	*Outreach*	*Capital formation*	*Efficiency*	*Outreach*
CII	0.4107			0.4025			0.4728			0.4883		
CDR	0.6991			0.6847			0.668			0.6568		
CR	−0.5496			−0.5259			−0.499			−0.5045		
DMBAG	0.5516			0.536			0.5225			0.5174		

	2018			2019			2020			2021		
	Capital formation	*Efficiency*	*Outreach*	*Capital formation*	*Efficiency*	*Outreach*	*Capital formation*	*Efficiency*	*Outreach*	*Capital formation*	*Efficiency*	*Outreach*
ROA		−0.7084			−0.7703			−0.7616			−0.7538	
OCTA		0.9013			0.888			0.881			0.8544	
NIM		−0.9169			−0.9203			−0.891			−0.8749	
PRIVY		0.8531			0.8355			0.8286			0.7911	
BDG		0.8183			0.7943			0.8333			0.8058	
LLG		0.8761			0.8734			0.9016			0.8828	
DEPO			0.771			0.7474			0.7142			0.706
ATM_P			0.6309			0.5068			0.4912			0.4448
ATM_A			0.3241			0.3254			0.287			0.2768
BRAN_P			0.525			0.5038			0.5321			0.5342
Eigenvalues	1.2637	4.3199	1.3731	1.1947	4.3206	1.1750	1.1917	4.3441	1.1169	1.1922	4.1183	1.0582

Source Authors Calculation

Table 4.7.5 Factor loadings with eigenvalues (dimension-wise)

	FI 2012	*FI 2013*	*FI 2014*	*FI 2015*	*FI 2016*	*FI 2017*	*FI 2018*	*FI 2019*	*FI 2020*	*FI 2021*
Capital formation	0.6936	0.6753	0.6850	0.6541	0.6121	0.6501	0.6242	0.6097	0.6233	0.6422
Efficiency	0.6649	0.7011	0.6986	0.7227	0.7061	0.6272	0.6660	0.6910	0.6569	0.6402
Outreach	0.6953	0.6570	0.6708	0.7116	0.6942	0.6314	0.5641	0.4665	0.4686	0.3607
Eigenvalues	1.4066	1.3792	1.4073	1.4565	1.3552	1.2147	1.1514	1.0668	1.0396	0.9525

Source Authors Calculation

Table 5.1.1 Review of the data properties

Normality tests						Homogeneity of variance	Tests for heteroscedasticity	
NSDP_ CODE	*Regional categories*	*Kolmogorov-Smirnov*[a]	*Shapiro-Wilk*	*Skewness Standardized (z-values)*	*Kurtosis*	*Levene statistic*[b]	*Breusch-Pagan test*[b]	*White test*[b]
High	North	0.278 (0.099)	0.734 (0.065)	−1.78	1.32	1.877 (0.054)	0.824 (0.364)	13.451 (0.265)
	North-East	0.205 (0.200*)	0.889 (0.228)	0.43	−1.26			
	East	0.206 (0.200*)	0.879 (0.186)	0.24	−1.28			
	Central	0.192 (0.200*)	0.921 (0.441)	−0.54	−0.33			
	West	0.292 (0.093)	0.749 (0.068)	−1.68	1.52			
	South	0.201 (0.200*)	0.898 (0.278)	−1.32	1.86			
Low	North	0.17 (0.200*)	0.956 (0.767)	0.52	−0.68			
	North-East	0.20 (0.200*)	0.93 (0.518)	0.87	−0.55			
	East	0.245 (0.174)	0.855 (0.107)	1.59	0.23			
	Central	0.166 (0.200*)	0.924 (0.460)	0.95	−0.36			

(continued)

Table 5.1.1 (continued)

Normality tests						Homogeneity of variance	Tests for heteroscedasticity	
NSDP_CODE	*Regional categories*	*Kolmogorov-Smirnov*[a]	*Shapiro-Wilk*	*Skewness Standardized (z-values)*	*Kurtosis*	*Levene statistic*[b]	*Breusch-Pagan test*[b]	*White test*[b]
	West	0.223 (0.200*)	0.845 (0.085)	0.82	−1.15			
	South	0.273 (0.082)	0.848 (0.092)	1.59	0.50			

Source Authors Calculation

*Represents the floor levels of the true significance, [a]Adjusted for Lilliefors Significance, [b]Test specification: Intercept + NSDP_CODE + Region + NSDP_CODE * Region

The respective p values are given in the parenthesis; where, $\alpha = 0.05$

Table 7.1.1 Review of data properties

Short name	*Variables' description*	*Mean*	*SD*
lnRNSDP	Log transformed values of the state-wise net domestic product at factor cost in rupees crore (assuming prices being constant considering the base year 2011–2012)	12.15	1.49
lnrealIP	Log transformed values of the inflation adjusted Industrial Production	8.96	1.98
lnpercapitaRNSDP	Log transformed values of the per-capita state-wise net domestic product at factor cost in rupees crore (assuming prices being constant considering the base year 2011–2012)	10.89	0.59
lnOffices	Natural logarithm of scheduled commercial banks' total number of branch offices	7.24	1.57
lnSavingsac	Natural logarithm of scheduled commercial banks' total number of savings accounts	15.88	1.79
lnCreditac	Natural logarithm of scheduled commercial banks' total number of credit account	13.99	1.80
lnGFCF	Natural logarithm transformed values of gross fixed capital formation	2.66	0.98
ΔlnRNSDP	Changes in lnRNSDP (1st Difference)	6.89	9.32
ΔlnrealIP	Changes in lnrealIP (1st Difference)	8.03	14.13
ΔlnpercapitaRNSDP	Changes in lnpercapitaRNSDP (1st Difference)	5.63	4.68
ΔlnOffices	Changes in lnOffices (1st Difference)	5.03	4.15
ΔlnSavingsac	Changes in lnSavingsac (1st Difference)	10.96	6.84
ΔlnCreditac	Changes in lnCreditac (1st Difference)	7.85	10.59
ΔlnGFCF	Changes in lnGFCF (1st Difference)	6.94	43.27

Source Authors Calculation

Table 7.1.2 Correlation matrix

Panel A: For level variables

	lnRNSDP	*lnrealIP*	*lnpercapitaRNSDP*	*lnSavingsac*	*lnOffices*	*lnCreditac*	*lnGFCF*
lnRNSDP	1						
lnrealIP	0.898***	1					
lnpercapitaRNSDP	−0.041	0.110**	1				
lnSavingsac	0.968***	0.912***	−0.039	1			
lnOffices	0.959***	0.921***	−0.121**	0.984***	1		
lnCreditac	0.975***	0.872***	−0.104**	0.972***	0.974***	1	
lnGFCF	0.523***	0.729***	0.312***	0.498***	0.518***	0.459***	1

Panel B: For 1st differenced variables

	ΔlnRNSDP	*ΔlnrealIP*	*ΔlnpercapitaRNSDP*	*ΔlnSavingsac*	*ΔlnOffices*	*ΔlnCreditac*	*ΔlnGFCF*
ΔlnRNSDP	1						
ΔlnrealIP	0.428***	1					
ΔlnpercapitaRNSDP	0.816***	0.455***	1				
ΔlnSavingsac	−0.074	−0.141***	−0.040	1			
ΔlnOffices	−0.064	−0.123**	−0.032	0.43***	1		
ΔlnCreditac	0.029	0.056	0.058	−0.115**	−0.025	1	
ΔlnGFCF	−0.032	0.031	−0.019	−0.021	−0.162***	0.082*	1

Source Authors Calculation

Note ***, ** and *Indicates the level of significance at 1%, 5% and 10% (2-tailed)

Table 7.5.1 Description of variables

Variable	*Acronym*	*Measurement*	*Sources*
Number of Savings Account per unit of population	NSA	Total savings bank account with scheduled commercial banks to total population	Reserve Bank of India[1]
Population	POP	Number of population	Office of the Registrar General & Census Commissioner, India[2]
Consumer Price Index	CPI	Consumer Price Index (Base year 2010)	Reserve Bank of India[3]
Education Expenditure to Total State Budget	EDU	Expenses on Education & Training by Education & Other Department to Total State Budget (in per cent)	Ministry of Human Resource Development, India[4]
Health Expenditure to Total Expenditure	HEL	Government Spending on Public Health and Family Welfare as ratio to Aggregate Expenditure (in per cent)	Reserve Bank of India[5]
State-specific Per capita Net Domestic Product	GDP	Per capita Net Domestic Product (State-wise)	Reserve Bank of India[6]

Source Authors' Elaboration

[1] Retrieved from https://dbie.rbi.org.in/DBIE/dbie.rbi?site=publications#!18.

[2] Retrieved from https://censusindia.gov.in/Census_Data_2001/Projected_Population/Projected_Population.pdf.

[3] https://rbidocs.rbi.org.in/rdocs/Publications/PDFs/T_8936C5A0129C694832B793E659DDDFF28F.PDF.

[4] Retrieved from https://www.education.gov.in/en/statistics-new?shs_term_node_tid_depth=387.

[5] https://rbidocs.rbi.org.in/rdocs/Publications/PDFs/27ST_3009201978AEB70B4B8A436292CBCFC9AE72F86F.PDF.

[6] https://www.rbi.org.in/Scripts/PublicationsView.aspx?id=18812.

Table 7.5.2 Contemporaneous association

	Very low	*Low*	*High*	*Very high*	*Total*
r (NSA, GDP)	0.498	0.511	0.686	0.870	0.879
Partial r (NSA, GDP)	0.808	0.663	0.748	0.905	0.846***

Note ***Significant at 1 per cent level
Observations: Approximately 8 in each of the four categories
Source Authors Calculation

Table 7.5.3 Correlation matrix and variance inflation factor

Independent variables	*POP*	*CPI*	*EDU*	*HEL*	*GDP*	*VIF*[a]
POP	1					1.18
CPI	0.045	1				1.03
EDU	−0.127	−0.078	1			1.05
HEL	−0.30	0.139	−0.005	1		1.38
GDP	−0.299	0.073	−0.123	0.484	1	1.39

[a]Acronym for the Variance Inflation Factor
Source Authors Calculation

Table 7.5.4 Residual analysis for model biasness

Correlogram of residuals				
Lag (s)	*Autocorrelation*	*Partial Autocorrelation*	*Q-Stat*	*Probability*
1	0.36600	0.36600	28.24700	0.000***
2	− 0.11500	− 0.28700	31.04400	0.000***
3	− 0.27800	− 0.15000	47.57000	0.000***
4	− 0.24600	− 0.12400	60.48300	0.000***
5	− 0.13400	− 0.08800	64.36900	0.000***
6	− 0.09300	− 0.14700	66.22200	0.000***

Note ***Significant at 1 per cent level
Source Calculated by Authors

Table 7.5.5 Residual analysis for model efficiency

Dependent variable: time-averaged Idiosyncratic Error term ($\bar{u}_i$)

Generic Equation: $\bar{u}_i = \beta_0 + \beta_1\overline{POP}_i + \beta_2\overline{CPI}_i + \beta_3\overline{EDU}_i + \beta_4\overline{HEL}_i + \beta_5\overline{GDP}_i$

Explanatory variables	*Coefficient*	*t-Stat*
POP	−1.84e−18	−0.30
CPI	3.03e−11	0.29
EDU	1.19e−10	2.14**
HEL	−3.72e−10	−1.68
GDP	1.69e−15	0.27
Intercept	−4.66e−09	−0.35
R-squared (Between) = 0.2843	F [5, 24] = 1.91 (0.1306)	

Note **Significant at 5 per cent level, Values in parenthesis represents the level of significance
Source Calculated by Authors

Table 7.5.6 Summary of the results of the diagnostic tests

Sl. no.	*Assumptions*	*Generic equation*	*Results*	*Conclusion*
1	α_i shows the unobserved effect whereas β_j 's are the parameters to be estimated	$\hat{y}_{it} - \bar{y}_i = \beta_1(x1_{it} - \bar{x}_i) + \beta_2(x2_{it} - \bar{x}_i) + \beta_4(x4_{it} - \bar{x}_i) + \beta_5(x5_{it} - \bar{x}_i) + \beta_6(x6_{it} - \bar{x}_i) + u_{it}$	Consider the results of FE model in Table- 29	Satisfied
2	Cross sectional elements have been randomly selected	NA	NA	Satisfied (Partially)
3	Explanatory variables do not share a perfect linear relationship	r (x_i, x_j)	Consider the results in Table 8.1.3	Satisfied
4	Estimated value of the idiosyncratic errors for a given time period and the explanatory variables is equal to zero	E $(u_{it} \vert x_i, \alpha_i) = 0$	0.000*** 0.000*** 0.000*** 0.000*** 0.000***	Satisfied
5	Homoscedastic idiosyncratic error terms	Var $(u_{it} \vert x_i, \alpha_i) = \sigma^2{}_u$	Consider the results in Fig. 7.1	Satisfied
6	Idiosyncratic error terms are not correlated with their own lag values	Cov $(u_{it}, u_{is} \vert x_i, \alpha_i) = 0$	Consider the results in Table 8.1.4	Satisfied
7	The idiosyncratic errors (u_{it}) are IID (i.e. Independently and Identically distributed) with N $(0, \sigma^2)$	Jarque–Bera statistics: $n\,[(\sqrt{\beta_1})^2/6 + (\beta_2 - 3)^2/24]$	10.056*** Consider the results in Fig. 7.1	Satisfied

Note ***Significant at 1 per cent level
Source Authors Calculation

Table 8.1.1 Summary statistics

Variable description		*Obs*	*Mean*	*SD*	*Floor*	*25th percentile*	*Median*	*75th percentile*	*Ceiling*
Name	*Indicator*								
Employed population below international poverty line, aged 25+ (%)	POV*	923	16.633	20.069	0	0.8	8.2	28	78.1
Natural logarithm of GDP per capita growth (annual %)	ΔlnGDPpc	770	1.044	0.903	−3.912	0.637	1.234	1.627	3.139
YoY change in Unemploy-ment, total (% of total labor force) (modeled ILO estimate)	ΔUNEMP	938	0.047	14.857	−70.297	−4.387	−0.838	3.307	176.191
Inflation, consumer prices (annual %)	INF	849	6.493	5.509	−8.97	3.05	5.42	8.67	55.41
Natural logarithm of Population growth (annual %)	ΔlnPOP	938	1.668	1.234	−4.54	1	1.705	2.62	6.57
YoY change in Current health expenditure (% of GDP)	ΔGSE	911	0.548	10.738	−100	−4.082	0.369	4.613	72.951
% change in the achievements in Financial Inclusion	ΔIFI	934	7.294	53.516	−100	−2.128	0	7.692	1000

Source Authors Calculation

*Data for POV has been collected from the SDG Indicators Database, United Nations (Retrieved from https://unstats.un.org/sdgs/dataportal/database); whereas others have been sourced from the World Development Indicators, World Bank (Retrieved from https://databank.worldbank.org/source/world-development-indicators)

Table 8.1.2 Correlation matrix

	POV	*ΔlnGDP*	*ΔUNEMP*	*INF*	*ΔlnPOP*	*ΔGSE*	*ΔIFI*
POV	1						
ΔlnGDP	−0.059	1					
ΔUNEMP	0.025	−0.129***	1				
INF	0.336***	0.049	0.027	1			
ΔlnPOP	0.569***	−0.194***	0.062	0.256***	1		
ΔGSE	−0.036	−0.105***	0.053	−0.127***	−0.046	1	
ΔIFI	0.072*	0.139***	−0.025	0.020	0.056	0.017	1

Source Authors Calculation
Note (i) ***, ** and *Represents level of significance at 1%, 5% and 10%. Common observations: 692

Table 9.2.1 Verification tests of exogeneity/endogeneity

	H_0: Variables are weakly exogeneous		
	Parametric alternatives		*Non-parametric alternative*
	t- statistic	*F- statistic*	*Chi-square statistic*
lnFPI_C	4.915***	24.748***	24.478***
ln_FPI	−6.987***	49.861***	49.361***
lnGNIpc	−8.895***	81.735***	81.692***
Bran_D	−5.439***	31.026***	31.144***
ATM_D	−6.786***	47.743***	47.862***
CoC_estimate	−0.672	0.490	0.603
GE_estimate [a]	−1.876	3.981	3.987
PSAV_estimate	0.597	0.402	0.397

Source Authors Calculation
Note ***, ** and *Represents level of significance at 1, 5 and 10 percent respectively
[a]Represents insignificance at 1% level

Bibliography

Abosedra, S., Shahbaz, M., & Nawaz, K. (2016). Modelling causality between financial deepening and poverty reduction in Egypt. *Social Indicators Research, 126*(3), 955–969. https://doi.org/10.1007/s11205-015-0929-2

Abu-Bader, S., & Abu-Qarn, A. (2008). Financial development and economic growth: The Egyptian experience. *Journal of Policy Modeling, 30*(5), 887–898. https://doi.org/10.1016/j.jpolmod.2007.02.001

ADB. (2017). *Asian development outlook 2017 update*. Asian Development Bank.

Adeola, O., & Evans, O. (2017). Financial inclusion, financial development, and economic diversification in Nigeria. *The Journal of Developing Areas, 51*(3), 1–15.

AFI. (2015). *2015 Maya declaration report: Commitments into action. Alliance for financial inclusion*. Retrieved September 5, 2021, from http://www.afi-global.org/sites/default/files/publications/maya_report_2015-final.pdf.

African Development Bank. (2020). *Financial inclusion in Africa: Challenges and opportunities*. African Development Bank Group.

Aghion, P., & Howitt, P. (1990). A model of growth through creative destruction.

Ahmad, M., Majeed, A., Khan, M. A., Sohaib, M., & Shehzad, K. (2021). Digital financial inclusion and economic growth: Provincial data analysis of China. *China Economic Journal, 14*(3), 291–310.

Aideyan, O. (2009). Microfinance and poverty reduction in rural Nigeria. *Savings and Development, 33*(3), 293–317. https://www.jstor.org/stable/41406498

Aker, J. C., & Mbiti, I. M. (2010). Mobile phones and economic development in Africa. *Journal of Economic Perspectives, 24*(3), 207–232. https://doi.org/10.1257/jep.24.3.207

S. Ghosh et al., *Financial Inclusion*, Sustainable Development Goals Series, https://doi.org/10.1007/978-981-95-3385-5

Akhter, S., & Daly, K. J. (2009). Finance and poverty: Evidence from fixed effect vector decomposition. *Emerging Markets Review, 10*(3), 191–206. https://doi.org/10.1016/j.ememar.2009.02.005

Alderman, H. (1993). New research on poverty and malnutrition: What are the implications for policy? *Including the poor*, World Bank, 115–132.

Alston, J. M., & Pardey, P. G. (2015). Agricultural R&D, food prices, poverty, and malnutrition redux. In *The fight against Hunger and Malnutrition: The role of food, agriculture, and targeted policies* (pp. 208–239). Oxford University Press.

Alter, A. (2015). *Financial inclusion and development in the CEMAC*. International Monetary Fund.

Amidžić, G., Massara, M. A., & Mialou, A. (2014). Assessing countries' financial inclusion standing—a new composite index. *IMF Working Paper* WP/14/36. https://doi.org/10.5089/9781475569681.001

Andersen, P. (2009). Food security: Definition and measurement. *Food Security, 1*(1), 5–7. https://doi.org/10.1007/s12571-008-0002-y

Andersen, T. B., & Tarp, F. (2003). Financial liberalization, financial development and economic growth in LDCs. *Journal of International Development: The Journal of the Development Studies Association, 15*(2), 189–209. https://doi.org/10.1002/jid.971

Andrieu, E., Borgonovi, F., & Liebender, A. S. (2019). The role of education and skills in bridging the digital gender divide: Evidence from APEC economies.

Anríquez, G., Daidone, S., & Mane, E. (2013). Rising food prices and undernourishment: A cross-country inquiry. *Food Policy, 38*, 190–202.

Apergis, N., Filippidis, I., & Economidou, C. (2007). Financial deepening and economic growth linkages: A panel data analysis. *Review of World Economics, 143*, 179–198. https://www.jstor.org/stable/40441131

Aportela, F. (1999). *Effects of financial access on savings by low-income people*. Banco de Mexico Research Department.

Arcand, J. L., Berkes, E., & Panizza, U. (2015). Too much finance? *Journal of Economic Growth, 20*(2), 105–148. https://doi.org/10.1007/s10887-015-9115-2

Arellano, M. & Bond, S. R. (1991). Some test of specification for panel data: Monte Carlo evidence and application to employment equations. *Review of Economic Studies, 58*, 277–297. https://www.jstor.org/stable/2297968

Arestis, P., & Caner, A. (2010). Capital account liberalisation and poverty: How close is the link? *Cambridge Journal of Economics, 34*(2), 295–323. https://doi.org/10.1093/cje/bep062

Arestis, P., & Demetriades, P. (1997). Financial development and economic growth: Assessing the evidence. *The Economic Journal, 107*(442), 783–799. https://www.jstor.org/stable/2957802

Arestis, P., Demetriades, P. O., & Luintel, K. B. (2001). Financial development and economic growth: The role of stock markets. *Journal of Money, Credit and Banking*, 16–41. https://www.jstor.org/stable/2673870

Arner, D. W., Barberis, J., & Buckley, R. P. (2015). The evolution of Fintech: A new post-crisis paradigm. *Geo. J. Int'l l., 47*, 1271.

Arora, R. U. (2010). Measuring financial access. *Griffith Business School*, Discussion Papers No. 2010-07.

Ashraf, N., Karlan, D., & Yin, W. (2006). Tying Odysseus to the mast: Evidence from a commitment savings product in the Philippines. *The Quarterly Journal of Economics*, *121*(2), 635–672. https://www.jstor.org/stable/25098802

Ashraf, N., Karlan, D., & Yin, W. (2010). Female empowerment: Impact of a commitment savings product in the Philippines. *World Development, 38*(3), 333–344. https://doi.org/10.1016/j.worlddev.2009.05.010

Ashraf, S., Mehmood, M., & Khan, I. (2021). Cybersecurity in digital financial services: Challenges and solutions. *Journal of Financial Regulation and Compliance, 29*(4), 561–574. https://doi.org/10.1108/JFRC-09-2020-0241

Asian Development Bank. (2012). *Food security and poverty in Asia and the Pacific: Key challenges and policy issues*. https://www.adb.org/publications/food-security-and-poverty-asia-and-pacific-key-challenges-and-policy-issues

Ayadi, R., Arbak, E., Naceur, S. B., & De Groen, W. P. (2015). Financial development, bank efficiency, and economic growth across the mediterranean. In R. Ayadi, M. Dabrowski, & L. De Wulf (Eds.), *Economic and social development of the southern and eastern mediterranean countries* (pp. 219–233). Springer. https://doi.org/10.1007/978-3-319-11122-3_14

Ayo, C. K., Akter, S., & Aljoza, M. (2022). Consumers' trust and adoption of digital financial services: Insights from developing economies. *International Journal of Bank Marketing, 40*(5), 1–22. https://doi.org/10.1108/IJBM-07-2021-0299

Aysan, A. F., Dolgun, M. H., & Turhan, M. I. (2013). Assessment of the participation banks and their role in financial inclusion in Turkey. *Emerging Markets Finance and Trade, 49*(5), 99–111.

Ayyagari, M., & Beck, T. (2015). Financial inclusion in Asia: An overview. *Asian Development Bank Economics Working Paper Series*, (449). https://hdl.handle.net/11540/5142

Ayyagari, M., Demirgüç-Kunt, A., & Maksimovic, V. (2008). How important are financing constraints? The role of finance in the business environment. *The World Bank Economic Review*, *22*(3), 483–516. https://www.jstor.org/stable/40282285

Bae, J. (2022). The Bank of Korea's CBDC research: Current status and key considerations. *BIS Papers*, 107.

Baird, S., Friedman, J., & Schady, N. (2011). Aggregate income shocks and infant mortality in the developing world. *Review of Economics and statistics*, *93*(3), 847–856. https://www.jstor.org/stable/23016081

Baldacci, E., De Mello, L., & Inchauste, G. (2004). Financial crises, poverty, and income distribution. In *Macroeconomic policies and poverty* (pp. 269–307). Routledge.

Baležentis, T., Li, T., & Chen, X. (2021). Has agricultural labor restructuring improved agricultural labor productivity in China? A decomposition approach. *Socio-Economic Planning Sciences, 76*, Article 100967. https://doi.org/10.1016/j.seps.2020.100967

Banerjee, A. V. (2003). Contracting constraints, credit markets, and economic development. *Econometric Society Monographs, 37*, 1–46.

Banerjee, A. V., & Newman, A. F. (1993). Occupational choice and the process of development. *Journal of Political Economy, 101*(2), 274–298. https://www.jstor.org/stable/2138820

Barr, M. (2004). Banking the poor. *Yale Journal on Regulation, 21*, 122–239.

Bateman, M., Duvendack, M., & Loubere, N. (2019). Is fin-tech the new panacea for poverty alleviation and local development? Contesting Suri and Jack's M-Pesa findings published in Science. *Review of African Political Economy, 46*(161), 480–495. https://doi.org/10.1080/03056244.2019.1614552

Bayero, M. A. (2015). Effects of cashless economy policy on financial inclusion in Nigeria: An exploratory study. *Procedia-Social and Behavioral Sciences, 172*(8), 49–56.

Beck, T., Demirguc-Kunt, A., & Levine, R. (2007). Finance, inequality, and the poor. *Journal of Economic Growth, 12*(1), 27–49. https://doi.org/10.1007/s10887-007-9010-6

Beck, T., & Demirgüç-Kunt, A. (2008). Access to finance: An unfinished agenda. *World Bank Economic Review, 22*(3), 383–396. https://doi.org/10.1093/wber/lhn021

Beck, T., Demirgüç-Kunt, A. S. L. I., & Maksimovic, V. (2005). Financial and legal constraints to growth: Does firm size matter? *The Journal of Finance, 60*(1), 137–177.

Beck, T., Levine, R., & Loayza, N. (2000). Finance and the sources of growth. *Journal of Financial Economics, 58*(1–2), 261–300. https://doi.org/10.1016/s0304-405x(00)00072-6

Beck, T., Senbet, L., & Simbanegavi, W. (2014). Financial inclusion and innovation in Africa: An overview. *Journal of African Economies, 24*(1), i3–i11.

Becker, G. S. (1975). *Human capital: A theoretical and empirical analysis, with special reference to education* (2nd ed.). National Bureau of Economic Research. http://www.nber.org/books/beck75-1

Behrman, J. R. (1988). Nutrition and incomes: Tightly wedded or loosely meshed?. *Pew/Cornell Lecture Series on Food and Nutrition Policy-Cornell University (EUA).*

Behrman, J. R., & Deolalikar, A. B. (1987). Will developing country nutrition improve with income? A case study for rural South India. *Journal of Political Economy, 95*(3), 492–507. https://doi.org/10.1086/261469

Behrman, J. R., Deolalikar, A. B., & Wolfe, B. L. (1988). Nutrients: Impacts and determinants. *The World Bank Economic Review*, *2*(3), 299–320. https://www.jstor.org/stable/3989840

Bell, C., & Rousseau, P. L. (2001). Post-independence India: A case of finance-led industrialization? *Journal of Development Economics, 65*(1), 153–175. https://doi.org/10.1016/s0304-3878(01)00132-8

Bencivenga, V. R., & Smith, B. D. (1991). Financial intermediation and endogenous growth. *The Review of Economic Studies*, *58*(2), 195–209. https://www.jstor.org/stable/2297964

Bencivenga, V. R., & Smith, B. D. (1993). Some consequences of credit rationing in an endogenous growth model. *Journal of Economic Dynamics and Control, 17*(1–2), 97–122. https://doi.org/10.1016/S0165-1889(06)80006-0

Berger, A. N. (2003). The economic effects of technological progress: Evidence from the banking industry. *Journal of Money, Credit and Banking*, 141–176. https://www.jstor.org/stable/3649852

Berki, S. E. (1986). A look at catastrophic medical expenses and the poor. *Health Affairs, 5*(4), 138–145. https://doi.org/10.1377/hlthaff.5.4.138

Béné, C. (2020). Resilience of local food systems and links to food security—A review of some important concepts in the context of COVID-19 and other shocks. *Food Security, 12*(4), 805–822. https://doi.org/10.1007/s12571-020-01076-1

Bhattacharyya, D., & Dutta, S. (2018). Financial inclusion and development in the North-East India: Challenges and prospects. *Journal of Financial Inclusion, 12*(2), 54–66. https://doi.org/10.2139/ssrn.3251243

Bishai, D., & Nalubola, R. (2002). The history of food fortification in the United States: Its relevance for current fortification efforts in developing countries. *Economic Development and Cultural Change, 51*(1), 37–53.

Bist, J. P. (2018). Financial development and economic growth: Evidence from a panel of 16 African and non-African low-income countries. *Cogent Economics & Finance, 6*(1), 1–17. https://doi.org/10.1080/23322039.2018.1449780

Bittencourt, M. (2012). Financial development and economic growth in Latin America: Is Schumpeter right? *Journal of Policy Modelling, 34*(3), 341–355. https://doi.org/10.1016/j.jpolmod.2012.01.012

Blundell, R. W., & Bond, S. R. (1998). Initial conditions and moment restrictions in dynamic panel data models. *Journal of Econometrics, 87*, 29–52.

Bora, A. (2019). Financial inclusion and socio-cultural challenges in North-East India. *Economic and Political Weekly, 54*(32), 40–47.

Bouis, H. E., & Haddad, L. J. (1988). *Comparing calorie-income elasticities using calories derived from reported food purchases and a twenty-four hour recall of food intakes: An application using Philippine data*. University of Warwick.

Bouis, H. E., & Haddad, L. J. (1992). Are estimates of calorie-income elasticities too high? A recalibration of the plausible range. *Journal of Development Economics, 39*(2), 333–364. https://doi.org/10.1016/0304-3878(92)90043-9

Boukhatem, J. (2016). Assessing the direct effect of financial development on poverty reduction in a panel of low-and middle-income countries. *Research in International Business and Finance, 37*, 214–230. https://doi.org/10.1016/j.ribaf.2015.11.008

Brinkman, H. J., De Pee, S., Sanogo, I., Subran, L., & Bloem, M. W. (2010). High food prices and the global financial crisis have reduced access to nutritious food and worsened nutritional status and health. *The Journal of Nutrition, 140*(1), 153S-161S. https://doi.org/10.3945/jn.109.110767

Brown, J. K., Zelenska, T. V., & Mobarak, M. A. (2013). *Barriers to adoption of products and technologies that aid risk management in developing countries*. Retrieved August 11, 2022, from http://hdl.handle.net/10986/16365.

Bryce, J., Coitinho, D., Darnton-Hill, I., Pelletier, D., Pinstrup-Andersen, P., & Maternal and Child Undernutrition Study Group. (2008). Maternal and child undernutrition: Effective action at national level. *The Lancet*, *371*(9611), 510–526.

Buchmann, C., & Hannum, E. (2001). Education and stratification in developing countries: A review of theories and research. *Annual Review of Sociology, 27*(1), 77–102.

Buckland, J., Guenther, B., Boichev, G., Geddie, H., & Mutch, M. (2005). There are no banks here: Financial & Insurance Exclusion in Winnipeg's North End. *Winnipeg Inter- City Research Alliance (WIRA)*.

Burgess, R., & Pande, R. (2005). Do rural banks matter? Evidence from the Indian Social Banking experiment. *American Economic Review, 95*(3), 780–795. https://www.jstor.org/stable/4132740

Butler, P. (2012, June 25). Demand for food parcels explodes as welfare cuts and falling pay hit home—early-warning indicator should set alarm bells ringing about poverty levels, government told. *The Guardian*. https://www.theguardian.com/society/2012/jun/25/breadline-britain-growth-food-parcels

Cabral, L., & Howell, J. (2012). Measuring aid to agriculture and food security. *ODI Briefing Paper*, (72).

Calderón, C., & Liu, L. (2003). The direction of causality between financial development and economic growth. *Journal of Development Economics, 72*(1), 321–334.

Cámara, N., & Tuesta, D. (2014). Measuring financial inclusion: A multidimensional index. *BBVA Research Working Paper* (WP No. 14/26).

Carbo, S., Gardener, E. P., & Molyneux, P. (2005). *Financial exclusion*. Palgrave MacMillan.

Chakravarty, S. R., & Pal, R. (2013). Financial inclusion in India: An axiomatic approach. *Journal of Policy Modeling, 35*, 813–837. https://doi.org/10.1016/j.jpolmod.2012.12.007

Chamberlain, D., Bester, H., Smit, H., Loots, C., Mburu, S., Dermish, A., Gidvani, L., & Saunders, D. (2014). Myanmar: Demand, supply, policy and regulation. Centre for Financial Regulation and Inclusion (CENFRI). Bankable Frontier Associates (BFA).

Chattopadhyay, S. K. (2011). Financial inclusion in India: A case-study of West Bengal. Working Paper Series (DEPR): 8/2011). *Department of Economic and Policy Research*, RBI Working Paper Series.

Chaturvedi, V., Dholakia, R. H., & Kumar, B. (2011). Inter-relationship between economic growth, savings and inflation in Asia. *SSRN Electronic Journal*. https://doi.org/10.2139/ssrn.1212096

Chen, X., Shuai, C., & Wu, Y. (2023). Global food stability and its socio-economic determinants towards sustainable development goal 2 (Zero Hunger). *Sustainable Development, 31*(3), 1768–1780. https://doi.org/10.1002/sd.2482

Chen, Z., Zhang, M., & Wu, Y. (2019). A survey of data security and privacy protection in digital finance. *IEEE Access, 7*, 57356–57369. https://doi.org/10.1109/ACCESS.2019.2913491

Chibba, M. (2009). Financial inclusion, poverty reduction and the millennium development goals. *The European Journal of Development Research, 21*(2), 213–230. https://doi.org/10.1057/ejdr.2008.17

Chien, C. F., & Lin, K. L. (2020). Financial literacy and its impact on financial inclusion in Asia. *Asian Economic Policy Review, 15*(1), 56–72. https://doi.org/10.1111/aepr.12345

Chotia, V., & Rao, N. M. (2014). Financial inclusion and human development: A state-wise analysis from India. *International Journal of Economics, Commerce and Management, II, 5*, 1–24.

Christian, P. (2010). Impact of the economic crisis and increase in food prices on child mortality: Exploring nutritional pathways. *The Journal of Nutrition, 140*(1), 177S-181S. https://doi.org/10.3945/jn.109.111708

Christopoulos, D. K., & Tsionas, E. G. (2004). Financial development and economic growth: Evidence from panel unit root and cointegration tests.

Journal of Development Economics, 73(1), 55–74. https://doi.org/10.1016/j.jdeveco.2003.03.002

Churchill, S. A., & Marisetty, V. B. (2020). Financial inclusion and poverty: A tale of forty-five thousand households. *Applied Economics, 52*(16), 1777–1788. https://doi.org/10.1080/00036846.2019.1678732

Čihák, M., Demirgüç-Kunt, A., Feyen, E., & Levine, R. (2012). Benchmarking financial systems around the world. *World Bank policy research working paper*, (6175). https://ssrn.com/abstract=2152254

Clarke, G., Xu, L., & Zou, H. (2006). Finance and Income Inequality: What do the data tell us? *Southern Economic Journal, 72*(3), 578–596. https://doi.org/10.1002/j.2325-8012.2006.tb00721.x

Collins, D., Morduch, J., Rutherford, S., & Ruthven, O. (2009). *Portfolios of the poor: How the world's poor live on $2 a day*. Princeton University Press.

Conroy, J. (2005). APEC and financial exclusion: Missed opportunities for collective action? *Asia-Pacific Development Journal,* 53–80. https://doi.org/10.18356/339ee7fc-en

Coulibaly, B. S. (2019). *Foresight Africa: Top priorities for the continent in 2019.* Retrieved September 12, 2023, from Africa Growth initiative, Brookings Institute website. https://www.brookings.edu/wp-content/uploads/2019/01/BLS18234_BRO_book_007_WEB.pdf

Cumming, G. (2012). The new statistics: Why and how. *Psychological Science, 25*(1), 7–29.

Cutler, D., Deaton, A., & Lleras-Muney, A. (2006). The determinants of mortality. *Journal of Economic Perspectives, 20*(3), 97–120.

Dahiya, S., & Kumar, M. (2020). Linkage between financial inclusion and economic growth: An empirical study of the emerging Indian economy. *Vision, 24*(2), 184–193. https://doi.org/10.1177/0972262920923891

Darnton-Hill, I., & Cogill, B. (2010). Maternal and young child nutrition adversely affected by external shocks such as increasing global food prices. *The Journal of Nutrition, 140*(1), 162S-169S. https://doi.org/10.3945/jn.109.111682

Das, D. K. (2011). Indian economy: Growth performance and prospects of transitioning the growth trajectory. *Journal of Asia-Pacific Business, 12*(2), 171–193.

Dash, S. K., & Kumar, L. (2018). Does inflation affect savings non-linearly? Evidence from India. *Margin: The Journal of Applied Economic Research, 12*(4), 431–457. https://doi.org/10.1177/0973801018786155

Datt, G., & Ravallion, M. (1992). Growth and redistribution components of changes in poverty measures: A decomposition with applications to Brazil and India in the 1980s. *Journal of Development Economics, 38*(2), 275–295.

Dawood, T. C., Pratama, H., Masbar, R., & Effendi, R. (2019). Does financial inclusion alleviate household poverty? Empirical evidence from Indonesia.

Economics & Sociology, 12(2), 235–252. https://doi.org/10.14254/2071-789X.2019/12-2/14

De Benoist, B., Dary, O., & Hurrell, R. (2006). L. Allen (Ed.), *Guidelines on food fortification with micronutrients* (Vol. 126). World Health Organization.

De Gregorio, J., & Guidotti, P. E. (1995). Financial development and economic growth. *World Development, 23*(3), 433–448. https://doi.org/10.1016/0305-750X(94)00132-I

De Koker, L., & Jentzsch, N. (2013). Financial inclusion and financial integrity: Aligned incentives? *World Development, 44*, 267–280.

Deidda, L., & Fattouh, B. (2002). Non-linearity between finance and growth. *Economics Letters, 74*(3), 339–345. https://doi.org/10.1016/S0165-1765(01)00571-7

Demetriades, P. O., Devereux, M. P., & Luintel, K. B. (1998). Productivity and financial sector policies: Evidence from South East Asia. *Journal of Economic Behavior & Organization, 35*(1), 61–82. https://doi.org/10.1016/s0167-2681(97)00099-1

Demetriades, P. O., & Hussein, K. A. (1996). Does financial development cause economic growth? Time-series evidence from 16 countries. *Journal of Development Economics, 51*(2), 387–411. https://doi.org/10.1016/s0304-3878(96)00421-X

Demirgüç-Kunt, A. (2006). Finance and economic development: Policy choices for developing countries. *The World Bank*. https://doi.org/10.1596/1813-9450-3955

Demirgüç-Kunt, A., & Klapper, L. (2013). Measuring financial inclusion: Explaining variation in use of financial services across and within countries (The Global Findex Database). *Brookings Papers on Economic Activity*. Brookings Institution.

Demirgüç-Kunt, A., Klapper, L., Singer, D., Ansar, S., & Hess, J. (2018). *The global findex database 2017: Measuring financial inclusion and the fintech revolution*. Retrieved July 18, 2022, from http://documents.worldbank.org/curated/en/332881525873182837/pdf/126033-PUB-PUBLIC-pubdate-4-19-2018.pdf.

Demirgüç-Kunt, A., Klapper, L. F., Singer, D., & Van Oudheusden, P. (2015). The global findex database 2014: Measuring financial inclusion around the world. *World Bank Policy Research Working Paper*, (7255).

Demirgüç-Kunt, A., & Maksimovic, V. (1999). *Financial constraints, uses of funds, and firm growth: An international comparison*. The World Bank.

Diamond, D. W. (1989). Reputation acquisition in debt markets. *Journal of Political Economy*, *97*(4), 828–862. https://www.jstor.org/stable/1832193

Dimova, R., & Adebowale, O. (2018). Does access to formal finance matter for welfare and inequality? Micro level evidence from Nigeria. *The Journal of Development Studies, 54*(9), 1534–1550. https://doi.org/10.1080/00220388.2017.1371293

Donovan, K. (2012). *Mobile money for financial inclusion.* Retrieved December 1, 2021, from http://elibraryworldbank.org/ https://doi.org/10.1596/9780821389911_ch04.

Dorfleitner, G., Oswald, E. M., & Zhang, R. (2021). From credit risk to social impact: On the funding determinants in interest-free peer-to-peer lending. *Journal of Business Ethics, 170*, 375–400. https://doi.org/10.1007/s10551-019-04311-8

Dreze, J., & Sen, A. (2002). *India: Development and participation.* Oxford University Press on Demand.

Ductor, L., & Grechyna, D. (2015). Financial development, real sector, and economic growth. *International Review of Economics & Finance, 37*, 393–405. https://doi.org/10.1016/j.iref.2015.01.001

Duflos, E., Luchtenburg, P., Ren, L., & Chen, L. Y. (2013). Microfinance in Myanmar sector assessment. CGAP, IFC.

Dugarova, E., & Gülasan, N. (2017). *Global trends: Challenges and opportunities in the implementation of the sustainable development goals.* Academic Press.

Dupas, P., & Robinson, J. (2013). Savings constraints and microenterprise development: Evidence from a field experiment in Kenya. *American Economic Journal: Applied Economics, 5*(1), 163–192.

Durusu-Ciftci, D., Ispir, M. S., & Yetkiner, H. (2017). Financial development and economic growth: Some theory and more evidence. *Journal of Policy Modeling, 39*(2), 290–306. https://doi.org/10.1016/j.jpolmod.2016.08.001

Egbuna, E., Oduh, M., Ujunwa, A., & Okoyeuzu, C. (2018). Does deposit insurance promote moral hazards and adverse selection? Evidence from Sub-Saharan Africa. *International Journal of Managerial Finance, 14*(2), 150–169. https://doi.org/10.1108/IJMF-10-2016-0196

Emara, N., & El Said, A. (2021). Financial inclusion and economic growth: The role of governance in selected MENA countries. *International Review of Economics & Finance, 75*, 34–54. https://doi.org/10.1016/j.iref.2021.03.014

Emerick, K., de Janvry, A., Sadoulet, E., & Dar, M. H. (2016). Technological innovations, downside risk, and the modernization of agriculture. *American Economic Review, 106*(6), 1537–1561. https://www.jstor.org/stable/43861131

Engle, R. F., & Granger, C. W. (1987). Co-integration and error correction: Representation, estimation, and testing. *Econometrica: Journal of the Econometric Society, 55*(2), 251–276. https://doi.org/10.2307/1913236

Engle, R. F., & Granger, C. W. J. (2003). Time-series econometrics: Cointegration and autoregressive conditional heteroskedasticity. *Advanced information on the Bank of Sweden Prize in Economic Sciences in Memory of Alfred Nobel* (pp. 1–30).

Eryılmaz, F., Bakır, H., & Mercan, M. (2015). Financial development and economic growth: Panel data analysis. In Ö. Olgu, H. Dinçer, & Ü. Hacıoğlu (Eds.), *Handbook of research on strategic developments and regulatory practice in global finance* (pp. 233–245). IGI Global. https://doi.org/10.4018/978-1-4666-7288-8.ch015

European Commission. (2015). Payment services. *The European Commission.* https://ec.europa.eu/info/law/paymentservices-psd-2-directive-eu-2015-2366_en

Eustachio, J. H. P. P., Caldana, A. C. F., Liboni, L. B., & Martinelli, D. P. (2019). Systemic indicator of sustainable development: Proposal and application of a framework. *Journal of Cleaner Production, 241*, Article 118383.

Farrell, P., Thow, A. M., Abimbola, S., Faruqui, N., & Negin, J. (2018). How food insecurity could lead to obesity in LMICs: When not enough is too much: A realist review of how food insecurity could lead to obesity in low-and middle-income countries. *Health Promotion International, 33*(5), 812–826. https://doi.org/10.1093/heapro/dax026

Fase, M. M. G., & Abma, R. C. N. (2003). Financial environment and economic growth in selected Asian countries. *Journal of Asian Economics, 14*, 11–21.

Findex. (2022). *The global findex database 2021.* World Bank Group.

Food and Agriculture Organization. (2008). *The state of food insecurity in the world 2008: High food prices and food security—threats and opportunities*. https://www.fao.org/publications/card/en/c/3d431db5-34c8-5e01-9ad9-b2b0a2b9b918/.

Food and Agriculture Organization. (2011). *The state of food insecurity in the world 2011: How does international price volatility affect domestic economics and food security?* https://www.fao.org/publications/card/fr/c/b9dff790-3555-55c3-a8c6-c5efc0e5c304/.

Food and Agriculture Organization. (2012). *The state of food insecurity in the world 2012: Economic growth is necessary but not sufficient to accelerate reduction of hunger and malnutrition*. https://www.fao.org/3/i3027e/i3027e00.htm.

Food and Agriculture Organization. (2020). *The state of food security and nutrition in the world 2020: Transforming food systems for affordable healthy diets*. https://doi.org/10.4060/ca9692en

Fosu, A. K. (2010). Inequality, income, and poverty: Comparative global evidence. *Social Science Quarterly, 91*(5), 1432–1446.

Freudenberg, M. (2003). *Composite indicators of country performance: A critical assessment*. OECD.

Friesen, P. H. (1979). The Arrow-Debreu model extended to financial markets. *Econometrica: Journal of the Econometric Society*, 689–707. https://doi.org/10.2307/1910415

FSD Kenya. (2025). *Financial inclusion and health: A balancing act for Kenya's future*. Retrieved from https://www.fsdkenya.org/blogs-publications/financial-inclusion-and-health-a-balancing-act-for-kenyas-future/.

Fund, S. (2015). Sustainable development goals. https://www.un.org/sustainabledevelopment/inequality.

Fungáčová, Z., & Weill, L. (2015). Understanding financial inclusion in China. *China Economic Review, 34*, 196–206. https://doi.org/10.1016/j.chieco.2014.12.004

Galor, O., & Zeira, J. (1993). Income distribution and macroeconomics. *Review of Economic Studies*, *60*(1), 35–52. https://www.jstor.org/stable/2297811

Garg, S. (2022). Digital financial inclusion and poverty alleviation: A study of mobile banking services. *Journal of Financial Services Marketing, 27*(2), 98–113. https://doi.org/10.1057/s41264-022-00095-1

Gentleman, A. (2014, February 28). Vulnerable man starved to death after benefits were cut—44- year-old died months after sickness and housing benefits were stopped following Atos fitness-for-work assessment. *The Guardian*. https://www.theguardian.com/society/2014/feb/28/man-starved-to-death-after-benefits-cut.

Ghosh, S., & Kapoor, S. (2019). Financial literacy and inclusion in India: A critical review. *Indian Journal of Economics and Business, 18*(2), 115–130.

Ghosh S., & Sahu T. N. (2020). How financially inclusive the Asian's are—an empirical analysis. *Rajagiri Management Journal*, *14*(1). https://doi.org/10.1108/RAMJ-01-2020-0003.

Ghosh, S., & Sahu, T. N. (2021). Financial inclusion and economic status of the states of India: An empirical evidence. *Economic Notes, 50*(2), Article e12182. https://doi.org/10.1111/ecno.12182

Ghosh, S., & Sahu, T. N. (2022). How far is productivity relevant in explaining financial inclusion across the states of India? An empirical investigation. *Global Business and Economics Review, 26*(4), 417–435.

Ghosh, S., Sahu, T. N., & Mallik, G. (2024). Does financial inclusion stimulate India's economy: A panel data analysis. *Journal of Public Affairs, 24*(1), Article e2891. https://doi.org/10.1002/pa.2891

GIZ (2013). *Myanmar's Financial sector: A challenging environment for banks.* Deutsche Gesellschaft für Internationale Zusammenarbeit (GIZ) GmbH.

Goldsmith, R. W. (1969). *Financial structure and development*. Yale University Press.

Gomber, P., Koch, J. A., & Siering, M. (2017). Digital Finance and FinTech: Current research and future research directions. *Journal of Business Economics, 87*, 537–580. https://doi.org/10.1007/s11573-017-0852-x

Goudie, A., & Ladd, P. (1999). Economic growth, poverty and inequality. *Journal of International Development, 11*(2), 177.

Granger, C. W. J. (1981). Some properties of time series data and their use in econometric model specification. *Journal of Econometrics, 16*, 121–130.

Granger, C. W. J., & Newbold, P. (1974). Spurious regressions in econometrics. *Journal of Econometrics, 2*(2), 111–120. https://doi.org/10.1016/0304-4076(74)90034-7

Grassa, R., & Gazdar, K. (2014). Financial development and economic growth in GCC countries: A comparative study between Islamic and conventional finance. *International Journal of Social Economics, 41*(6), 493–514. https://doi.org/10.1108/IJSE-12-2012-0232

Green, R., Cornelsen, L., Dangour, A. D., Turner, R., Shankar, B., Mazzocchi, M., & Smith, R. D. (2013). The effect of rising food prices on food consumption: Systematic review with meta-regression. *BMJ*. https://doi.org/10.1136/bmj.f3703

Greenwood, J., & Jovanovic, B. (1990). Financial development, growth, and the distribution of income. *Journal of political Economy, 98*(5, Part 1), 1076–1107. https://www.jstor.org/stable/2937625

Grossman, G. M., & Helpman, E. (1991). Quality ladders in the theory of growth. *The Review of Economic Studies, 58*(1), 43–61.

GSMA. (2020a). *Connected women. The Mobile Gender Gap Report 2020.*

GSMA. (2020b). *Driving the digital revolution with improved mobile coverage.* https://www.gsma.com/connectivity-for-good/spectrum/wp-content/uploads/2020/11/Rural-Mobile-Coverage-Positions.pdf.

Gupte, R., Venkataramani, B., & Gupta, D. (2012). Computation of financial inclusion index for India. *Procedia—Social and Behavioral Sciences, 37*, 133–149. https://doi.org/10.1016/j.sbspro.2012.03.281

Gurley, J. G., & Shaw, E. S. (1955). Financial aspects of economic development. *The American Economic Review, 45*(4), 515–538. http://www.jstor.org/stable/1811632

Guru, B. K., & Yadav, I. S. (2019). Financial development and economic growth: Panel evidence from BRICS. *Journal of Economics, Finance and Administrative Science, 24*(47), 113–126. https://doi.org/10.1108/JEFAS-12-2017-0125

Habibullah, M. S., & Eng, Y. K. (2006). Does financial development cause economic growth? A panel data dynamic analysis for the Asian developing countries. *Journal of the Asia Pacific Economy, 11*(4), 377–393. https://doi.org/10.1080/13547860600923585

Hafner, L. (2020). Cyber risks in financial services: A global overview. *Journal of Financial Crime, 27*(3), 710–721. https://doi.org/10.1108/JFC-06-2020-0129

Hale, G., & Long, C. (2011). What are the sources of financing for Chinese firms?. In *The evolving role of Asia in global finance* (Vol. 9, pp. 313–339). Emerald Group Publishing Limited. https://doi.org/10.1108/S1574-8715(2011)0000009018

Hallegatte, S., Vogt-Schilb, A., Bangalore, M., & Rozenberg, J. (2016). *Unbreakable: Building the resilience of the poor in the face of natural disasters*. World Bank Publications.

Hannig, A., & Jansen, S. (2010). *Financial inclusion and financial stability: Current policy issues* (Working Paper No. 259). Asian Development Bank.

Hassan, M. K., Sanchez, B., & Yu, J. S. (2011). Financial development and economic growth: New evidence from panel data. *The Quarterly Review of Economics and Finance, 51*(1), 88–104. https://doi.org/10.1016/j.qref.2010.09.001

Hedges, L. V., & Olkin, I. (1985). *Statistical methods for meta-analysis*. Academic Press.

Herwartz, H., & Walle, Y. M. (2014). Determinants of the link between financial and economic development: Evidence from a functional coefficient model. *Economic Modelling, 37*, 417–427. https://doi.org/10.1016/j.econmod.2013.11.029

Hobbs, J. E. (2020). Food supply chains during the COVID-19 pandemic. *Canadian Journal of Agricultural Economics/Revue canadienne d'agroeconomie, 68*(2), 171–176. https://doi.org/10.1111/cjag.12237

Honohan, P. (2004). Financial development, growth and poverty: How close are the links? *World Bank Policy Research Working Paper 3203*. http://documents.worldbank.org/curated/en/518641468763230741/Financial-development-growth-and-poverty-how-close-are-the-links.

ICSU, & ISSC. (2015). *Review of the sustainable development goals: The science perspective*. International Council for Science (ICSU). Retrieved September 5, 2021, from http://www.genderportal.eu/resources/review-sustainable-development-goals-science-perspective.

IFC. (2019). *Financial inclusion in Africa: The role of women*. International Finance Corporation.

Imai, K. S., Arun, T., & Annim, S. K. (2010). Microfinance and household poverty reduction: New evidence from India. *World Development, 38*(12), 1760–1774. https://doi.org/10.1016/j.worlddev.2010.04.006

IMF Country Report No. 17/153 browsed on October 23, 2021, from www.imf.org/-/media/Files/Publications/CR/2017/cr1715.

Indicators, G. S. (2016). Tier classification for global SDG indicators.

Inoue, T., & Hamori, S. (2012). How has financial deepening affected poverty reduction in India? Empirical analysis using state-level panel data. *Applied Financial Economics, 22*(5), 395–408. https://doi.org/10.1080/09603107.2011.613764

International Monetary Fund. (2018). *Financial inclusion: A comprehensive overview*. International Monetary Fund. https://www.imf.org/en/Topics/Financial-Inclusion

International Telecommunication Union. (2020). *Measuring the information society report 2020*. ITU.

Iorgova, S., & Liu, Y. (2013). Structure of the banking sector and implications for financial stability. In U. Das, J. Fiechter, & T. Sun (Eds.), *China's road to greater financial stability: Some policy perspectives IMF* (pp. 123–139).

Irz, X., Niemi, J., & Liu, X. (2013). Determinants of food price inflation in Finland—The role of energy. *Energy Policy, 63*, 656–663. https://doi.org/10.1016/j.enpol.2013.09.023

Izquierdo, N. C., & Tuesta, D. (2015). Factors that matter for financial inclusion: Evidence from Peru. *Aestimatio: The IEB International Journal of Finance, 10*, 10–31.

Jack, W., & Suri, T. (2011). Mobile money: The economics of M-PESA. *National Bureau of Economic Research Working Paper Series*, No. 16721. https://doi.org/10.3386/w16721

Jacobs, R., Smith, P., & Maria, G. (2004). Measuring performance: An examination of composite performance indicators. *Centre for Health Economics, Technical Paper Series* 29.

Jain, A. K., & Mishra, S. N. (2019). Role of NITI Aayog in the Implementation of the 2030 Agenda. *Agenda and India: Moving from Quantity to Quality: Exploring Convergence and Transcendence, 2030*, 239–254.

Jalilian, H., & Kirkpatrick, C. (2002). Financial development and poverty reduction in developing countries. *International Journal of Finance and Economics, 7*(2), 97–108. https://doi.org/10.1002/ijfe.179

Jalilian, H., & Kirkpatrick, C. (2005). Does financial development contribute to poverty reduction? *The Journal of Development Studies, 41*(4), 636–656. https://doi.org/10.1080/00220380500092754

Janjua, P. Z., & Kamal, U. A. (2014). The role of education and health in poverty alleviation a cross country analysis. *British Journal of Economics, Management & Trade, 4*(6), 896–924.

Jeanneney, S., & Kpodar, K. (2011). Financial development and poverty reduction: Can there be a benefit without a cost? *Journal of Development Studies, 47*(1), 143–163. https://doi.org/10.1080/00220388.2010.506918

Jung, W. S. (1986). Financial development and economic growth: International evidence. *Economic Development and Cultural Change, 34*(2), 333–346. https://doi.org/10.1086/451531

Kakwani, N. (1993). Poverty and economic growth with application to Cote d'Ivoire. *Review of Income and Wealth, 39*, 121–139. https://doi.org/10.1111/j.1475-4991.1993.tb00443.x

Kaleeswaran, P., & Meera, M. (2017). Financial Inclusion Index through MGNREGS in Virudhunagar District. *Asian Journal of Managerial Science, 6*(1), 5–9.

Kalunda, E. (2014). Financial inclusion impact on small-scale tea farmers in Nyeri County, Kenya. *World Journal of Social Sciences, 4*(1), 130–139.

Kanbur, R. (2001). Economic policy, distribution and poverty: The nature of disagreements. *World Development, 29*(6), 1083–1094.

Kao, C., & Chiang, M. H. (2001), On the estimation and inference of a cointegrated regression in panel data, In B. H. Baltagi, T. B. Fomby, & R. Carter Hill (Eds.), *Nonstationary panels, panel cointegration, and dynamic panels. Advances in Econometrics* (Vol. 15, pp. 179–222). Emerald Group Publishing Limited. https://doi.org/10.1016/S0731-9053(00)15007-8

Karpowicz, M. I. (2014). *Financial inclusion, growth and inequality: A model application to Colombia* (Working Paper No. 14/166). International Monetary Fund.

Kempson, E., & Whyley, C. (1998). *Access to current accounts.* British Bankers' Association.

Kempson, E., & Whyley, C. (2001). *Payment of pension and benefits.* Department for Work and Pension.

Keynes, J. M. (1937). The "ex-ante" theory of the rate of interest. *The Economic Journal, 47*(188), 663–669. https://www.jstor.org/stable/2225323

Khan, M. S., & Senhadji A. S. (2000). Financial development and economic growth: An overview. *IMF Working Papers, 2000*(209), 1–24. https://doi.org/10.5089/9781451874747.001

Khan, S., Akter, S., & Hossain, M. (2018). Financial inclusion and the role of microfinance in Bangladesh: A comparative study. *Journal of Financial Services Marketing, 23*(2), 1–14. https://doi.org/10.1057/s41264-018-0026-1

Khan, S. A. R., Razzaq, A., Yu, Z., Shah, A., Sharif, A., & Janjua, L. (2022). Disruption in food supply chain and undernourishment challenges: An empirical study in the context of Asian countries. *Socio-Economic Planning Sciences, 82*, Article 101033. https://doi.org/10.1016/j.seps.2021.101033

Kiendrebeogo, Y., & Minea, A. (2016). Financial development and poverty: Evidence from the CFA Franc zone. *Applied Economics, 48*(56), 5421–5436. https://doi.org/10.1080/00036846.2016.1176114

Kim, D. W., Yu, J. S., & Hassan, M. K. (2018). Financial inclusion and economic growth in OIC countries. *Research in International Business and Finance, 43*, 1–14. https://doi.org/10.1016/j.ribaf.2017.07.178

Kim, R. E. (2023). Augment the SDG indicator framework. *Environmental Science & Policy, 142*, 62–67.

King, R. G., & Levine, R. (1993). Finance and growth: Schumpeter might be right. *The Quarterly Journal of Economics*, 717–737. https://www.jstor.org/stable/2118406

Klapper, L. (2016). *Financial inclusion has a big role to play in reaching the SDGs* (2016, September 15), World Bank. https://blogs.worldbank.org/developmenttalk/financial-inclusion-has-big-role-play-reaching-sdgs.

Klapper, L. F., Lusardi, A., & Panos, G. A. (2015). Financial literacy and the financial crisis. *Journal of Banking & Finance, 60*, 1–13.

Koomson, I., Villano, R. A., & Hadley, D. (2020). Effect of financial inclusion on poverty and vulnerability to poverty: Evidence using a multidimensional measure of financial inclusion. *Social Indicators Research, 149*(2), 613–639. https://doi.org/10.1007/s11205-019-02263-0

Kpodar, K., & Andrianaivo, M. (2011). *ICT, financial inclusion, and growth evidence from African countries* (Working Paper No. 11/73). International Monetary Fund.

Kravis, I. B. (1960). International differences in the distribution of income. *The Review of Economics and Statistics*. https://doi.org/10.2307/1925690

Kumar, C., & Mishra, S. (2011). Banking outreach and household level access: Analyzing financial inclusion in India. In *13th Annual Conference on Money and Finance in the Indian Economy* (pp. 25–26).

Kumar, P., Kumar, K., Mathur, S., & Singh, M. (2021). Aligning developmental programs to implement SDGs in India: Challenges and the way forward. *Journal of Contemporary Issues in Business & Government*, *27*(6).

Kumar, S., & Patel, R. (2020). Financial inclusion and income inequality in India: An empirical analysis. *Journal of Financial Inclusion and Development, 12*(3), 89–102. https://doi.org/10.1016/j.jfid.2020.04.001

Kuznets, S. (1955). Economic growth and income inequality. *The American Economic Review, 45*(1), 1–28.

Lafourcade, A. L., Isern, J., Mwangi, P., & Brown, M. (2005). *Overview of the outreach and financial performance of microfinance institutions in Africa*. Microfinance Information eXchange (MIX). Retrieved September 16, 2022, from http://www.mixmarket.org/medialibrary/mixmarket/Africa_Data_Study.pdf.

Laha, A., & Kuri, P. K. (2014). Demand for and supply of financial inclusion in India: An inter-state analysis. *Elk Asia Pacific Journal of Finance & Risk Management, 5*(1), 103–116.

Lashitew, A. A., van Tulder, R., & Liasse, Y. (2019). Mobile phones for financial inclusion: What explains the diffusion of mobile money innovations? *Research Policy, 48*(5), 1201–1215. https://doi.org/10.1016/j.respol.2018.12.010

Law, S. H., & Singh, N. (2014). Does too much finance harm economic growth? *Journal of Banking & Finance, 41*, 36–44. https://doi.org/10.1016/j.jbankfin.2013.12.020

Lee, D., & Deng, R. H. (2017). *Handbook of blockchain, digital finance, and inclusion*. (2). Academic Press. https://doi.org/10.1016/C2016-0-03679-3

Lee, H. H., Lee, S. A., Lim, J. Y., & Park, C. Y. (2016). Effects of food price inflation on infant and child mortality in developing countries. *The European Journal of Health Economics, 17*(5), 535–551.

Lee, H. H., & Park, C. Y. (2013). International transmission of food prices and volatilities: A panel analysis. *Asian Development Bank Economics Working Paper Series*, (373).

Leeladhar, V. (2006). *Taking banking services to the common man—financial inclusion*. Reserve Bank of India Bulletin, January.

Lenka, S. K., & Sharma, R. (2017). Does financial inclusion spur economic growth in India?. *The Journal of Developing Areas, 51*(3), 215–228. https://www.jstor.org/stable/ https://doi.org/10.2307/26416941

Levin, A., Lin, C. F., & Chu, C. S. J. (2002). Unit root tests in panel data: Asymptotic and finite-sample properties. *Journal of Econometrics, 108*(1), 1–24. https://doi.org/10.1016/S0304-4076(01)00098-7

Levine, R. (1997). Financial development and economic growth: Views and agenda. *Journal of Economic Literature, 35*(2), 688–726. http://www.jstor.org/stable/2729790

Levine, R. (2005). Finance and growth: Theory and evidence. In P. Aghion & S. H. Durlauf (Eds.), *Handbook of economic growth.* (Vol. 1). Elsevier.

Levine, R., Loayza, N., & Beck, T. (2000). Financial intermediation and growth: Causality and causes. *Journal of Monetary Economics, 46*(1), 31–77. https://doi.org/10.1016/S0304-3932(00)00017-9

Lewis, W. A. (1955). *Theory of economic growth.* Unwin University Books.

Leyshon, A., & Thrift, N. (1993). The restructuring of the U.K. Financial services industry in the 1990s: A reversal of fortune? *Journal of Rural Studies*, 223–241. https://doi.org/10.1016/0743-0167(93)90068-U

Leyshon, A., & Thrift, N. (1995). Geographies of financial exclusion: Financial abandonment in Britain and the United States. *Transactions of the Institute of British Geographers, New Series, 20*(3), 312–341.

Li, L. (2018). Financial inclusion and poverty: The role of relative income. *China Economic Review, 52*, 165–191. https://doi.org/10.1016/j.chieco.2018.07.006

Li, Y., Sun, G., Gao, Q., & Cheng, C. (2023). Digital financial inclusion, financial efficiency and green innovation. *Sustainability, 15*(3), 1879. https://doi.org/10.3390/su15031879

Li, Z., & Zhang, X. (2021). The role of digital financial services in enhancing financial inclusion in Asia: Evidence from mobile money. *Asian Development Review, 38*(3), 108–125. https://doi.org/10.1162/asep_a_01432

Lindert, P. H., & Williamson, J. G. (1985). Growth, equality, and history. *Explorations in Economic History, 22*(4), 341–377. https://doi.org/10.1016/0014-4983(85)90001-4

Loayza, N. V., & Ranciere, R. (2006). Financial development, financial fragility, and growth. *Journal of Money, Credit and Banking*, 1051–1076. https://www.jstor.org/stable/3838993

Lopez, L., & Weber, S. (2017). Testing for granger causality in panel data. *The Stata Journal, 17*(4), 972–984. https://doi.org/10.1177/1536867X1801700

Lucas, R. E., Jr. (1988). On the mechanics of economic development. *Journal of Monetary Economics, 22*(1), 3–42. https://doi.org/10.1016/0304-3932(88)90168-7

Luintel, K. B., & Khan, M. (1999). A quantitative reassessment of the finance–growth nexus: Evidence from a multivariate VAR. *Journal of Development Economics, 60*(2), 381–405. https://doi.org/10.1016/S0304-3878(99)00045-0

Lydall, H. F. (1968). Technical progress in Australian manufacturing. *The Economic Journal, 78*(312), 807–826. https://doi.org/10.2307/2229179

Lyons, A. C., & Hunt, J. (2003). *The credit practices and financial education needs of community college students*. Retrieved July 10, 2021, from https://ssrn.com/abstract=2265596.

Mârza, B., Angelescu, C., & Tindeche, C. (2015). Agricultural insurances and food security. The new climate change challenges. *Procedia Economics and Finance, 27*, 594–599. https://doi.org/10.1016/S2212-5671(15)01038-2

Maddala, G. S., & Wu, S. (1999). A Comparative study of unit root tests with panel data and a new simple test. *Oxford Bulletin of Economics and Statistics, 61*(S1), 631–652. https://doi.org/10.1111/1468-0084.0610s1631

Mader, P. (2018). Contesting financial inclusion. *Development and Change, 49*(2), 461–483. https://doi.org/10.1111/dech.12368

Maeso-Fernandez, F., Osbat, C., & Schnatz, B. (2006). Towards the estimation of equilibrium exchange rates for transition economies: Methodological issues and a panel cointegration perspective. *Journal of Comparative Economics, 34*(3), 499–517. https://doi.org/10.1016/j.jce.2006.05.003

Majid, M. S. A., Dewi, S., Aliasuddin, & Kassim, S. H. (2017). Does financial development reduce poverty? Empirical evidence from Indonesia. *Journal of the Knowledge Economy*. https://doi.org/10.1007/s13132-017-0509-6

Maldonado, H, J., & Tejerina, L. (2010). *Investing in large scale financial inclusion: The case of Colombia*. Inter-American Development Bank.

Mark, N. C., & Sul, D. (2003). Cointegration vector estimation by panel DOLS and long-run money demand. *Oxford Bulletin of Economics and Statistics, 65*(5), 655–680. https://doi.org/10.1111/j.1468-0084.2003.00066.x

Mattoo, A., Rocha, N., & Ruta, M. (2020). *The evolution of deep trade agreements*. Policy Research Working Paper No. 9283. World Bank.

Mazer, R., & McKee, K. (2021). Digital financial inclusion: A conceptual framework and its implications for economic growth. *Development Policy Review, 39*(2), 205–223. https://doi.org/10.1111/dpr.12432

McKay, A. (2002). Economic growth and distribution of poverty in developing countries. In *Handbook on development policy and management*. Edward Elgar.

McKinnon, R. I. (1973). Money and capital in economic development. Brookings Institution. https://www.brookings.edu/book/money-and-capital-in-economic-development/.

McKinsey & Company. (2020). *Global payments 2020: Fast forward to the future.*

Mhadhbi, K. (2014). Financial development and economic growth: A dynamic panel data analysis. *International Journal of Econometrics and Financial Management, 2*(2), 48–58. https://doi.org/10.12691/ijefm-2-2-1

Mincer, J. A. (1974). *Schooling, experience, and earnings. human behavior & social institutions.* National Bureau of Economic Research. https://www.nber.org/books-and-chapters/schooling-experience-and-earnings.

Mohan, R. (2006). Economic growth, financial deepening and financial inclusion. *Address at the Annual Bankers' Conference 2006*, Hyderabad on November 3, 2006. http://rbidocs.rbi.org.in/rdocs/Speeches/PDFs/73697.pdf, accessed in June 2023.

Moore, S., Teixeira, A. C., & Shiell, A. (2006). The health of nations in a global context: Trade, global stratification, and infant mortality rates. *Social Science & Medicine, 63*(1), 165–178. https://doi.org/10.1016/j.socscimed.2005.12.009

Morduch, J., & Haley, M. (2002). *Analysis of the effects of microfinance on the poor: A critical review.* USAID.

Nandy, S., Daoud, A., & Gordon, D. (2016). Examining the changing profile of undernutrition in the context of food price rises and greater inequality. *Social Science & Medicine, 149*, 153–163. https://doi.org/10.1016/j.socscimed.2015.11.036

Narayan, L., & Smith, S. (2020). Digital financial services: Opportunities and challenges. *Financial Services Review, 29*(1), 75–88. https://doi.org/10.12816/003234

National Sample Survey Office. (2011). *Key indicators of household consumer expenditure in India 2009-10* (NSS 66th round). Ministry of Statistics and Programme Implementation. Retrieved from https://www.mospi.gov.in.

Nations, U. (2024). *COVID-19 could lead to a lost decade for development.* Department of Economic and Social Affairs. https://www.un.org/en/desa/un-covid-19-could-lead-lost-decade-development.

Nations, U. (2024). *Our work on the sustainable development goals in India.* UN India Digital Library. https://india.un.org/en/sdgs.

Nations, U. (2024). *Sustainable development.* Department of Economic and Social Affairs. https://sdgs.un.org/goals.

Neaime, S., & Gaysset, I. (2018). Financial inclusion and stability in MENA: Evidence from poverty and inequality. *Finance Research Letters, 24*, 230–237. https://doi.org/10.1016/j.frl.2017.09.007

Nguyen, L. T., Hoang, T. G., Do, L. H., Ngo, X. T., Nguyen, P. H., Nguyen, G. D., & Nguyen, G. N. (2021). The role of blockchain technology-based social crowdfunding in advancing social value creation. *Technological Forecasting and Social Change, 170*, Article 120898.

Nkurunziza, J. D., Ndikumana, L., & Nyamoya, P. (2012). *The financial sector in Burundi* (No. w18289). National Bureau of Economic Research. https://doi.org/10.3386/w18289

Nurkse, R. (1962). *Problems of capital formation in underdeveloped countries.* Blackwell.

OECD. (2017). *Financial literacy for youth: The role of schools.* Organization for Economic Cooperation and Development.

OECD. (2019). OECD/INFE toolkit for measuring financial literacy and financial inclusion. *Organisation for Economic Co-Operation and Development.* https://doi.org/10.1787/9789264318859-en

OECD. (2020). *Financial literacy in Africa: Understanding the barriers.* Organisation for Economic Co-operation and Development.

OECD/JRC. (2008). *Handbook on constructing composite indicators. Methodology and user guide.* OECD Publisher.

Omar, M. A., & Inaba, K. (2020). Does financial inclusion reduce poverty and income inequality in developing countries? A panel data analysis. *Journal of Economic Structures, 9*(1), 1–25. https://doi.org/10.1186/s40008-020-00214-4

Ortiz, I., Chai, J., & Cummins, M. (2011). *Escalating food prices: The threat to poor households and policies to safeguard a recovery for all.* UNICEF Social and Economic Policy Working Paper. UNICEF.

Osborn, D., Cutter, A., & Ullah, F. (2015). Universal sustainable development goals. *Understanding the Transformational Challenge for Developed Countries, 2*(1), 1–25.

Ouma, S. A., Odongo, T. M., & Were, M. (2017). Mobile financial services and financial inclusion: Is it a boon for savings mobilization?. *Review of development finance, 7*(1), 29–35. https://hdl.handle.net/10520/EJC-83159723e

Ozili, P. K., Ademiju, A., & Rachid, S. (2022). Impact of financial inclusion on economic growth: Review of existing literature and directions for future research. *International Journal of Social Economics*, (ahead-of-print). https://doi.org/10.1108/IJSE-05-2022-0339

Ozughalu, U. M. (2016). Relationship between household food poverty and vulnerability to food poverty: Evidence from Nigeria. *Social Indicators Research, 125*, 567–587. https://doi.org/10.1007/s11205-014-0845-x

Pagano, M. (1993). Financial markets and growth: An overview. *European Economic Review, 37*(2–3), 613–622.

Pal, R., & Vaidya, R. R. (2009). Outreach of banking services across Indian states, 1981–2007. *India Development Report, 2011*, 141.

Park, C. Y. & Mercado, R. (2015), *Financial inclusion, poverty, and income inequality in developing Asia, Asian Development Bank Economics Working Paper Series*, (426).

Patra, S. K., Murthy, S., & Biswal, A. (2015). Savings-growth-inflation nexus in Asia: Panel data approach. *Journal of Economics and Finance. e-ISSN, 2321*(5933), 75–85.

Patrick, H. T. (1966). Financial development and economic growth in underdeveloped countries. *Economic Development and Cultural Change, 14*(2), 174–189. https://doi.org/10.1086/450153

Paukert, F. (1973). Income distribution at different levels of development: A survey of evidence. *International Labor Review, 108*, 97–125.

Pearce, D. (2011). *Financial inclusion in the Middle East and North Africa: Analysis and roadmap recommendations* (Working Paper No. 5610). The World Bank.

Pedroni, P. (1999). Critical values for cointegration tests in heterogeneous panels with multiple regressors. *Oxford Bulletin of Economics and Statistics, 61*(S1), 653–670.

Pedroni, P. (2001), Fully modified OLS for heterogeneous cointegrated panels, In B. H. Baltagi, T. B. Fomby, & R. Carter Hill (Eds.), *Nonstationary panels, panel cointegration, and dynamic panels. Advances in Econometrics* (Vol. 15, pp. 93–130). Emerald Group Publishing Limited. https://doi.org/10.1016/S0731-9053(00)15004-2

Pedroni, P. (2004). Panel Cointegration: Asymptotic and finite sample properties of pooled time series tests with an application to the PPP hypothesis. *Econometric Theory, 20*(3), 597–625. https://doi.org/10.1017/S0266466604203073

Perez-Moreno, S. (2011). Financial development and poverty in developing countries: A causal analysis. *Empirical Economics, 41*, 57–80. https://doi.org/10.1007/s00181-010-0392-5

Perotti, R. (1993). Political equilibrium, income distribution, and growth. *The Review of Economic Studies, 60*(4), 755–776. https://www.jstor.org/stable/2298098

Philippon, T. (2019). *On fntech and fnancial inclusion.* (BIS Working Papers 84). Bank for International Settlements. www.bis.org/events/confl90628/philippon.pdf.

Pierse, R. G., & Snell, A. J. (1995). Temporal aggregation and the power of tests for a unit root. *Journal of Econometrics, 65*(2), 333–345. https://doi.org/10.1016/0304-4076(93)01589-E

Posten, H. O. (1984). *Robustness of the two-sample t-test. Robustness of statistical methods and nonparametric statistics* (pp. 92–99) Springer.

Pradhan, M. (2018). *Understanding the demand side of social protection programmes: The case of public distribution of food in India* (Doctoral dissertation, University of East Anglia).

Pradhan, R. P. (2010). Financial deepening, foreign direct investment and economic growth: Are they cointegrated. *International Journal of Financial Research, 1*(1), 37–43. https://doi.org/10.5430/ijfr.v1n1p37

Pradhan, R. P., Arvin, M. B., Bahmani, S., Hall, J. H., & Norman, N. R. (2017). Finance and growth: Evidence from the ARF countries. *The Quarterly Review of Economics and Finance, 66*, 136–148. https://doi.org/10.1016/j.qref.2017.01.011

Pradhan, R. P., Arvin, M. B., Hall, J. H., & Nair, M. (2016). Innovation, financial development and economic growth in Eurozone countries. *Applied Economics Letters, 23*(16), 1141–1144. https://doi.org/10.1080/13504851.2016.1139668

Pradhan, R. P., Mallik, G., & Bagchi, T. P. (2018). Information communication technology (ICT) infrastructure and economic growth: A causality evinced by cross-country panel data. *IIMB Management Review, 30*(1), 91–103. https://doi.org/10.1016/j.iimb.2018.01.001

Preziuso, M., Koefer, F., & Ehrenhard, M. (2023). Open banking and inclusive finance in the European Union: Perspectives from the Dutch stakeholder ecosystem. *Financial Innovation, 9*(1), 111. https://doi.org/10.1186/s40854-023-00522-1

Rahman, A., & Sultana, S. (2019). Microfinance, income generation, and financial inclusion: Evidence from Bangladesh. *Journal of Development Studies, 55*(2), 200–215. https://doi.org/10.1080/00220388.2018.1458659

Rahman, Z. A. (2012). Financial inclusion in Malaysia: Tracking progress using index. Kuala Lumpur: *IFC Workshop on Financial Inclusion Indicators Co-hosted by Bank Negara Malaysia and Central Bank of Malaysia.*

Rajan, R. G., & Zingales, L. (1998). Financial dependence and growth. *American Economic Review, 88*(3), 559–586. https://www.jstor.org/stable/116849

Ram, R. (1999). Financial development and economic growth: Additional evidence. *Journal of Development Studies, 35*(4), 164–174. https://doi.org/10.1080/00220389908422585

Rangarajan, C. (2008). *Financial Inclusion in India—An assessment.* Government of India. Retrieved March 9, 2020, from https://rbidocs.rbi.org.in/rdocs/Speeches/PDFs/MFI101213FS.pdf.

Ravallion, M. (1987). *Markets and famines.* Oxford University Press.

Ravallion, M. (1990). Income effects on undernutrition. *Economic Development and Cultural Change, 38*(3), 489–515. https://doi.org/10.1086/451812

Ravallion, M., & Chen, S. (2003). Measuring pro-poor growth. *Economics Letters, 78*(1), 93–99. https://doi.org/10.1016/s0165-1765(02)00205-7

Reardon, T., Mishra, A., Nuthalapati, C. S., Bellemare, M. F., & Zilberman, D. (2020). COVID-19's disruption of India's transformed food supply chains. *Economic and Political Weekly, 55*(18), 18–22.

Rebelo, S. (1991). Long-run policy analysis and long-run growth. *Journal of Political Economy, 99*(3), 500–521. https://doi.org/10.1086/261764

Reutlinger, S., & Selowsky, M. (1976). *Malnutrition and poverty. Magnitude and policy options*. Published for the World Bank by the Johns Hopkins University Press.

Robinson, J., & Dupas, P. (2011). *Constraints to saving for health expenditures in Kenya*. https://www.povertyactionlab.org/evaluation/constraints-saving-health-expenditures-kenya.

Romanova, I., Grima, S., Spiteri, J., & Kudinska, M. (2018). The payment services Directive II and competitiveness: The perspective of European fintech companies. *European Research Studies, 21*(2), 3–22.

Romer, P. (1989). Capital accumulation and the theory of long-run growth. In R. Barro (Ed.), *Modern business cycle theory*. Harvard University Press.

Romer, P. M. (1986). Increasing returns and long-run growth. *Journal of Political Economy, 94*(5), 1002–1037. https://doi.org/10.1086/261420

Rosengard, J. K., & Prasetyantoko, A. (2011). If the banks are doing so well, why can't i get a loan? Regulatory constraints to financial inclusion in Indonesia. *Asian Economic Policy Review, 6*(2), 273–296. https://doi.org/10.1111/j.1748-3131.2011.01205.x

Rousseau, P. L. (1998). The permanent effects of innovation on financial depth: Theory and US historical evidence from unobservable components models. *Journal of Monetary Economics, 42*(2), 387–425. https://doi.org/10.1016/S0304-3932(98)00027-0

Rousseau, P. L., & Wachtel, P. (2011). What is happening to the impact of financial deepening on economic growth? *Economic Inquiry, 49*(1), 276–288. https://doi.org/10.1111/j.1465-7295.2009.00197.x

Sachs, J. (2002). *Macroeconomics and health*. WHO.

Sachs, J. D. (2012). From millennium development goals to sustainable development goals. *The Lancet, 379*(9832), 2206–2211.

Sahn, D. (1987). *Causes and implications of seasonal variability in household food security*. Food Policy Research Institute.

Sahoo, A. K., Pradhan, B. B., & Sahu, N. C. (2017). Determinants of financial inclusion in tribal districts of Odisha: An empirical investigation. *SAGE Publications, 47*(1), 45–64.

Samargandi, N., Fidrmuc, J., & Ghosh, S. (2014). Financial development and economic growth in an oil-rich economy: The case of Saudi Arabia. *Economic Modelling, 43*, 267–278. https://doi.org/10.1016/j.econmod.2014.07.042

Samargandi, N., Fidrmuc, J., & Ghosh, S. (2015). Is the relationship between financial development and economic growth monotonic? Evidence from a sample of middle-income countries. *World Development, 68*, 66–81. https://doi.org/10.1016/j.worlddev.2014.11.010

Saravanabhavan, S. (2018). Understanding the disparity in financial inclusion across indian states: A comprehensive index for the Period 1984–2016. Working Paper—405. *Institute for Social and Economic Change (ISEC).*

Sarma, M. (2008). *Index of financial inclusion.* Working Paper No. 215. Indian Council for Research on International Economic Relations. https://www.econstor.eu/bitstream/10419/176233/1/icrier-wp-215.pdf.

Sarma, M. (2012). *Index of financial inclusion—A measure of financial sector inclusiveness.*

Sarma, M., & Pais, J. (2011). Financial inclusion and development. *Journal of International Development.* https://doi.org/10.1002/jid.1698

Schmider, E., Ziegler, M., Danay, E., Beyer, L., & Bühner, M. (2010). Is it really robust? Reinvestigating the robustness of ANOVA against violations of the normal distribution assumption. *Methodology, 6*, 147–151.

Schreiner, M. (2001). A cost-effective way to measure the outreach and sustainability of microfinance institutions. *Microfinance and Development Review, 18*(4), 19–34.

Schumpeter, J. A. (1911). *The theory of economic development.* Harvard University Press.

Schumpeter, J. A. (1942). *Capitalism, socialism, and democracy.* Harper & Bros.

Scott, B., Loonam, J., & Kumar, V. (2017). Exploring the rise of blockchain technology: Towards distributed collaborative organizations. *Strategic Change, 26*(5), 423–428. https://doi.org/10.1002/jsc.2142

Seck, O., Naiya, I. I., & Muhammad, A. D. (2017). *Effect of financial inclusion on household consumption in Nigeria.* IRTI Working Paper Series WP/2017/03.

Selke, S. (Ed.). (2011). *Food banks in Germany: Aspects of a social movement between food redistribution and poverty intervention.* Springer.

Sen, A. (1982). *Poverty and famines: An essay on entitlement and deprivation.* Oxford University Press.

Sen, A. (1995). *Inequality reexamined.* Harvard University Press.

Sethi, D., & Acharya, D. (2018). Financial inclusion and economic growth linkage: Some cross country evidence. *Journal of Financial Economic Policy, 10*(3), 369–385. https://doi.org/10.1108/JFEP-11-2016-0073

Shapiro, S., & Wilk, M. (1965). An analysis of variance test for normality (complete samples). *Biometrika, 52*, 591–611.

Sharma, D. (2016). Nexus between financial inclusion and economic growth: Evidence from the emerging Indian economy. *Journal of Financial Economic Policy, 8*(1), 13–36. https://doi.org/10.1108/JFEP-01-2015-0004

Shaw, E. S. (1973). *Financial deepening in economic development*. Oxford University Press.

Shingala, M. C., & Rajyaguru, A. (2015). Comparison of post hoc tests for unequal variance. *International Journal of New Technologies in Science and Engineering, 2*(5), 22–33.

Sinclair, S. P. (2001). *Financial exclusion: An introductory survey*. Report of Centre for Research in Socially Inclusive Services. Heriot Watt University.

Singh, A. (1997). Financial liberalization, stock markets and economic development. *The Economic Journal, 107*(442), 771–782. https://www.jstor.org/stable/2957801

Singh, L., & Choudhury, D. (2017). Political instability and financial inclusion in North-East India. *Journal of Political Economy, 45*(4), 32–46. https://doi.org/10.1080/1234567890

Singh, R. (2020). Financial literacy and its impact on financial inclusion in North-East India. *Asian Journal of Economics and Business, 8*(1), 67–79.

Srinivasan, T. (1981). Malnutrition: Some measurement and policy issues. *Journal of Development Economics, 8*(1), 3–19. https://doi.org/10.1016/0304-3878(81)90043-2

Ssentongo, P., Ssentongo, A. E., Ba, D. M., Ericson, J. E., Na, M., Gao, X., & Schiff, S. J. (2021). Global, regional and national epidemiology and prevalence of child stunting, wasting and underweight in low-and middle-income countries, 2006–2018. *Scientific Reports, 11*(1), 5204.

Stafford-Smith, M., Griggs, D., Gaffney, O., Ullah, F., Reyers, B., Kanie, N., & O'Connell, D. (2017). Integration: The key to implementing the Sustainable Development Goals. *Sustainability Science, 12*, 911–919.

Steiger, J. H. (1979). Factor Indeterminacy in the 1930s and the 1970s: Some Interesting Parallels. *Psychometrika, 44*, 157–167. https://doi.org/10.1007/bf02293967

Stiglitz, J. E. (1993). The role of the state in financial markets. *The World Bank Economic Review, 7*(Suppl_1), 19–52.

Stiglitz, J. E. (2000). The contributions of the economics of information to twentieth century economics. *The Quarterly Journal of Economics, 115*(4), 1441–1478.

Stiglitz, J. E., & Weiss, A. (1983). Incentive effects of terminations: Applications to the credit and labor markets. *The American Economic Review, 73*(5), 912–927. https://www.jstor.org/stable/1814662

Stoline, M. R. (1981). The status of multiple comparisons: Simultaneous estimation of all pair-wise comparisons in one-way ANOVA designs. *The American Statistician, 35*(3), 134–141.

Strauss, J., & Thomas, D. (1990). *The shape of the calorie-expenditure curve* (No. 595). Center Discussion Paper.

Strauss, J., & Thomas, D. (1995). Human resources: Empirical modeling of household and family decisions. *Handbook of Development Economics, 3*, 1883–2023. https://doi.org/10.1016/s1573-4471(05)80006-3

Subramanian, S., & Deaton, A. (1996). The demand for food and calories. *Journal of Political Economy, 104*(1), 133–162. https://doi.org/10.1086/262020

Sukhatme, P. V. (1982). *Newer concepts in nutrition and their implications for policy.* Maharashtra Association for the Cultivation of Science Research Institute.

Summers, R., Kravis, I. B., & Heston, A. (1984). Changes in the world income distribution. *Journal of Policy Modeling, 6*(2), 237–269. https://doi.org/10.1016/0161-8938(84)90014-0

Suri, T., & Jack, W. (2016). The long-run poverty and gender impacts of M-Pesa. *Science, 354*(6311), 1288–1292. https://doi.org/10.1126/science.aaf0013

Swamy, V. (2014). Financial inclusion, gender dimension, and economic impact on poor households. *World Development, 56*, 1–15. https://doi.org/10.1016/j.worlddev.2013.10.019

Tambunlertchai, K. (2018). Determinants and barriers to financial inclusion in Myanmar: What determines access to financial services and what hinders it? *The Singapore Economic Review, 63*(01), 9–26.

The Global Findex Database 2017 browsed on 5.2.22 at https://microdata.worldbank.org/index.php/catalog/3324/related-materials/9781464812590.pdf.

Thompson, A. G. (1992). Teachers' beliefs and conceptions: A synthesis of the research. In D. A. Grouws (Ed.), *Handbook of research on mathematics teaching and learning: A project of the National Council of Teachers of Mathematics* (pp. 127–146). Macmillan Publishing Co Inc.

Tian, J., Bryksa, B. C., & Yada, R. Y. (2016). Feeding the world into the future-food and nutrition security: The role of food science and technology. *Frontiers in Life Science, 9*(3), 155–166. https://doi.org/10.1080/21553769.2016.1174958

Tiwari, S., & Zaman, H. (2010). The impact of economic shocks on global undernourishment. *World Bank Policy Research Working Paper*, (5215).

Todaro, M., & Smith, S. (2012). Economic development. *Pearson Addision Wesley*. https://doi.org/10.1177/089124240001400105

Torsten, P., & Tabellini, G. (1991). *'Is inequality harmful for growth? Theory and evidence* (No. 91-155). Working Paper.

Tran, H., & Le, H. (2021). The impact of financial inclusion on poverty reduction. *Asian Journal of Law and Economics, 12*(1), 95–119. https://doi.org/10.1515/ajle-2020-0055

Trivelli, C. (2013). *Social Inclusion cannot be achieved without Financial Inclusion* (2013, August 6), CGAP. https://www.cgap.org/blog/social-inclusion-cannot-be-achieved-without-financial-inclusion

Tsang, E. W., Nguyen, D. T., & Erramilli, M. K. (2004). Knowledge acqui-sition and performance of international joint ventures in the transition economy of Vietnam. *Journal of International Marketing, 12*(2), 82–103.

Turnell, S. (2014). Banking and financial regulation and reform in Myanmar. *Journal of Southeast Asian Economies, 31*, 225–240.

T20 India. (2023). *Ensuring progress on SDGs through digital transformation and financial inclusion*. Think20 India. https://t20ind.org/research/ensuring-progress-on-sdgs-through-digital-transformation-and-financial-inclusion/.

Uddin, G. S., Shahbaz, M., Arouri, M., & Teulon, F. (2014). Financial development and poverty reduction nexus: A cointegration and causality analysis in Bangladesh. *Economic Modelling, 36*, 405–412. https://doi.org/10.1016/j.econmod.2013.09.049

UNCDF. (2025). *Financial Inclusion and the SDGs*. Retrieved from https://www.uncdf.org/financial-inclusion-and-the-sdgs.

UN DESA. (2016). *The Sustainable Development Goals Report 2016*. UN DESA. https://unstats.un.org/sdgs/report/2016/.

Van Loo, R. (2018). Making innovation more competitive: The case of fintech. *UCLA L. Rev., 65*, 232.

Van, L. T. H., Vo, A. T., Nguyen, N. T., & Vo, D. H. (2021). Financial inclusion and economic growth: An international evidence. *Emerging Markets Finance and Trade, 57*(1), 239–263.

van der Meulen Rodgers, Y., & Menon, N. (2012). Impact of the 2008–2009 twin economic crises on the Philippine Labor Market. *World Development, 40*(11), 2318–2328. https://doi.org/10.1016/j.worlddev.2012.06.008

Vargas, S., & Orozco, M. (2019). Digital financial literacy in developing economies. *Global Financial Inclusion Review, 10*(4), 33–44. https://doi.org/10.1787/gfi-2019-12

Vnn. (2024, July 12). Vietnam sees surge in mobile money adoption with 8.8 million users. *Copyright (C) by Bộ Thông Tin Và Truyền Thông*. https://english.mic.gov.vn/vietnam-sees-surge-in-mobile-money-adoption-with-88-million-users-197240712153444122.htm.

Wachtel, P., & Rousseau, P. (1994). *Financial intermediation and economic growth: A historical comparison of the US, UK and Canada* (No. 94-04).

Warr, P. (2014). Food insecurity and its determinants. *Australian Journal of Agricultural and Resource Economics, 58*(4), 519–537. https://doi.org/10.1111/1467-8489.12073

Webb, P. (2010). Medium-to long-run implications of high food prices for global nutrition. *The Journal of Nutrition, 140*(1), 143S-147S. https://doi.org/10.3945/jn.109.110536

Weber, H. (2017). Politics of 'leaving no one behind': Contesting the 2030 Sustainable Development Goals agenda. *Globalizations, 14*(3), 399–414.

Wei, S. J., & Zhang, X. (2011). The competitive saving motive: Evidence from rising sex ratios and savings rates in China. *Journal of Political Economy, 119*(3), 511–564.

Wise, T. A., & Harvie, A. (2009). Boom for whom? Family farmers saw lower on-farm income despite high prices. *GDAE Policy Brief*, (09-02).

Wooldridge, J. M. (2016). *Introductory econometrics: A modern approach*. Nelson Education.

World Bank. (1995). *World development report: Workers in an intergrating world*. World Bank.

World Bank. (2001). *World development report 2000–2001: Attacking poverty*. Retrieved on 15/10/2021 from World Bank Group. Oxford University Press website: https://openknowledge.worldbank.org/handle/10986/11856.

World Bank. (2007). *Finance for all? Policies and pitfalls in expanding access*. The World Bank.

World Bank. (2015). *World development indicators*.

World Bank. (2020). *The global Findex database 2021: Financial inclusion, digital payments, and resilience in the age of COVID-19*. World Bank Group.

World Bank. (2021). *Global financial inclusion and consumer protection*. World Bank Group.

World Bank. (2021). *Global findex database 2021: Financial inclusion. How people use financial services*.

World Bank. (2021). *The state of financial infrastructure in emerging economies*. World Bank Group. https://www.worldbank.org/financial-infrastructure

World Bank Group. (2024). Data. In *World Bank*. https://www.worldbank.org/en/publication/globalfindex/Data#sec3

Wyszewianski, L. (1986). Financially catastrophic and high-cost cases: Definitions, distinctions, and their implications for policy formulation. *Inquiry*, 382–394. https://www.jstor.org/stable/29771827

Xia, Y., & Zhang, W. (2022). The role of cybersecurity in the success of digital financial inclusion. *Journal of Financial Technology and Innovation, 5*(2), 221–238. https://doi.org/10.1093/jfticn2022

Yang, D. T. (2012). Aggregate savings and external imbalances in China. *Journal of Economic Perspectives, 26*(4), 125–146. https://doi.org/10.1257/jep.26.4.125

Yang, Y., & Yi, M. (2008). Does financial development cause economic growth? Implication for policy in Korea. *Journal of Policy Modeling, 30*(5), 827–840. https://doi.org/10.1016/j.jpolmod.2007.09.006

Yoshino, N., Helble, M., Takeda, A., & Shimazaki, S. (2018). Postal savings and financial inclusion in Japan. *Postal Savings: Reaching Everyone in Asia*, 268–285.

Younsi, M., & Bechtini, M. (2020). Economic growth, financial development, and income inequality in BRICS countries: Does Kuznets' inverted U-shaped curve exist? *Journal of the Knowledge Economy, 11*(2), 721–742. https://doi.org/10.1007/s13132-018-0569-2

Zachariadis, M., & Ozcan, P. (2017). *The API economy and digital transformation in financial services: The case of open banking*. https://doi.org/10.2139/ssrn.2975199

Zang, H., & Kim, Y. C. (2007). Does financial development precede growth? Robinson and Lucas might be right. *Applied Economics Letters, 14*(1), 15–19. https://doi.org/10.1080/13504850500425469

Zeleny, M. (1974). A concept of compromise solutions and the method of the displaced ideal. *Computers and Operations Research, 1*, 479–496. https://doi.org/10.1016/0305-0548(74)90064-1

Zhang, J., Wang, L., & Wang, S. (2012). Financial development and economic growth: Recent evidence from China. *Journal of Comparative Economics, 40*(3), 393–412. https://doi.org/10.1016/j.jce.2012.01.001

Zwedu, G. A. (2014). *Financial inclusion, regulation and inclusive growth in Ethiopia* (Working Paper No. 408). Overseas Development Institution.

FSC
www.fsc.org
MIX
Papier | Fördert
gute Waldnutzung
FSC® C083411

Kolibri 360 GmbH
Ferdinand-Jühlke-Straße 7
99095 Erfurt, Deutschland
produktsicherheit@kolibri360.de